Molecular Robotics II

Satoshi Murata · Takashi Nakakuki · Taro Toyota ·
Shin-ichiro Nomura · Akinori Kuzuya
Editors

Molecular Robotics II

Toward Chemical AI

Springer

Editors
Satoshi Murata
Department of Robotics
Graduate School of Engineering, Tohoku University
Sendai, Japan

Taro Toyota
Graduate School of Arts and Sciences
University of Tokyo
Tokyo, Japan

Akinori Kuzuya
Department of Chemistry and Materials Engineering
Kansai University
Suita, Japan

Takashi Nakakuki
Department of Intelligent and Control Systems
Faculty of Computer Science and Systems Engineering, Kyushu Institute of Technology
Kitakyushu, Japan

Shin-ichiro Nomura
Department of Robotics
Graduate School of Engineering, Tohoku University
Sendai, Japan

ISBN 978-981-95-2393-1 ISBN 978-981-95-2394-8 (eBook)
https://doi.org/10.1007/978-981-95-2394-8

This Springer imprint is published by the registered company Springer Nature Singapore Pte Ltd.
The registered company address is: 152 Beach Road, #21-01/04 Gateway East, Singapore 189721, Singapore

Preface

This book is designed as a sequel to "Molecular Robotics—An Introduction," published by Springer Singapore, 2022, where the previous subject was the robotization of a "single" artificial cell defined by a compartment (or simply vesicle). Microsized robots composed of biomolecules are intelligent systems that integrate various nanotechnologies. Over the past two decades, the research and development of individual elemental technologies for molecular robots for capsules, sensors, actuators, and molecular computing have been addressed worldwide. Based on this background, a previous book summarized the techniques and theories intended for the system integration of molecular robots.

Although various evolutionary scenarios can be envisioned for molecular robots, the most promising and straightforward road map might be "multicellularity." These multicellular systems can realize more complex and sophisticated information processing than unicellular systems, leading to chemically synthesized artificial intelligence (chemical AI). However, various technical and theoretical problems remain to be solved for linking individual artificial cells and integrating them into a single system. Typical challenges include geometric network structures formed by multiple artificial cells, molecular communication technology between artificial cells, molecular computing systems as operating systems mounted on multicellular systems, and the dynamic deformation technology of multicellular systems.

This book outlines the technical and theoretical topics of these challenges in chemical AI, reviewing the remarkable research results reported in recent years. We hope that this book will facilitate the development of highly systematic molecular systems and progress toward the realization of chemical AI.

Fukuoka, Japan

Takashi Nakakuki
nakakuki@ics.kyutech.ac.jp

Acknowledgements I would like to express my deep gratitude to all the people involved in writing, compiling, and arranging this book. This book contains a great deal of results obtained from Grant-in-Aid for Transformative Research Area (A) "Molecular Cybernetics" (2020–2024) in Japan. I also appreciate the contributions of all the researchers involved in this project.

Contents

Chapter 1
Introduction

Satoshi Murata

Abstract In recent years, attention has been focused on the bottom-up approach, in which the assembly of molecules spontaneously creates a desired system by programming the properties of the molecules themselves. The emergence of this methodology, in which molecules are designed and self-assembled to create artifacts with molecular or atomic resolution, is a historic turning point in manufacturing methodology. Once artifacts have molecular/atomic precision, it will be possible not only to mimic the advanced functions of living organisms but also to realize functions that surpass those of living organisms. Molecular robotics, the subject of this book, is at the forefront of this paradigm shift in manufacturing. In the previous book, "Molecular Robotics—An Introduction," we introduced the frontiers of single-cell molecular robotics research, and this book, "Molecular Robotics II—Toward Chemical AI," focuses on attempts to extend this to multicellular molecular robots.

This book is a sequel to Molecular Robotics—An Introduction (2022) [1]. In "An Introduction," the concept of molecular robotics and its application areas are introduced. Then, it is organized into various aspects, such as design theory, systemization technology, actuators, materials, medical applications, and social acceptance. This book not only covers subsequent advances in the field of molecular robotics but also details recent developments toward multicellularization in molecular robotics.

Although the idea of nanosized robots was mentioned in Feynman's famous lecture [2], the first example of synthesizing robots with actual molecules was probably the work of Shinkai in 1981 on light-responsive molecular tweezers [3]. This research subsequently led to synthesizing molecules such as rotaxanes and catenanes. In 2016, Sauvage et al. were awarded the Nobel Prize in Chemistry for their work on molecular machines [4]. The discovery of the DNA double helix by Watson and Crick led to unraveling molecular mechanisms in living organisms [5]. Once the operating principles of biomolecules are understood in detail, it becomes possible

S. Murata (✉)
Department of Robotics, Graduate School of Engineering, Tohoku University, Sendai, Japan
e-mail: satoshi.murata.a4@tohoku.ac.jp

S. Murata et al. (eds.), *Molecular Robotics II*,
https://doi.org/10.1007/978-981-95-2394-8_1

to modify them to add artificial functions or even give them functions completely different from their original functions.

Although various molecular devices that modify biomolecules have been developed, integrating them into a system has been difficult. This is because combining the actions (e.g., chemical reactions) of different molecular devices in the same space (e.g., in the same test tube) is difficult. When many different chemical reactions occur at the same place and simultaneously, it is difficult to ensure that only the desired reaction occurs and to create a meaningful network by chaining such reactions together.

DNA computing uses the sequence specificity of DNA hybridization to perform various calculations [6]. For example, in DNA logic operations, a DNA molecule is given to a DNA logic element as an input signal, and another DNA molecule corresponding to the result of the operation is the output. A circuit can be created by connecting several logic elements as input to another DNA logic element. In "structural DNA nanotechnology," technologies have been developed to create various nanostructures using specific hybridization of many DNA strands to construct nanostructures called DNA tiles or DNA origamis [7]. The size and complexity of these DNA nanostructures will soon reach levels comparable to those of natural proteins.

Molecular robotics is a field of research based on DNA computing and structural DNA nanotechnology; however, it also uses molecular materials other than DNA, such as peptides, engineered proteins, and synthetic polymers, to create autonomous molecular systems with sensors, processors, and actuators [1]. The first molecular robot based on DNA molecules was the molecular spider, which achieved guided locomotion on DNA origami through enzymatic reactions with DNA [8]. As this type of robot is essentially a single molecule, it is difficult to add additional functionalities.

An evolutionary scenario for molecular robots has been proposed to determine how their functionality can be enhanced [9]. To overcome the limitations of single-molecule molecular robots, the "Molecular Robotics" project (2012–2017, Grant-in-Aid for Scientific Research on Innovative Areas, MEXT, Japan) conducted by the authors' group artificially synthesized a single micron-sized container (compartment) and integrated a set of molecular devices for information processing and locomotion inside it. Thus, the randomness of single-molecule reactions can be avoided, and functions can be designed deterministically based on reaction kinetics [10]. These are known as first-generation molecular robots (amoeba-type molecular robots). Furthermore, by scaling up the reaction field using functional polymer gels and operating a set of molecular devices on them, it is possible to express functions that require spatial anisotropy, such as chemotaxis. We call this the second generation of molecular robots (slime-type molecular robots). In the previous book, we discussed the first- and second-generation molecular robots.

The subject of this book is third-generation molecular robots. It is a multicellular molecular robot that combines single-cell molecular robots of the amoeba type. Each cell has a different function, and more complex systems can be constructed by connecting them. The authors have been working on a research and development project for multicellular molecular robots[1] (2020–2025, Grant-in-Aid for Scientific Research on Transformative Research Areas "Molecular Cybernetics," MEXT, Japan). This project aims to create a multicellular molecular robot that performs information processing as a "minimal artificial brain" [11, 12].

This book is structured as follows. In Chapter 2, "From Molecular Robotics to Molecular Cybernetics," we describe in detail how the concept of molecular cybernetics is derived from molecular robotics. Part I, "Molecular Devices to Control Molecular System," introduces various molecular devices constituting a minimal artificial brain. This chapter describes methods to provide the desired functions to molecules, accelerate intermolecular reactions, give functions to DNA origamis, and amplify molecular concentrations using artificial nucleic acids. Part II, "Importance of Artificial Membrane Devices in Molecular Robotics," introduces methods for realizing molecular interactions across lipid membrane compartments, which correspond to the chassis of an amoeba-type molecular robot. Transducers for the molecular communication between compartments are essential for creating multicellular systems. Examples of such transducers, including channels that allow various molecules and ions to pass through membranes, devices that transmit DNA signals across membranes, and applications of functionalized membranes in various sensing tasks, are described. Part III, "Cellular Compartments for Cell-Mimetic Molecular Robots: Giant Vesicles," describes techniques for creating, maintaining, manipulating, and observing multicellular systems. The central focus is on systems that use microfluidic devices to create, sort, and align artificial cells (liposomes) to which various physicochemical stimuli are applied, and their responses are observed. This chapter also includes methods for deforming an artificial cell by encapsulating a molecular motor. In Part IV, "Intelligent Molecular Computation Systems," the immediate goal of molecular cybernetics is discussed: How to implement the learning circuit of Pavlov's dog? Various implementation methods are shown, ranging from simple counter circuits to relatively complex learning circuits with plasticity. This chapter also examines the limitations of the learning ability and discusses the theory of transmission efficiency when molecular circuits communicate via molecular diffusion. Part V introduces the activities of journalists and ethicists in molecular cybernetics. It discusses how, even at the current fundamental stage, researchers consider the risks of their studies, which may have a significant impact on society. This chapter also describes several advanced attempts to link these considerations to responsible research practices.

[1] The project was named after Winner's Cybernetics. "Cybernetics," written by Norbert Wiener [13], discussed the mechanisms of communication and control common to living organisms and artificial objects and became the starting point for the later development of control engineering, systems engineering, and robotics. He introduced control structures, such as feedback, and a neural network prototype. Winner assembled actual systems using electrical and mechanical components available at the time of the demonstration, but molecular cybernetics attempted to construct systems by synthesizing biomolecules, the materials of living organisms themselves.

References

1. S Murata Eds 2022 Molecular robotics-an introduction Springer
2. RP Feynman 1992 There's plenty of room at the bottom J. Microelectromechanical Syst. 1 1
3. S Shinkai 1981 Photoresponsive crown ethers. 2. Photocontrol of ion extraction and ion transport by a Bis(crown ether)with a butterfly-like motion J. Am. Chem. Soc. 103 111 115
4. The Nobel Prize in Chemistry 2016 is awarded to Jean-Pierre Sauvage, Sir J. Fraser Stoddart and Bernard L. Feringa for their design and production of molecular machines. https://www.nobelprize.org/prizes/chemistry/2016/press-release/
5. JD Watson FHC Crick 1953 Molecular structure of nucleic acids Nature 4356 737 738
6. DNA computing. https://en.wikipedia.org/wiki/DNA_computing
7. NC Seeman 2015 Structural DNA nanotechnology MRS Cambridge
8. K Lund 2010 Molecular robots guided by prescriptive landscapes Nature 465 7295 206 210
9. S Murata 2013 Molecular robotics: a new paradigm for artifacts N. Gener. Comput. 31 27 45
10. Sato, Y., et al.: Micrometer-sized molecular robot changes its shape in response to signal molecules. Sci. Robot. **2**(4):eaal3735, (2017)
11. S Murata 2022 Molecular cybernetics: challenges toward cellular chemical artificial intelligence Adv. Funct. Material 32 37
12. A Kuzuya 2023 From molecular robotics to molecular cybernetics: the first step toward chemical artificial intelligence IEEE Trans. Mol. Bio. Multi-Scale Commun 9 3 354 363
13. N Wiener 1961 Cybernetics, second edition: or control and communication in the animal and the machine MIT Press

Chapter 2
From Molecular Robotics to Molecular Cybernetics

Akinori Kuzuya

"Molecular Robotics [1]" is a research field that aims to construct intelligent molecular systems involving all of the three basic elements of robotics: sensing, planning, acting, by using functional molecules [2]. This field was particularly inspired by the development of "DNA Computing [3]" and "Structural DNA Nanotechnology [4]," and its goal is to construct molecular systems by combining functional molecules, most of which are biomolecules such as proteins and lipids, together with precisely programmable DNA. This chapter briefly overviews major achievement of molecular robotics and its successor field of "molecular cybernetics."

2.1 Evolutionary Scenario of Molecular Robotics

Most primitive molecular robots, zeroth-generation molecular robots, are usually considered to be independent supramolecules composed of DNA and other functional biomolecules (Fig. 2.1). A typical example is the "molecular spider" proposed by Lund et al. [5], which "walks" along "DNA track" placed on DNA origami based on Brownian motion induced by random walks and thermal fluctuations. To achieve directional motion, it is necessary to introduce "fuel" to drive the molecular motion. The "molecular swarm robots" developed by Kakugo, Kuzuya et al. are a typical example that combines DNA and microtubule/kinesin molecular motor system using chemical energy as the fuel [6].

Amoeba-type robots employing compartments to overcome the limitations of the zeroth generation are referred to as the first-generation molecular robots. Amoeba-type robots can be regarded as a single artificial cell that encapsulates molecular systems. Compartments can be made of liposomes with lipid bilayers, synthetic

A. Kuzuya (✉)
Department of Chemistry and Materials Engineering, Kansai University, Osaka, Japan
e-mail: kuzuya@kansai-u.ac.jp

S. Murata et al. (eds.), *Molecular Robotics II*,
https://doi.org/10.1007/978-981-95-2394-8_2

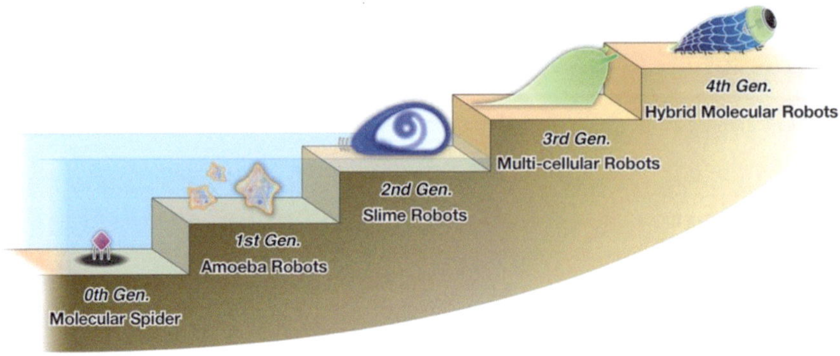

Fig. 2.1 Evolutionary scenario of molecular robotics. (Courtesy of Dr. Shogo Hamada of Science Tokyo)

polymer vesicles, DNA nanostructures, etc., which are flexible and can be deformed like an amoeba by an encapsulated molecular motor systems. The first amoeba robot was reported by Nomura and Murata [7]. In liposomes, they encapsulated microtubules together with DNA-modified kinesin that can selectively attach to and detach from lipid membranes in response to light signals sensed by photo-responsive DNA, and successfully controlled liposome deformation.

The second generation of molecular robots, called slime robots, utilizes polymer gels and other materials as reaction fields and actuators themselves. This expands their size to a few millimeters and resolves the size limitations of the first-generation molecular robots. One of the most notable slime robots is DASH (DNA-based assembly and synthesis of hierarchical materials) developed by S Hamada [8]. They created dynamic biomaterials powered by artificial metabolism, and demonstrated programmed pattern formation, etc.

The third generation of molecular robots are multicellular robots. They can realize multiple functions between different cells and are expected to realize more complex systems, such as natural organs and tissues. The fourth generation is a hybrid of electronic devices and molecular systems. This may enable the realization of functions that overcome the limitations of molecular and chemical systems, and this is where molecular robots enter a stage where they transcend natural systems.

2.2 Molecular Cybernetics

It is natural that the realization of single-cell molecular robots (amoeba robots) encourages the development of multicellular, third-generation molecular robots. "Molecular Cybernetics [9]" aims to construct a molecular-based artificial intelligence, or "Chemical AI," and to develop communication methods between molecular robots divided into multiple liposomes.

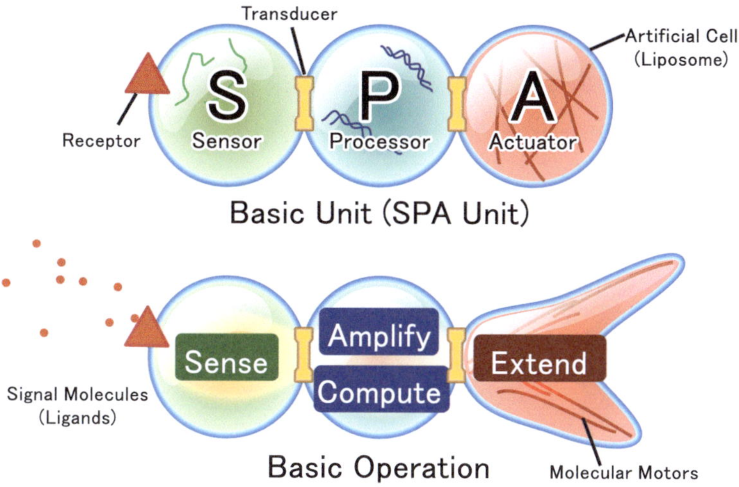

Fig. 2.2 A model of chemical AI composed of three artificial cells communicate to each other; sensor (S), processor (P), and actuator (A) cells

Figure 2.2 shows a possible set of chemical AI. It consists of three artificial cells, corresponding to the three layers of the nervous system in nature. Each of them is responsible for sensing chemical input (sensor = S), memory and learning (processor = P), and output such as transformation (actuator = A). Chemical information processing is initiated by the input of various molecular stimuli to the SPA unit, sensed by receptors on S cell, mediated by chemical reactions corresponding to the calculations in the P cell, and terminated by the output of specific molecular signals from A cell. Communication between each artificial cell is maintained by "transducers" or "artificial channels" that pass molecular information through two lipid bilayers. New connections between SPA units, which mimic brain plasticity based on dynamic reconnection of neurons, are controlled by programmed a cell behavior.

In Chapter 3 and beyond, various techniques that may contribute to the development of molecular cybernetics are introduced.

References

1. M Hagiya 2014 Molecular robots with sensors and intelligence Acc. Chem. Res. 47 1681 1690
2. N Winner 1948 Cybernetics: or control and communication in the animal and the machine MIT Press Cambridge, USA
3. G Paun G Rothenberg A Salomaa 1998 DNA computing Springer-Verlag Berlin, Heidelberg, Germany
4. NC Seeman 2016 Structural DNA nanotechnology Cambridge University Press Cambridge, UK
5. K Lund 2010 Molecular robots guided by prescriptive landscapes Nature 465 206 210

6. JJ Keya 2018 DNA-assisted swarm control in a biomolecular motor system Nature Commun. 9 453
7. Y Sato 2017 Micrometer-sized molecular robot changes its shape in response to signal molecules. Science Robot. 2 eaal3735
8. S Hamada 2019 Dynamic DNA material with emergent locomotion behavior powered by artificial metabolism. Science Robot. 4 eaaw3512
9. S Murata 2022 Molecular cybernetics: challenges toward cellular chemical artificial intelligence Adv. Funct. Mater. 32 2201866

Part I
Molecular Devices to Control Molecular Systems

Chapter 3
Functional Expansion of DNA Molecule: Stimulus-Responsive Nucleic Acids

Keiji Murayama, Keiichi Imato, and Yusuke Takezawa

3.1 Stimulus-Responsive Nucleic Acids for Molecular Robot

To realize the molecular robot using nucleic acids, a molecular device that can switch the function and behavior by responding to external stimulus is strongly required. Most effective but simple method is to modify the DNA with stimuli-responsive molecules to facilitate functional switching triggered by external stimuli such as pH, heat, DNA signaling, light, or metal–ion binding. In this section, we introduce designs and successful applications of light-responsive nucleic acids and metal-responsive DNAs.

3.2 Photo-Responsive Nucleic Acids

Among potential external stimuli, photo-stimulation is an ideal trigger because it can achieve spatiotemporal control without contaminating the solution [1]. Photoremovable protecting (caging) groups easily allow to regulate structures, assemblies, and functions of DNA with light and have been extensively installed on DNA [2,

K. Murayama (✉)
Nagoya University, Nagoya, Japan
e-mail: murayama@chembio.nagoya-u.ac.jp

K. Imato
Hiroshima University, Hiroshima, Japan
e-mail: kimato@hiroshima-u.ac.jp

Y. Takezawa
The University of Tokyo, Tokyo, Japan
e-mail: takezawa@chem.s.u-tokyo.ac.jp

S. Murata et al. (eds.), *Molecular Robotics II*,
https://doi.org/10.1007/978-981-95-2394-8_3

3]. While a wide variety of photocleavable protecting groups are known, *ortho*-nitrobenzyl groups are most commonly used for caging DNA (Fig. 3.1a) [4]. Another representative example is a family of coumarins, which dissociate by photoirradiation in the presence of a nucleophile (Fig. 3.1b). Initially, these caging groups have been inserted into DNA backbones to cleave the chains and break the strands with light and then activate or deactivate their functions (Fig. 3.1c) [5]. Alternatively, the cages have been integrated into nucleobases, blocking and unlocking hydrogen bonding between them. Even photoregulation of gene expression in vivo and microscopic operation of DNA-based materials have been already achieved by using these photoreleasable-protecting groups. Although the parent (unmodified) *ortho*-nitrobenzyl group requires UV light around 365 nm for the deprotection, red and near-infrared light, which effectively penetrates into biological tissues, is also available in recently developed systems [6].

Since these uncaging reactions of photo-caged compounds are irreversible, applicable designs as photo-reactive molecular systems are limited in some cases. In contrast, photochromic molecules can reversibly isomerize the structure by the irradiation with different wavelengths of light (Fig. 3.2). For example, *trans*-azobenzene isomerizes to *cis*-form by the irradiation of UV light, whereas blue light irradiation causes *cis*-to-*trans* isomerization. By incorporation of these photochromic molecules into DNA, reversibly photoregulatable DNA-based nanomachines have been developed to extend the flexibility of the applications.

Asanuma et al. designed azobenzene-modified DNA (azo-DNA) using acyclic D-threoninol linker that can incorporate various functional molecules into DNA without destabilization of the duplex (Fig. 3.3) [7, 8]. The duplex formation and dissociation of azo-DNA can be controlled by the irradiation of light. *Trans*-Azobenzene tethered by DNA can intercalate into base pairs, stabilizing the DNA duplex through the π–π stacking interaction, whereas *cis*-azobenzene has sterically balky structure that disturbs the neighbor base pairing, resulting in dissociation of the duplex. Upon the irradiation with UV light (310–370 nm), *cis*-rich state is induced, dissociating the duplex. The irradiation with blue light (400–500 nm) recovers *trans*-rich state, facilitating reformation of the duplex (Fig. 3.3). Azo-DNA with multiple azobenzenes

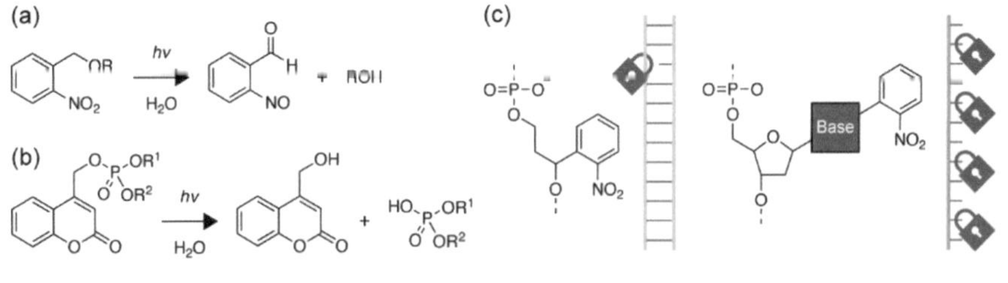

Fig. 3.1 Typical structures and photocleavage reactions of photoresponsive DNAs. **a** *ortho*-nitrobenzyl groups. **b** coumarin groups. **c** DNA photo-caged by *ortho*-nitrobenzyl groups

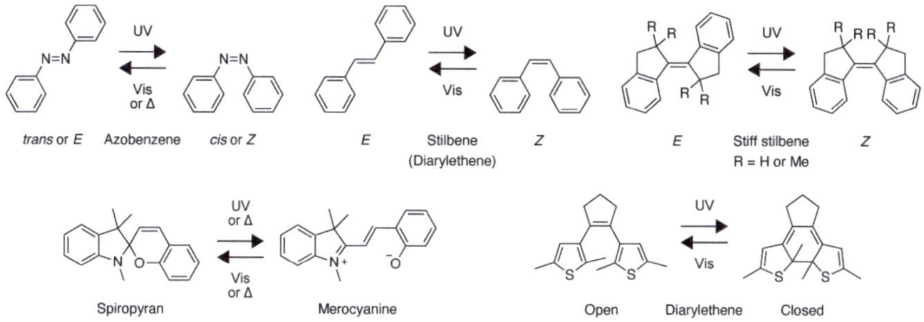

Fig. 3.2 Chemical structures and isomerization of photochromic molecules. Incorporation of these molecules into DNA can establish reversely photoresponsive DNAs

achieved almost complete photoregulation between duplex formation and dissociation at 37 °C [7]. The photoregulatable azo-DNAs have successfully been applied to various nanomaterials and biological tools with photo-controllability [8].

As an example, Asanuma's group has reported photoswitching of RNA digestion by azobenzene-modified DNAzyme (Fig. 4a) [9]. Both termini of the recognition arms of 10–23 DNAzyme were designed as overhang composed of azo-DNA complementary each other. In *trans*-state, the overhang azo-DNA forms duplex, which resulted in stem-loop hairpin structure that suppresses the RNA cleavage activity due to topological distortion of catalytic loop structure (OFF). Upon UV irradiation, the stem duplex of azo-DNA is dissociated, and reactivates the RNA cleavage by taking a correct conformation of the catalytic loop (ON). The real-time photoswitching of RNA digestion was demonstrated in the presence of excess amount of substrate RNA under multiple-turnover conditions with alternative irradiation with UV and visible light. The UV light clearly turned on the RNA digestion and visible light almost completely turned it off (Fig. 4b). The newly constructed nanomachine realized clear-cut and reversible photoregulation of the RNA digestion, which promises photoswitching the mechanical movement of a molecular robot.

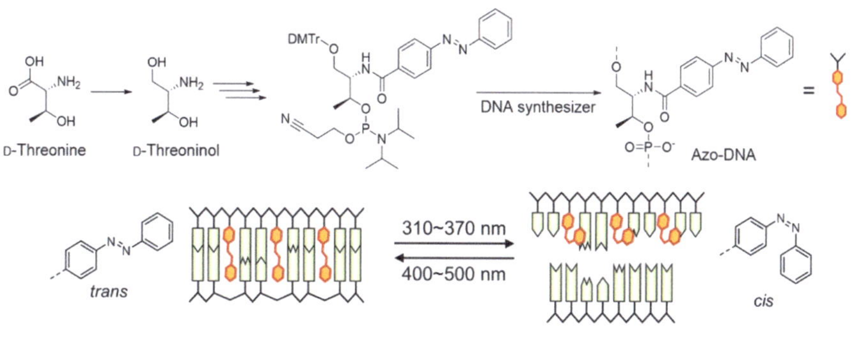

Fig. 3.3 Design of Azo-DNA. Introduction of azobenzene into DNA via D-threoninol linker enables photoregulation of duplex formation and dissociation by the irradiation with light

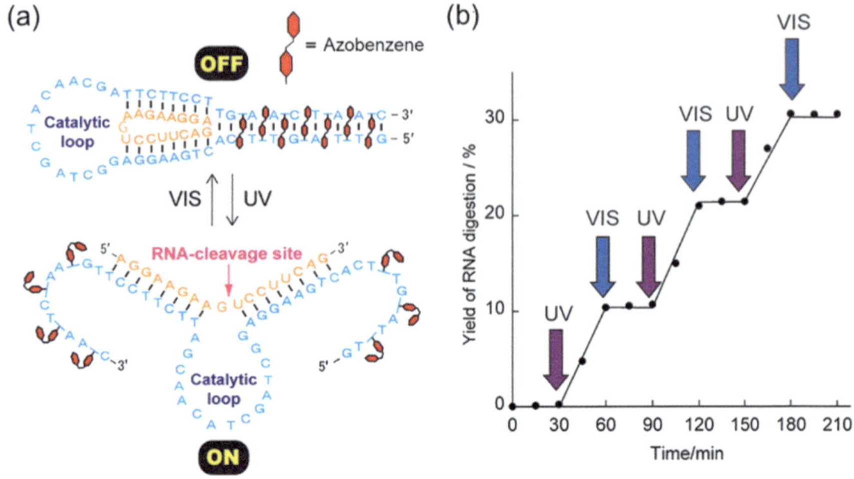

Fig. 3.4 Photoresponsive DNAzyme composed of Azo-DNA. **a** Design and **b** photoregulation of RNA-clevage activity of photoresponsive DNAzyme. Adapted with permission from [8] with partial modifications. Copyright 2014 American Chemical Society

To expand the applications of the photoregulation system, chemical modification on azobenzene can change the responsive wavelength. S-DM-Azo contains alkylthio modification on para-position of azobenzene, with the 2,6-dimethyl substitution to suppress thermal back reaction in *cis* state [10]. Upon the irradiation with 350–400 nm light, S-DM-Azo isomerizes to generate *cis*-rich state, dissociating the duplex, whereas 430–550 nm or 340 nm light induces *trans*-rich state with reformation of the duplex. Normal azobenzene and S-DM-Azo can be orthogonally isomerized via corresponding wavelength of light (Fig. 3.5).

The orthogonal photoregulation system enables complex manipulation of the DNA-based nanomachine. Famulok and Asanuma et al. established the orthogonally photocontrollable DNA walker [11] using DM-Azo [12] and S-DM-Azo. The walker was designed as a dimer of two legs (Fig. 3.6). The path is assembled from the scaffold strand and four foothold strands: F1, F2, F3, and F4. At the initial state, the walker binds to F1 and F2 (state 1). The irradiation with 385 nm isomerizes S-DM-Azo to *cis*-form, resulting in hybridization between the walker and F3 after dissociation from F1 (state 2). Then, 365 nm irradiation causes *cis*-isomerization of DM-Azo, enables move of the walker leg to F4 from F2 (state 3). After that, the irradiation with 385 nm regenerates state 2 via toehold-mediated strand displacement. Finally, the initial state 1 was regenerated through the irradiation with 450 nm light. Thus, the movement of the walker can be operated only by the irradiation of corresponding wavelength of light [11].

One of the most commonly used photochromic (photoisomerizable) molecules following azobenzene is spiropyran (Fig. 3.2). Spiropyran is characterized by large polarity changes in the photoisomerization with a planar merocyanine [13, 14] and has been predominantly utilized as additives to bind with DNA strands through

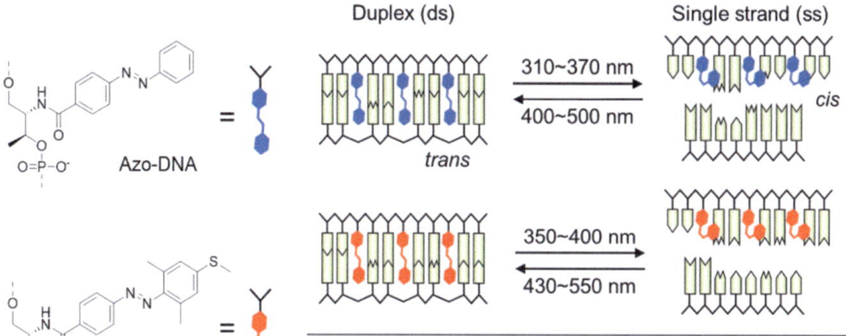

	340 nm	365 nm	385 nm	450 nm
Azo-DNA	ss	ss	ds	ds
S-DM-Azo-DNA	ds	ss	ss	ds

Fig. 3.5 Photoregulation of Azo-DNA and S-DM-Azo-DNA. [10] These two azobenzenes can be isomerized with different wavelengths of light, i.e., orthogonal photo-control is possible

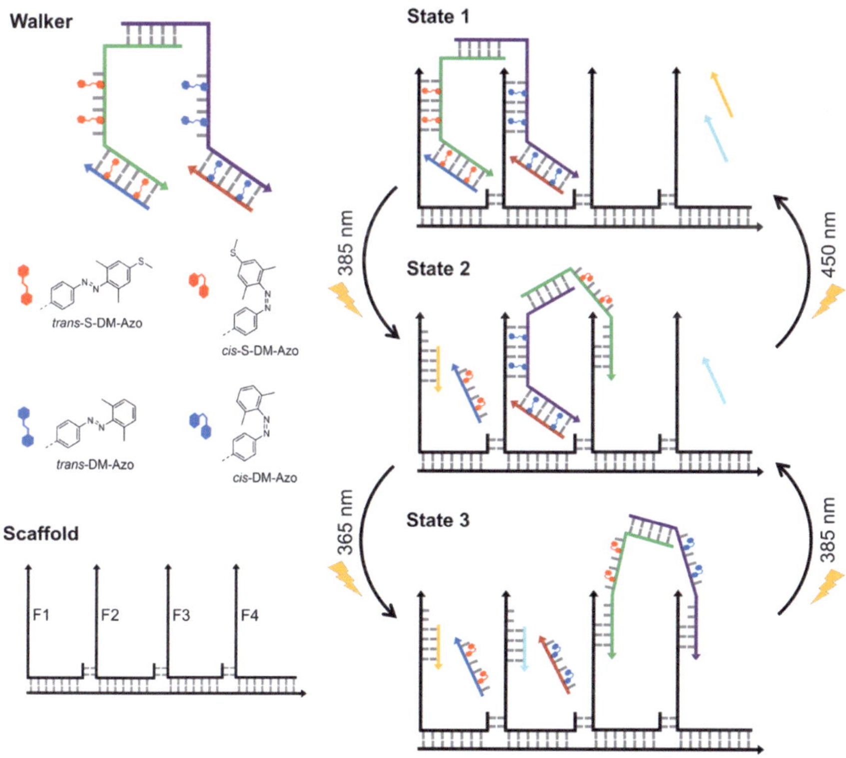

Fig. 3.6 Photoregulatable DNA walker modified with DM-Azo and S-DM-Azo. [11] Non-autonomous move of DNA walker can be achieved through the orthogonal photo-control of two different azo-DNAs

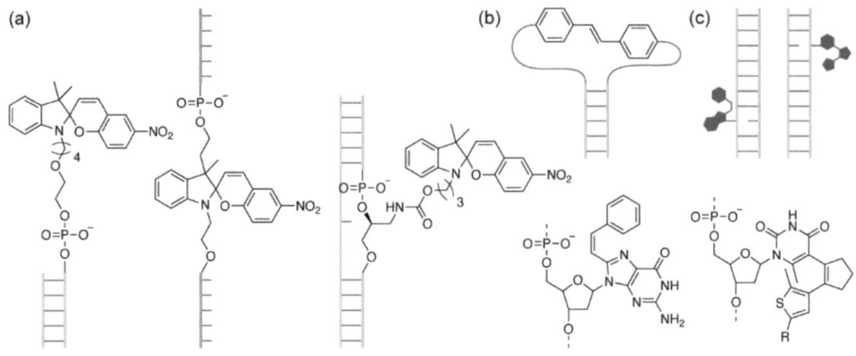

Fig. 3.7 **a** Spiropyran- and **b** stilbene-incorporated DNA. **c** Typical structures of diarylethene-conjugated nucleobases incorporated into DNA

supramolecular interactions or induce proton transfer, controlling structures and functions of DNA. In a few reports, spiropyran has been incorporated into DNA structures as a terminal pendant, a linker, or a part of a building block like a nucleotide (Fig. 3.7a) [15–17]. The incorporation into the terminal was demonstrated to decrease the melting temperature (T_m) for duplex formation from 21.6 °C to 18.0 °C upon photoisomerization from merocyanine to spiropyran by visible light irradiation, i.e., spiropyran can control the duplex stability with light [15]. However, any regulation of DNA structures and functions has never been achieved in the other systems due to inhibited photoisomerization in the strands or hydrolysis of merocyanine [16, 17], which is thermally equilibrated with spiropyran and partially generated in a polar solvent.

A drawback of the above azobenzene- and spiropyran-based systems is rapid thermal isomerization even at room temperature, precluding precise photo-control of DNA. The metastable *cis* isomer of azobenzene thermally isomerizes to the *trans* isomer, and the equilibrium between spiropyran and merocyanine is thermally restored after photoisomerization in both directions (Fig. 3.2). On the other hand, the two types of photoisomerization in diarylethenes, between the open and closed forms and between the *E* and *Z* isomers, can exclude thermal isomerization (Fig. 3.2) [18], and therefore have been also installed on DNA. Initially, stilbene, which undergoes reversible *E*–*Z* photoisomerization, has been employed as a bridge for hairpin conformations of DNA (Fig. 3.7b) [19–21]. The DNA hairpins were destabilized by *E*-to-*Z* photoisomerization of the stilbene linker with T_ms decreased by < 30 °C, although the hairpin conformations and specific neighboring bases completely quenched the *E*-to-*Z* photoisomerization by electron transfer in some cases. Then, diarylethenes have been conjugated with nucleobases (Fig. 3.7c) [22–29]. Compared to the *E*–*Z* photoisomerization [22], the photoinduced ring-opening and closing reactions of the diarylethene-conjugated nucleobases in DNA strands have been well studied [23–29]. Structures of the integrated nucleobases strongly affected the performance such as thermal stability of the metastable closed forms, conversions, quantum yields, reversibility (durability) of the two-way photoisomerization, and fluorescence properties. Based on the investigations, diarylethene-nucleotides offering these properties at a high level in DNA duplexes have been developed. Additionally, the photoresponsive DNA strands have been applied in transcription [24, 27, 28] and all-optical excitonic devices [25, 29]. However, the changes in T_ms were as small as less than 10 °C, and the mechanism for stabilization and destabilization of the DNA strands has not been clearly described.

The *E*–*Z* photoisomerization of stilbene-based molecules generally involves irreversible cyclization in the *Z* isomer to form a six-membered ring. In this context, stiff-stilbene-based photochromic molecules exclude the cyclization and are promising (Fig. 3.2) [30–32] but have been predominantly employed as additives to control DNA structures and functions [33, 34]. Only one paper on DNA with a stiff-stilbene-based motor, which rotates in only one direction through alternating *E*–*Z* photoisomerization and thermal helix inversion, has been reported [35]. The motor was incorporated into a DNA hairpin and the *E* isomer was found to stabilize the hairpin structure with a higher T_m compared to the *Z* isomer, although the thermal helix inversion was inhibited by the hairpin conformation.

3.3 Metal-Responsive Nucleic Acids

Metal ions are also versatile chemical stimuli to construct stimuli-responsive molecular systems. Metal ions have been extensively utilized to operate non-biological molecular switches and machines in supramolecular chemistry research [36], and this principle can be applied in molecular robotics and DNA nanotechnology. A promising approach for creating metal-responsive DNA systems lies in interstrand metal complexation between two opposing nucleobases (or nucleobase analogs) and a bridging metal ion, known as metal-mediated artificial base pairing (Fig. 8a) [37, 38]. The formation of metal-mediated base pairs generally stabilizes DNA duplexes, allowing for metal-dependent DNA hybridization.

Natural pyrimidine nucleobases, thymine (**T**) and cytosine (**C**), form metal-mediated base pairs (**T**–Hg^{2+}–**T** and **C**–Ag^{+}–**C**) in the presence of Hg^{2+} and Ag^{+} ions, respectively [39]. These metallo-base pairs have been used to develop Hg^{2+}- and Ag^{+}-responsive DNA molecular machines. For example, Willner et al. developed Hg^{2+}-responsive DNA tweezers by incorporating **T**–**T** mismatches into the sequence [40]. The designed DNA tweezers are opened and closed by adding and removing Hg^{2+} ions through the formation and dissociation of **T**–Hg^{2+}–**T** base pairs. Hg^{2+}-driven bipedal DNA walkers [41] and DNA catenane rotary motors [42] have also been constructed. Although various DNA systems responsive to Hg^{2+} and Ag^{+} ions have been devised based on similar working mechanisms, designing their base sequences is often troublesome due to the binding of these metal ions to **T** and **C** bases at unintended positions.

Metal-mediated base pairs consisting of artificial ligand-type nucleobase analogs and metal ions other than Hg^{2+} or Ag^{+} have recently been utilized for developing

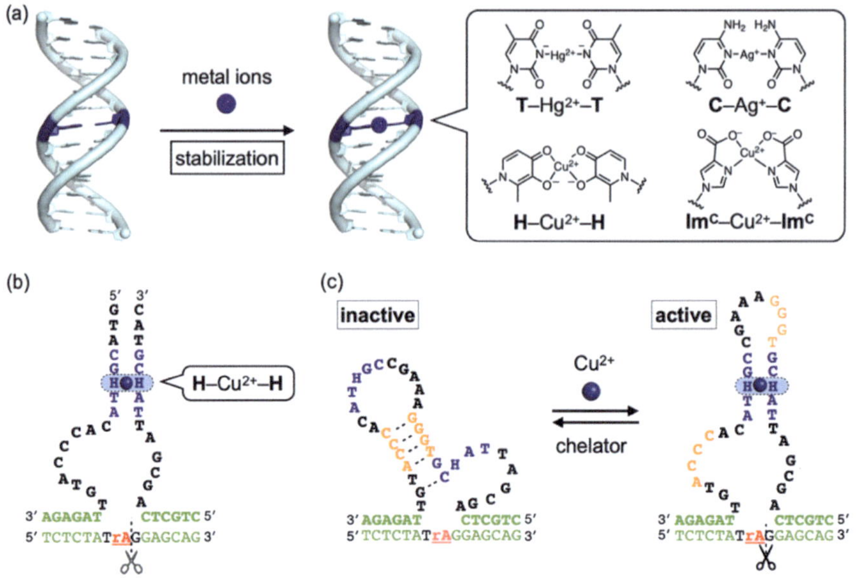

Fig. 3.8 Metal-responsive DNA molecular systems with metal-mediated artificial base pairs. **a** Metal-dependent stabilization of DNA duplexes by metal-mediated artificial base pairing. **b** Base sequence of a Cu^{2+}-responsive split DNAzyme with an **H**–Cu^{2+}–**H** base pair. "**rA**" in the substrate strand represents an adenine ribonucleotide at the cleavage site. **c** Base sequence of a Cu^{2+}-responsive single-stranded DNAzyme with an **H**–Cu^{2+}–**H** base pair. Both catalytically inactive and active secondary structures are shown. Adapted from ref. 45, CC BY 4.0

metal-responsive DNA systems [43, 44]. Over the past two decades, significant efforts have been dedicated to creating synthetic ligand-type nucleobases [37, 38]. Typically, metal ions with square planar (such as Cu^{2+}) or linear coordination geometry (such as Ag^{+}) are employed to ensure an optimal fit with planar natural base pairs. For instance, Tanaka and Shionoya et al. synthesized 3-hydroxy-4-pyridone-type nucleobase (**H**), which forms a stable Cu^{2+}-mediated base pair (**H**–Cu^{2+}–**H**) within DNA duplexes [45]. The addition of an equivalent amount of Cu^{2+} ions increases the thermal stability of a DNA duplex containing an **H**–**H** mismatch pair ($\Delta T_m = +\ 13.1$ °C) due to the formation of an **H**–Cu^{2+}–**H** base pair. Metal-mediated base pairing was initially explored for constructing one-dimensional metal arrays [46, 47], but has recently been applied to developing metal-responsive functional DNAs [43, 44].

Takezawa and Shionoya developed a series of metal-responsive DNAzymes using artificial metal-mediated base pairs. DNAzymes are DNA molecules that catalyze chemical reactions, including RNA cleavage, and are widely used in DNA-based sensors and molecular machines. The metal-responsive DNAzymes were rationally designed to be activated by the formation of metal-mediated base pairs. One of the simplest strategies is to split a known DNAzyme into two fragments and introduce ligand-type nucleobases into the stem duplex [48, 49]. This approach allows metal-mediated base pairing to induce the hybridization of the split strands, thereby reconstructing the original DNAzyme structure. For example, a Cu^{2+}-responsive RNA-cleaving DNAzyme was developed through the incorporation of **H** nucleobases (Fig. 8b) [48]. The activity of this DNAzyme was suppressed in the absence of Cu^{2+} ions but enhanced upon addition of one equivalent of Cu^{2+}, showing a 5.5-fold increase in the reaction rate. A more efficient activation (12-fold) was achieved with a split DNAzyme containing 4-carboxyimidazole-type nucleobases (**Im**$^{\mathbf{C}}$), which form stable **Im**$^{\mathbf{C}}$–Cu^{2+}–**Im**$^{\mathbf{C}}$ base pairs [49]. The catalytic activity of these DNAzymes was reversibly regulated by the addition, removal, and reduction of Cu^{2+}. Additionally, metal-dependent orthogonal activation of two different DNAzymes was demonstrated using both **H**–Cu^{2+}–**H** and **T**–Hg^{2+}–**T** base pairs [48].

Metal-responsive allosteric DNAzymes can be designed without splitting the original DNAzymes [50]. This approach involves an intrastrand structural transformation induced by metal-mediated base pairing. An **H**–**H** mismatch pair was incorporated into the stem region, and a sequence complementary to the catalytic domain was inserted into the loop to form an inactive structure (Fig. 8c). The addition of Cu^{2+} ions induced the catalytically active form with an **H**–Cu^{2+}–**H** base pair, resulting in a 6.8-fold increase in the DNAzyme activity [50]. Notably, one of the smallest DNAzymes, 8–17 DNAzyme, was also successfully reshaped into a Cu^{2+}-responsive allosteric DNAzyme using an **Im**$^{\mathbf{C}}$–Cu^{2+}–**Im**$^{\mathbf{C}}$ base pair [51]. This strategy was further applied to develop a DNAzyme that exhibits an AND logic-gate response to both Cu^{2+} and Ag^{+} ions [50].

Bifacial nucleobases capable of forming both hydrogen-bonded and metal-mediated base pairs have emerged as a new class of metal-responsive units to control DNA structures and functions. For example, 5-hydroxyuracil (**U**$^{\mathbf{OH}}$) [52] and 5-carboxyuracil (**caU**) [53] can form both metal-mediated base pairs (**U**$^{\mathbf{OH}}$–Gd^{3+}–**U**$^{\mathbf{OH}}$ and **caU**–Cu^{2+}–**caU**) and Watson–Crick-like hydrogen-bonded pairs (**U**$^{\mathbf{OH}}$–A and **caU**–A) (Fig. 9a). The addition of specific metal ions can switch the pairing partners of these bifacial bases, thereby imparting metal responsiveness to DNA structures. Takezawa and Shionoya et al. applied bifacial **U**$^{\mathbf{OH}}$ bases to induce DNA strand displacement reactions (SDRs) using Gd^{3+} ions as input signals (Fig. 9b) [54]. SDRs involve the displacement of one strand in a pre-hybridized duplex with an input single-stranded DNA, being recognized as one of the fundamental principles in DNA-based molecular robotics. Initially, a DNA strand containing four **U**$^{\mathbf{OH}}$ bases (**3**) hybridized with its complementary strand (**2**) via **U**$^{\mathbf{OH}}$–A pairing. The addition of Gd^{3+} ions (4 equiv.) caused the invader strand containing **U**$^{\mathbf{OH}}$ bases (**1**) to displace the incumbent strand (**2**) through the formation of **U**$^{\mathbf{OH}}$–Gd^{3+}–**U**$^{\mathbf{OH}}$ base pairs.

The bifacial **U**$^{\mathbf{OH}}$ base was further applied to the development of metal-responsive DNA tweezers as a simple model for DNA molecular devices (Fig. 9c) [54]. Four **U**$^{\mathbf{OH}}$ bases were introduced at both ends of the stopper strand **d**, which closes the tweezers via hydrogen-bonded **U**$^{\mathbf{OH}}$–A base pairing. The addition of Gd^{3+} ions released strand **d** by forming **U**$^{\mathbf{OH}}$–Gd^{3+}–**U**$^{\mathbf{OH}}$ base pairs, causing the DNA tweezers to open. The tweezers were closed again by adding EGTA, a chelating agent that removes Gd^{3+} ions. The opening and closing of the tweezers were repeated by adding Gd^{3+} ions and EGTA alternately. Additionally, a Gd^{3+}-responsive allosteric DNAzyme was developed by using **U**$^{\mathbf{OH}}$ nucleobases [54]. The **U**$^{\mathbf{OH}}$-modified DNAzyme was designed

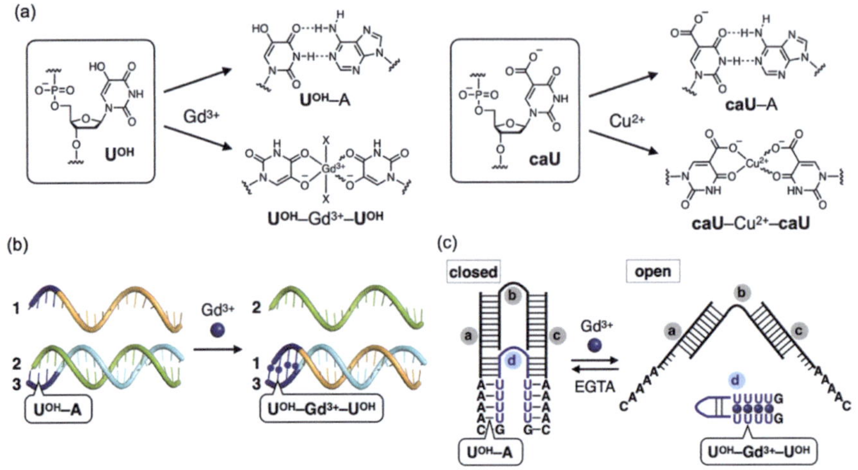

Fig. 3.9 Hydrogen-bonded and metal-mediated base pairing of bifacial nucleobases and their applications. **a** 5-Hydroxyuracil (**U^OH^**) and 5-carboxyuracil (**caU**) nucleobases. "**X**" represents additional coordinating ligands such as water molecules and neighboring nucleobases. **b** Gd^{3+}-mediated DNA strand displacement reactions (SDRs) using DNA strands containing **U^OH^** bases. **U^OH^** nucleotides in the DNA strands are highlighted. **c** Schematic representation of Gd^{3+}-responsive DNA tweezers. **U** represents **U^OH^** nucleotides. Adapted from ref. 55, CC BY 4.0

so that Gd^{3+} addition transforms it from a catalytically inactive structure containing **U^OH^**–A base pairs into an active structure containing **U^OH^**–Gd^{3+}–**U^OH^** base pairs. Similarly, a Cu^{2+}-responsive DNAzyme was strategically developed by incorporating **caU** bases, which form both **caU**–A and **caU**–Cu^{2+}–**caU** base pairs [55].

The molecular design principle described above can be applied to create various types of metal-responsive functional DNAs and DNA nanoarchitectures. In fact, Müller et al. reported metal-responsive DNA aptamers whose binding affinity to their targets can be regulated through metal-mediated base pairing [56]. Ligand-type nucleobases can be introduced into DNA oligomers using conventional automated DNA synthesizers or by enzymatic reactions. Consequently, metal-responsive DNA systems with metal-mediated base pairs will undoubtedly expand the toolbox of DNA nanotechnology and molecular robotics.

References

1. C Brieke 2012 Light-controlled tools Angew. Chem. Int. Ed. 51 8446 8476
2. P Klán 2013 Photoremovable protecting groups in chemistry and biology: reaction mechanisms and efficacy Chem. Rev. 113 119 191
3. Q Liu 2014 Optochemical control of Deoxyoligonucleotide function via a nucleobase-caging approach Acc. Chem. Res. 47 45 55
4. MP O'Hagan 2023 Photocleavable Ortho-Nitrobenzyl-protected DNA architectures and their applications Chem. Rev. 123 6839 6887

5. Ordoukhanian, P., et al.: Design and synthesis of a versatile photocleavable DNA building block. Application to phototriggered hybridization. J. Am. Chem. Soc. 117, 9570–9571 (1995)
6. R Weinstain 2020 Visible-to-NIR-Light activated release: from small molecules to nanomaterials Chem. Rev. 120 13135 13272
7. H Asanuma 2007 Synthesis of Azobenzene-Tethered DNA for reversible photo-regulation of DNA Nat. Protoc. 2 203 212
8. Y Kamiya 2014 Light-driven DNA nanomachine with a photoresponsive molecular engine Acc. Chem. Res. 47 1663 1672
9. MG Zhou 2010 A light-driven DNA nanomachine for the efficiently photoswitching RNA digestion Angew. Chem. Int. Ed. 49 2167 2170
10. H Nishioka 2012 A Photon-fueled DNA nanodevice that contains two different photoswitches Angew Chem. Int. Ed. 51 1165 1168
11. M Skugor 2019 Orthogonally photocontrolled non-autonomous DNA walker Angew. Chem. Int. Ed. 58 6948 6951
12. H Nishioka 2007 2',6'-Dimethylazobenzene as an efficient and thermo-stable photo-regulator for the photoregulation of DNA hybridization Chem. Commun. 7 4354 4356
13. R Klajn 2014 Spiropyran-based dynamic materials Chem. Soc. Rev. 43 148 184
14. L Kortekaas 2019 The evolution of spiropyran: fundamentals and progress of an extraordinarily versatile photochrome Chem. Soc. Rev. 48 3406 3424
15. H Asanuma 2001 Spiropyran as a regulator of DNA hybridization with reversed switching mode to that of Azobenzene Chem. Lett. 30 108 109
16. C Beyer 2010 Synthesis of DNA with spirobenzopyran as an internal covalent modification Synlett 2010 1371 1376
17. C Brieke 2013 Spiropyran photoswitches in the context of DNA: synthesis and photochromic properties Chem. Eur. J. 19 15726 15734
18. M Irie 2014 Photochromism of Diarylethene molecules and crystals: memories, switches, and actuators Chem. Rev. 114 12174 12277
19. RL Letsinger 1995 Use of a Stilbenedicarboxamide bridge in stabilizing, monitoring, and photochemically altering folded conformations of Oligonucleotides J. Am. Chem. Soc. 117 7323 7328
20. FD Lewis 1999 Phototriggered DNA hairpin formation in a Stilbenediether-Linked Bis(oligonucleotide) conjugate J. Am. Chem. Soc. 121 11928 11929
21. FD Lewis 2002 Synthesis, structure, and photochemistry of exceptionally stable synthetic DNA hairpins with stilbene diether linkers J. Am. Chem. Soc. 124 12165 12173
22. S Ogasawara 2008 Straightforward and reversible photoregulation of hybridization by using a photochromic nucleoside Angew. Chem. Int. Ed. 47 8839 8842
23. S Barrois 2012 Covalent modification of 2′-deoxyuridine with two different molecular switches Synlett 23 711 716
24. H Cahová 2013 Nucleoside-based diarylethene photoswitches and their facile incorporation into photoswitchable DNA Angew. Chem. Int. Ed. 52 3186 3190
25. DL Kellis 2019 An all-optical excitonic switch operated in the liquid and solid phases ACS Nano 13 2986 2994
26. T Kolmar 2021 Development of red-shifted and fluorogenic nucleoside and oligonucleotide diarylethene photoswitches Chem. Eur. J. 27 17386 17394
27. SM Büllmann 2021 Optochemical control of transcription by the use of 7-deaza-adenosine-based diarylethenes Chem. Commun. 57 6596 6599
28. T Kolmar 2021 Development of high-performance pyrimidine nucleoside and oligonucleotide diarylethene photoswitches Angew. Chem. Int. Ed. 60 8164 8173
29. SM Büllmann 2022 A DNA-based two-component excitonic switch utilizing high-performance diarylethenes Angew. Chem. Int. Ed. 61 e202117735
30. D Villarón 2020 Stiff-Stilbene photoswitches: from fundamental studies to emergent applications Angew. Chem. Int. Ed. 59 13192 13202
31. K Imato 2022 Sterically hindered stiff-stilbene photoswitch offers large motions, 90% two-way photoisomerization, and high thermal stability J. Org. Chem. 87 15762 15770

32. K Imato 2023 Thermally stable photomechanical molecular hinge: sterically hindered stiff-stilbene photoswitch mechanically isomerizes JACS Au. 3 2458 2466
33. MP O'Hagan 2019 A photoresponsive stiff-stilbene ligand fuels the reversible unfolding of G-Quadruplex DNA Angew. Chem. Int. Ed. 58 4334 4338
34. MP O'Hagan 2020 Enhanced sampling molecular dynamics simulations correctly predict the diverse activities of a series of stiff-stilbene G-quadruplex DNA ligands Chem. Sci. 12 1415 1426
35. AS Lubbe 2018 Photoswitching of DNA hybridization using a molecular motor J. Am. Chem. Soc. 140 5069 5076
36. B Champin 2007 Transition metal complexes as molecular machine prototypes Chem. Soc. Rev. 36 358 366
37. Y Takezawa 2017 Artificial DNA base pairing mediated by diverse metal ions Chem. Lett. 46 622 633
38. Y Takezawa 2017 Self-assemblies based on metal-mediated Artificial Nucleobase Pairing JL Atwood Eds Comprehensive Supramolecular Chemistry II 4 Elsevier Oxford 259 293
39. Y Tanaka 2015 Structures, physicochemical properties, and applications of T-Hg^{II}-T, C-Ag^{I}-C, and other metallo-base-pairs Chem. Commun. 51 17343 17360
40. ZG Wang 2010 All-DNA finite-state automata with finite memory Proc. Natl. Acad. Sci. U.S.A. 107 21996 22001
41. ZG Wang 2011 DNA machines: bipedal walker and stepper Nano Lett. 11 304 309
42. C-H Lu 2013 A three-station DNA catenane rotary motor with controlled directionality Nano Lett. 13 2303 2308
43. Y Takezawa 2023 Recent advances in the development of metal-responsive functional DNAs based on metal-mediated Artificial Base Pairing J Müller B Lippert Eds Modern Avenues in Metal-Nucleic Acid Chemistry (Metal Ions in Life Science 25 CRC Boca Raton 257 289
44. Y Takezawa 2024 Rational design of metal-responsive functional DNA supramolecules. J. Incl. Phenom. Macrocycl. Chem. 104 349 369.
45. K Tanaka 2002 Efficient incorporation of a copper hydroxypyridone base pair in DNA J. Am. Chem. Soc. 124 12494 12498
46. K Tanaka 2003 A discrete self-assembled metal array in artificial DNA Science 299 1212 1213
47. K Tanaka 2006 Programmable self-assembly of metal ions inside artificial DNA duplexes Nat. Nanotechnol. 1 190 194
48. Y Takezawa 2019 Enzymatic synthesis of Cu(II)-responsive deoxyribozymes through polymerase incorporation of artificial ligand-type nucleotides J. Am. Chem. Soc. 141 19342 19350
49. Y Takezawa 2020 Sharp switching of DNAzyme activity through the formation of a Cu^{II}-mediated carboxyimidazole base pair Angew. Chem. Int. Ed. 59 21488 21492
50. T Nakama 2020 Allosteric regulation of DNAzyme activities through intrastrand transformation induced by Cu(II)-mediated artificial base pairing J. Am. Chem. Soc. 142 10153 10162
51. Y Takezawa 2024 Metal-dependent activity control of a compact-sized 8–17 DNAzyme based on metal-mediated unnatural base pairing Chem. Commun. 60 288
52. Y Takezawa 2015 Bifacial base-pairing behaviors of 5-hydroxyuracil DNA bases through hydrogen bonding and metal coordination Chem. Eur. J. 21 14713 14716
53. Y Takezawa 2020 Metal-dependent DNA base pairing of 5-carboxyuracil with itself and all four canonical nucleobases J. Am. Chem. Soc. 142 21640 21644
54. Y Takezawa 2023 Metal-mediated DNA strand displacement and molecular device operations based on base-pair switching of 5-hydroxyuracil nucleobases Nat. Commun. 14 4759
55. Y Takezawa 2024 Ligase-mediated synthesis of Cu^{II}-responsive allosteric DNAzyme with bifacial 5-carboxyuracil nucleobases Chem. Sci. 15 2365 2370
56. Heddinga 2022 Modulating aptamer function by copper(II)-mediated base pair formation Org. Biomol. Chem. 20 4787 4793
57. AS Lubbe 2017 Recent developments in reversible photoregulation of oligonucleotide structure and function Chem. Soc. Rev. 46 1052 1079

Chapter 4
Boosters for DNA Reactions

Naohiko Shimada and Atsushi Maruyama

Toehold-mediated strand displacement (TMSD) reaction is a process by which single-stranded invader DNA binds to single-strand region, called toehold, of partially double-stranded DNA, causing strand exchange. DNAzymes are enzymes consisting of DNA, and some of them cleave nucleic acids as a substrate. These reactions are fundamental to operation of DNA nanodevices such as DNA robots or DNA computers. Boosting of the reactions can lead to speed up of the operation of the nanodevices. In this chapter, we introduce two types of polymers, cationic comb-type copolymers and ureido polymers, for boosting the DNA reactions.

4.1 Structure and Property of Cationic Comb-Type Copolymers

We have reported that cationic comb-type copolymers composed of cationic main chain such as polylysine and abundant hydrophilic side chains such as dextran or polyethylenglycol (> 90 wt%) form a water-soluble complex with anionic DNA. DNA hybrids including duplex [1, 2], triplex [3] and quadruplex [4] complexed with the copolymers are thermodynamically stabilized. The copolymers also increase DNA hybridization rate by two orders [5]. The properties allow us to enhance DNA strand displacement reaction [6, 7], which is exchange reaction between duplex and its complementary single-stranded DNA, resulting in boosting DNAzyme reaction under multiple turnover conditions [8, 9]. Thus, cationic comb-type copolymers

N. Shimada (✉) · A. Maruyama
Institute of Science Tokyo, Yokohama, Japan
e-mail: nshimada@life.isct.ac.jp

A. Maruyama
e-mail: maruyama@life.isct.ac.jp

S. Murata et al. (eds.), *Molecular Robotics II*,
https://doi.org/10.1007/978-981-95-2394-8_4

act as a booster to enhance the DNA reactions. These were summarized in *Molecular Robotics: An Introduction* at Chapter 6.6 [10]. Topics other than those already described are presented in this chapter.

4.2 Cationic Comb-Type Copolymers Boost DNA Reactions Based on Toehold-Mediated Strand Displacement

As we previously introduced in *Molecular Robotics: An Introduction* [10], a cationic comb-type copolymer, poly(L-lysine)-*graft*-dextran (PLL-*g*-Dex, Fig. 4.1a), boosts TMSD reaction, resulting in significant improvement of operation time of DNA logic gates from the order of hours to minutes [11]. In this section, the effect of the copolymers on other TMSD-based reactions is introduced. DNA hybridization chain reaction (HCR), which is also based on TMSD reaction, has been utilized for nonenzymatic isothermal signal amplification [12]. HCR is started by addition of an initiator DNA which triggers the alternative opening of two hairpin DNAs through TMSD to generate a longer double-stranded (ds) DNA (Fig. 4.1b). We evaluated effect of the copolymer on the HCR [13]. In the absence of the copolymer, a large fraction of unreacted hairpins remained (Fig. 4.2a). In contrast, only a small fraction of unreacted hairpins was observed in the presence of PLL-*g*-Dex. As shown in Fig. 4.2b, the addition of the copolymer resulted in 40-fold faster reaction. These results indicated that HCR was enhanced by cationic comb-type copolymers.

DNA catalytic hairpin assembly (CHA) [14] based on TMSD reaction is also an enzyme-free amplification circuit. Two hairpins are alternatively opened by the addition of initiator to from ds DNA (Fig. 4.3b). We prepared three types of dextran-grafted copolymers with different backbones (Fig. 4.3a) to evaluate their effect on CHA. All copolymers accelerated the CHA reaction (Fig. 4.3c) [15]. Interestingly,

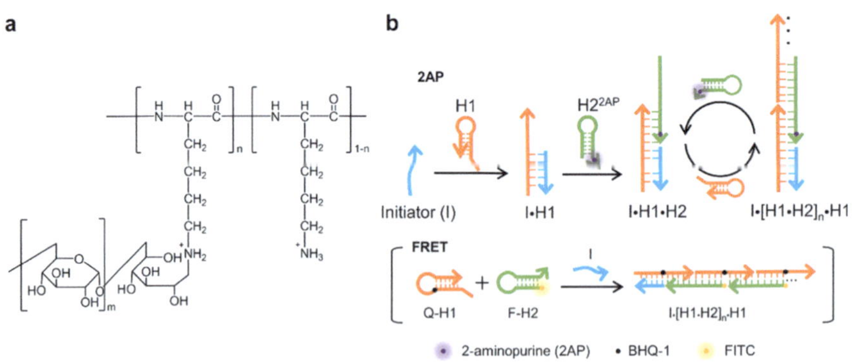

Fig. 4.1 **a** Chemical formula of PLL-*g*-Dex. **b** Schematic of HCR. Arrowheads indicate the 3′ ends of oligonucleotides. Reproduced with permission from Ref. [13]. Copyright 2022 American Chemical Society

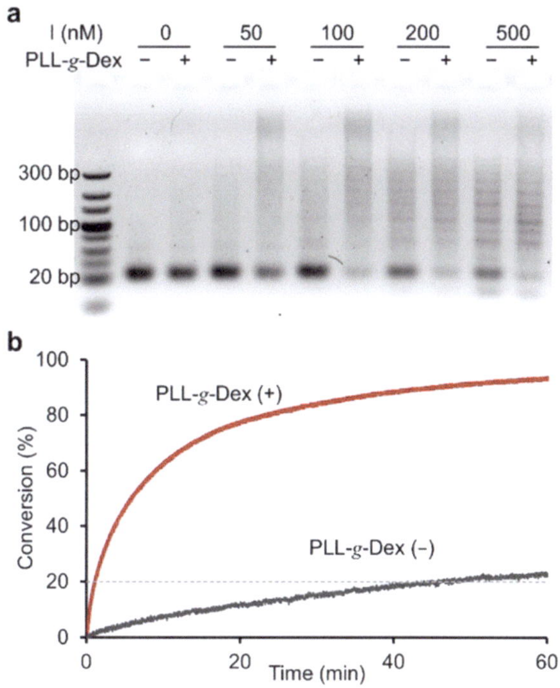

Fig. 4.2 PLL-*g*-Dex accelerates HCR. **a** Gel electrophoresis of HCR in the absence (−) and presence (+) of PLL-*g*-Dex. **b** Percent conversion of hairpins with (red line) or without (black line) PLL-*g*-Dex. Reproduced with permission from Ref. [13]. Copyright 2022 American Chemical Society

a copolymer, PLD composed of polylysine, which had most hydrophilic backbone among polymers tested, showed most effective acceleration. In addition, the PLD markedly improved signal and background ratio (Fig. 4.3d). Thus, cationic comb-type copolymers boosted TMSD-based signal amplification circuit such as HCR and CHA as well as TMSD-based DNA logic gates.

4.3 Structure and Solution Property of Ureido Polymers

Poly(allylamine-*co*-allylurea) (PAU, Fig. 4.4), a kind of ureido polymers, is easily prepared by the addition of potassium cyanate to poly(allylamine). Interestingly, PAU dissolved in the aqueous solution precipitates upon cooling. The precipitate dissolves again upon heating above a certain temperature called as an upper critical solution temperature (UCST). Thus, ureido polymers show UCST-type solution behavior which is a rare phenomenon for water-soluble polymers [16]. Moreover, the precipitate of PAU is liquid droplets containing abundant water molecules resulting from liquid–liquid phase separation. The formation of liquid droplet of ureido polymer is called simple coacervation, which is formed by one type of polymer. The simple coacervates of ureido polymers can be used in media containing many biomolecules. For example, the cooling-inducible simple coacervates of PAU change

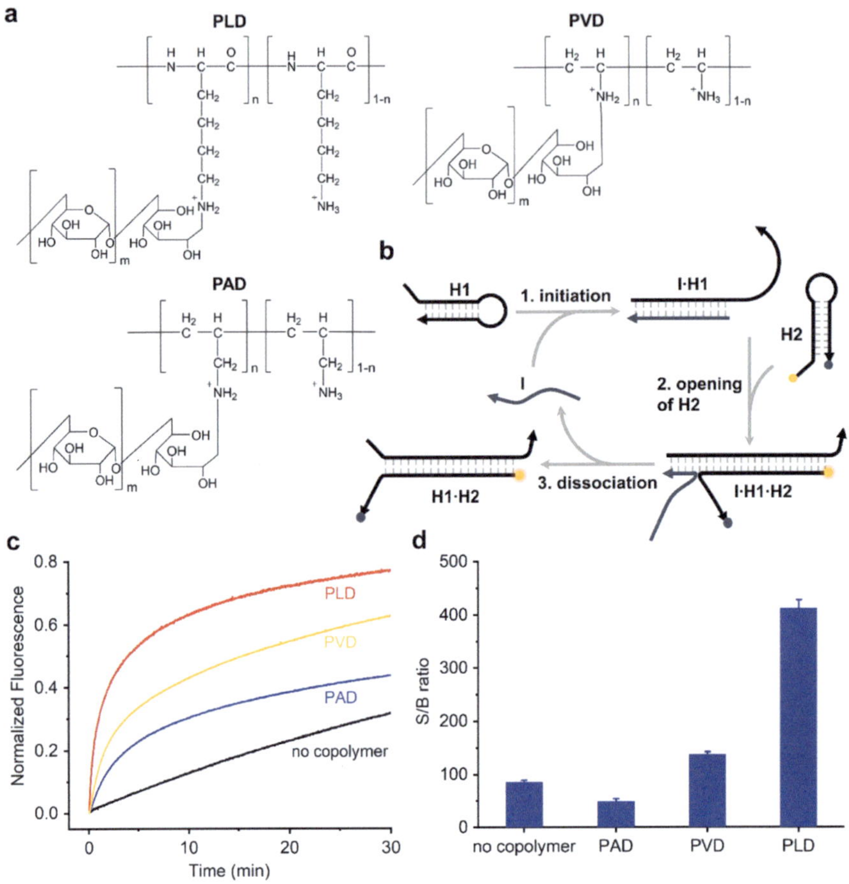

Fig. 4.3 Cationic copolymers accelerate operation of the CHA circuit. **a** Chemical structures of cationic copolymers used in this study. **b** Scheme of CHA circuit. **c** Effects of different copolymers on CHA circuit. **d** Signal-to-background (S/B) ratio of CHA circuit without or with copolymers. Reproduced with permission from Ref. [15]. Copyright 2023 Wiley–VCH GmbH

cell morphology from elongated to rounded shape in cell culture media containing serum biomolecules [17]. Thus, ureido polymer showing simple coacervation could be used as temperature response liquid biomaterials.

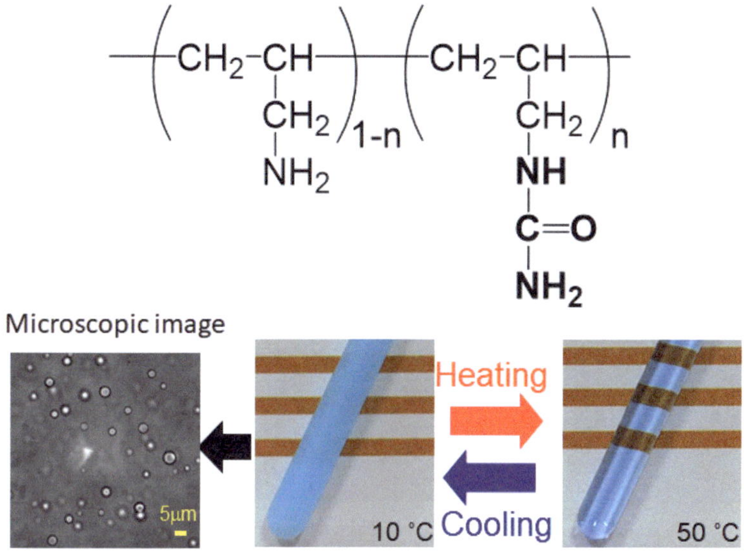

Fig. 4.4 Chemical formula of PAU (upper) and UCST-type phase change of PAU (lower)

4.4 Cooling-Induced Liquid Droplet of Ureido Polymers Boosts DNA Enzyme Reaction

Some proteins are encapsulated into the droplets of ureido polymers [18]. β-galactosidase (β-gal), a protein enzyme, was encapsulated into the droplets induced by cooling [19]. The activity of β-gal in the droplet decreased by 3.7-fold than that in the absence of the polymer. When increasing in temperatures at which the droplets dissolved, the activity was recovered to the same level as that in the absence of the polymer. The results indicated that environment in the liquid droplet of the ureido polymer changed activities of protein enzymes. DNA enzymes could have changed their activity in the droplets as well as protein enzymes. We evaluated effect of the droplets of ureido polymers on activity of DNAzymes (Fig. 4.5) [20]. Activity of DNAzyme was significantly enhanced by encapsulation into droplets of two PAUs (PAU93 and PAU95 which have 93 and 95 mol% ureido groups and 7 and 5 mol% amino groups, respectively), while other PAUs did not enhance the activity because DNAzyme was not encapsulated into droplets of the PAUs (Fig. 4.4B). For example, at 35 °C, encapsulation in the droplets of PAU93 boosted the DNAzyme reaction by 7.2-fold. At 55 °C, the reaction profiles in the presence of PAUs were very similar with that in the absence of PAU (Fig. 4.5a), indicating that the boosting effect was lost by heating to temperatures at which the droplets dissolved. The observed rate constant for DNAzyme reaction with PAU93 at 35 °C was similar to that for DNAzyme alone at 55 °C which is close to optimal temperature of the DNAzyme. This indicated that the cooling-inducible droplet of ureido polymers increased activity of DNAzyme at near physiological temperature to close to that observed at the optimal

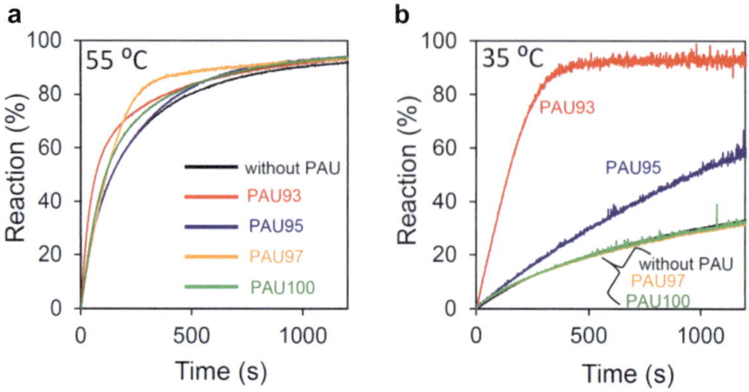

Fig. 4.5 Reaction profiles of 2 nM DNAzyme with 100 nM substrate in the absence or the presence of 1 mg mL^{-1} PAUs at **a** 55 °C and **b** 35 °C. Reproduced with permission from Ref. [20]. Copyright 2023 The Royal Society of Chemistry

temperature. Thus, the cooling-inducible liquid droplets of ureido polymers provide environment to boost the activity of DNAzymes. It should be noted that the temperature stimulus for boosting activity of DNAzymes could be altered to other stimuli including pH change [18] or light irradiation [21] by chemical modification to the ureido polymer. Ureido polymers could be utilized as stimuli-responsive materials for boosting activity of DNAzymes in the field of DNA nanotechnology.

References

1. Sato, Y., et al.: Spectroscopic investigation of cationic comb-type Copolymers/DNA interaction: interpolyelectrolyte complex enhancement synchronized with DNA hybridization. Langmuir **23**, 65–69 (2007)
2. Maruyama, A., et al.: Characterization of interpolyelectrolyte complexes between double-stranded DNA and polylysine comb-type copolymers having hydrophilic side chains. Bioconjug. Chem. **9**, 292–299 (1998)
3. Maruyama, A., et al.: Comb-type polycations effectively stabilize DNA triplex. Bioconjug. Chem. **8**, 3–6 (1997)
4. Moriyama, R., et al.: The role of cationic comb-type copolymers in chaperoning DNA annealing. Biomaterials **32**, 7671–7676 (2011)
5. Wu, L., et al.: Poly(l-lysine)-graft-dextran copolymer accelerates DNA hybridization by two orders. Soft Matter **4**, 744–747 (2008)
6. Kim, W.J., et al.: DNA strand exchange stimulated by spontaneous complex formation with cationic comb-type copolymer. J. Am. Chem. Soc. **124**, 12676–12677 (2002)
7. Kim, W.J., et al.: Cationic comb-type copolymers for DNA analysis. Nat. Mater. **2**, 815–820 (2003)
8. Gao, J., et al.: Enhancement of deoxyribozyme activity by cationic copolymers. Biomater. Sci. **3**, 308–316 (2015)
9. Hanpanich, O., et al.: Cationic copolymer-chaperoned DNAzyme sensor for microRNA detection. Biomaterials **225**, 119535 (2019)

10. Kuzuya, A.: Molecular robotics: an introduction. In: Murata S. (ed.) pp. 215–245. Springer Nature Singapore, Singapore (2022)
11. Shimada, N., et al.: DNA computing boosted by a cationic copolymer. Adv. Func. Mater. **28**, 1707406 (2018)
12. Chai, H., et al.: Recent progress in DNA hybridization chain reaction strategies for amplified biosensing. ACS Appl. Mater. Interfaces **13**, 38931–38946 (2021)
13. Wang, J., et al.: Cationic copolymer-augmented DNA hybridization chain reaction. ACS Appl. Mater. Interfaces **14**, 39396–39403 (2022)
14. Yin, P., et al.: Programming biomolecular self-assembly pathways. Nature **451**, 318–322 (2008)
15. Wang, J., et al.: A cationic copolymer enhances responsiveness and robustness of DNA circuits. Small **19**, 2304091 (2023)
16. Shimada, N., et al.: Ureido-derivatized polymers based on both poly(allylurea) and poly(L-citrulline) exhibit UCST-type phase transition behavior under physiologically relevant conditions. Biomacromolecules **12**, 3418–3422 (2011)
17. Shimada, N., et al.: Reversible monolayer/spheroid cell culture switching by ucst-type thermoresponsive ureido polymers. ACS Appl. Mater. Interfaces **8**, 31524–31529 (2016)
18. Shimada, N., et al.: Design of UCST polymers for chilling capture of proteins. Biomacromolecules **14**, 1452–1457 (2013)
19. Komachi, T., et al.: Evaluation of cooling-induced liquid-liquid phase separation of ureido polymers as a cold-shock stress granules model. Macromol. Biosci. **21**, 2000345 (2021)
20. Komachi, T., et al.: Encapsulation in cooling-induced coacervates augments DNA enzyme activity. Polym. Chem. **14**, 4743–4747 (2023)
21. Ikeuchi, N., et al.: Light-regulated liquid–liquid phase separation for spatiotemporal protein recruitment and cell aggregation. ACS Appl. Mater. Interfaces

Chapter 5
Design and Application of DNA Origami Actuators

Yuki Suzuki

DNA forms canonical double helical structures through complementary base pairing. Owing to this unique property, DNA is widely used as a programmable nanomaterial for constructing artificially designed nanostructures via self-assembly [1]. In a method called "DNA origami," long single-stranded DNA is folded into a designed nanoscale shape through hybridization with many synthetic short DNA strands, called staple strands [2]. The most innovative aspect of the DNA origami method is that it can achieve a wide variety of shapes depending on how the scaffold strands are folded [2–5]. The nanostructures produced using this method are not limited to static nanostructures. The emergence of this technique has accelerated the development of various dynamic DNA nanostructures that imitate basic mechanical motions [6–9].

A popular design approach for such DNA origami nanomachines is based on connecting relatively rigid parts composed of double-helix bundles via flexible ssDNA joints. For example, in the DNA box developed by Andersen et al., a specific edge of the grid is connected to the box via several single-stranded DNA strands, thereby enabling its open/close hinge-like motion [10]. Isomerization of the crossover joint has also been utilized as a mechanism for reconfiguration. The DNA nanopliers developed by Kuzuya et al. comprise two structural elements that are connected by a central crossover [6] (Fig. 5.1a). The antiparallel-to-parallel (and vice versa) isomerization of this crossover enables rotation against each pincer (Fig. 5.1b). Similarly, in the rotary device shown in Fig. 5.1c, the planar base and T-shaped rotor are connected by a scaffold ssDNA strand that crosses over these two parts [11].

Two popular mechanisms control reconfiguration of these dynamic DNA origami nanostructures: (1) strand displacement reaction [12], which is induced by adding external signal DNA strands (often referred to as "fuel" strands), and (2) formation of non-canonical DNA structures [13]. For the strand displacement reaction, the target DNA to be displaced is generally designed to have a protruding single-stranded

Y. Suzuki (✉)
Division of Applied Chemistry, Graduate School of Engineering, Mie University, Tsu, Mie, Japan
e-mail: ysuzuki@chem.mie-u.ac.jp

S. Murata et al. (eds.), *Molecular Robotics II*,
https://doi.org/10.1007/978-981-95-2394-8_5

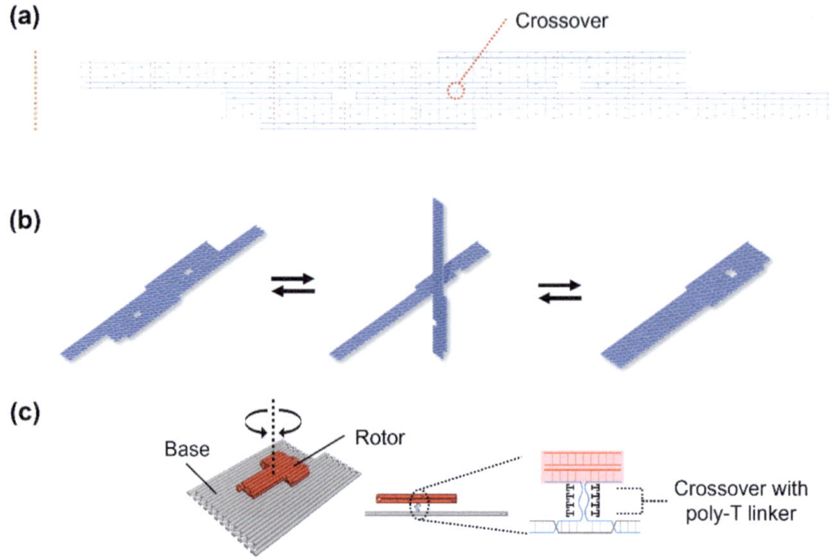

Fig. 5.1 Reconfigurable DNA origami nanostructures **a** CaDNAno diagram of the DNA origami pliers. Upper and bottom parts are connected by the scaffold DNA crossing over the two pliers. **b** Schematics of the reconfiguration of the DNA origami pliers. The antiparallel-to-parallel (and vice versa) isomerization of the crossover enables rotation against each pincer. **c** Design of the rotary DNA origami device. The base sequence of the crossing over connection is designed to have only thymine bases. Reproduced in part from [11] with permission.

region called a "toehold." When a fuel strand fully complementary to the target DNA is added, a strand displacement reaction occurs in the toehold region. The second mechanism is exemplified by the formation of triplexes and tetraplexes. The most common method for realizing pH-induced reconfiguration is through parallel triplex or i-motif formation, whereas G-quadruplex formation is used to implement K^+ and Na^+ responsivity.

5.1 Multi-reconfigurable Actuator

Multi-reconfigurable DNA nanodevices can be realized by implementing multiple connections and mechanisms to operate them in a combinatorial and reversible manner. Figure 5.2 shows the dynamic nanolattice whose overall shape is determined by the combination of orthogonal signal DNAs [14]. Each frame of this nanolattice comprises four rigid four-helix bundles (4HBs) with four flexible ssDNA joints (10 nt) that connect them and carry the latch DNAs, each protruding from each 4HB. Reconfiguration of each frame is achieved by controlling the angle between the 4HBs using a set of external "fuel" DNA strands termed "trigger DNAs." Each trigger DNA can bridge a pair of latch DNAs to suppress the flexible motion of the joints and hold

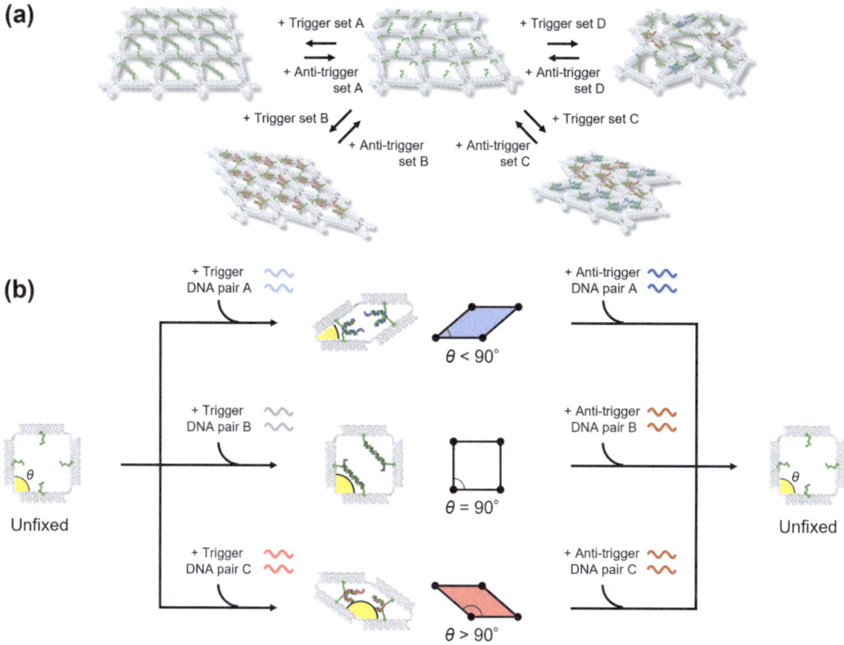

Fig. 5.2 Multi-reconfigurable DNA origami lattice actuator. **a** Schematic representation of the multi-reconfigurable DNA origami nanolattice. **b** Schematic representation of the transformation mechanism of a frame in the lattice. Pairs of trigger DNAs are designed to specifically bind to latch DNAs to suppress the flexible motion of the single-stranded joints and hold the 4-helix bundles at predetermined relative angles. Trigger DNAs are designed to be displaced by the addition of the corresponding set of anti-trigger DNAs. Reproduced in part from [14] with permission

bundles at predetermined angles. Three different bridging types were employed with three orthogonal pairs of trigger DNAs to transform each frame into a right-angled rhombus, left-angled rhombus, or square (Fig. 5.2b). A toehold sequence was added to each trigger DNA to allow its displacement from the frame by adding the corresponding fully complementary ssDNA (anti-trigger DNA). The orthogonal sets of these signals regulated the shape of each of the nine frames, thus enabling the reconfiguration of the nanolattice into various shapes, the interconversions of which were achieved by a toehold-mediated strand displacement reaction.

5.2 Actuators Exhibiting Bending and Coiling Transformation

Design strategies for realizing soft and lifelike motions have been proposed in addition to the conventional approach to nanomechanical DNA origami devices [15–17]. The bending nanoactuator shown in Fig. 5.3 contains repeated functional modules.

The module comprises a stem, piers, and bridge strands that apply tension to induce bending. The tension applied to each module can be tuned by varying the length of the linker inserted in the bridge strand, i.e., the module exhibits different degrees of bending depending on the linker length. A large deformation of the entire shape was induced by the cumulative effect of bending, which could be adjusted by varying the length of the linker. Furthermore, the curved shape can be tuned by adjusting the number and location of bridge strands. The modularity of this design also allowed the linker sequence to be replaced by a stimulus-responsive sequence, thereby leading to reversible actuation via chemical cues. For example, K^+ responsivity can be achieved by introducing G-quadruplex (GQ) DNA strands into deformable modules. In the absence of K^+, these DNA strands relax and remain single stranded. However, in the presence of K^+, they are folded into a compact quadruplex structure, which induces bending deformation by pulling the adjacent piers inward. Through the cumulative effect of this actuation, the entire shape is deformed from its original linear shape into an arched shape in a signal-dependent and reversible manner (Fig. 5.3).

A nanoactuator that undergoes three-dimensional transformation can be constructed [17]. This actuator also consists of K^+-responsive repeated modules but is designed as a six-helix duplex bundle with an intrinsic right-handed twist. Therefore, the cumulative actuation of the module results in the right-handed coiling of the structure (Fig. 5.4). A mechanism to switch the direction of bending deformation of the modules was proposed using structural changes in the local DNA strand between the double-helix and G-quadruplex structures. This was achieved by designing an anti-GQ strand with a toehold. The anti-GQ strand forms duplex DNA with a GQ-forming bridge, which imposes stress and pushes the piers outward. This double helix can be dissociated by the addition of a releaser strand via the toehold-mediated strand displacement reaction. Consequently, this shrinkable DNA actuator has dual responsivity against K^+ and specific DNA strands.

An experimental comparison of the coiled structure formed in the presence of K^+ and that formed after the addition of anti-strands revealed a clear difference in the coil radii, coil pitches, and spring constants, and that the inner and outer sides of the coil were inverted (Fig. 5.4b). This demonstrated that the topology and mechanical properties of the device could be switched depending on the type of external stimulus.

5.3 Applications of DNA Origami Actuators

The highly addressable nature of a DNA origami actuator derived from staple sequences allows the functionalization of its surface with a multitude of molecules and enables the distance between them to change in accordance with the actuation induced by external stimuli. This feature has enabled the manipulation of nanomaterials such as proteins and inorganic nanoparticles to alter their spatial arrangements. Representative studies include switching the relative orientation of gold nanorods [18, 19] and changing the distance between two enzymes [20]. Its application has recently been expanded to regulate cell morphology and motility by controlling

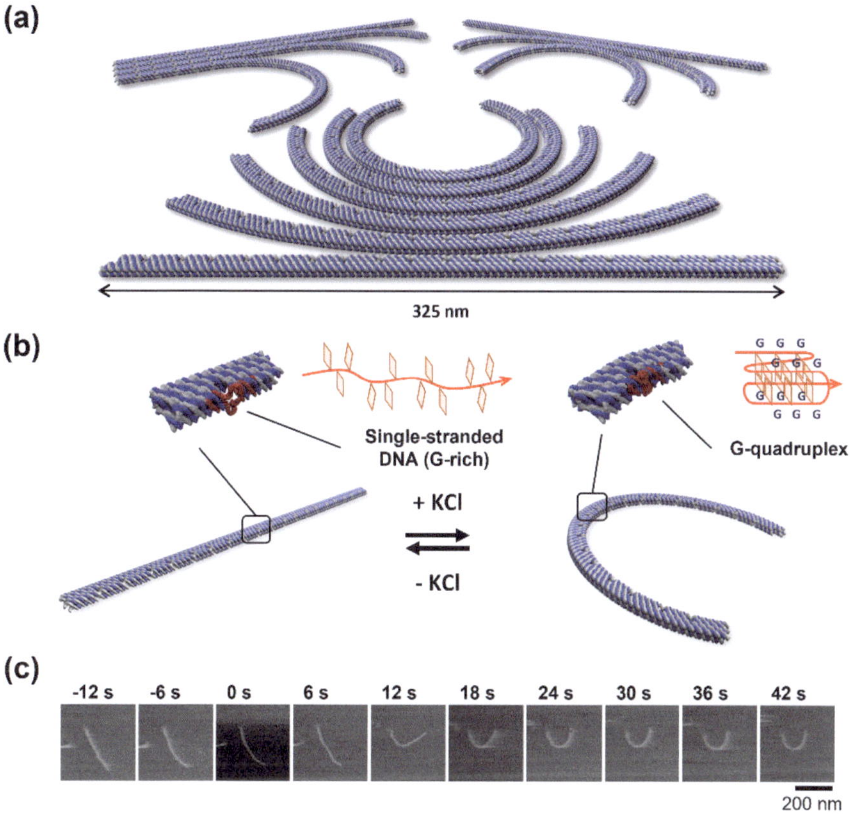

Fig. 5.3 DNA origami nanoactuator exhibiting bending transformation **a** Schematics of DNA origami actuators that bend into various shapes. **b** Example of deformation in response to potassium ions. **c** High-speed atomic force microscope image of the deformation. Potassium chloride solution was injected at time 0 s. Reproduced in part from [15] with permission

the clustering and declustering of membrane proteins such as integrins, which are responsible for the adhesion of cells to the extracellular matrix. The binding of integrin to its ligand containing the arginyl-glycyl-aspartic acid (RGD) motif leads to clustering of integrin molecules, which activate downstream pathways to modulate cell mechanics. Based on this mechanism, DNA origami actuators are functionalized with several RGD peptides, whose spacing is altered in concert with the contraction of the actuator triggered by environmental or external stimuli, including pH changes [16], nucleic acid strands, or light [21]. This leads to an alteration in the clustering state of the integrin molecules at the nanoscale level, thereby enabling the modulation of cell morphology and migration (Fig. 5.5).

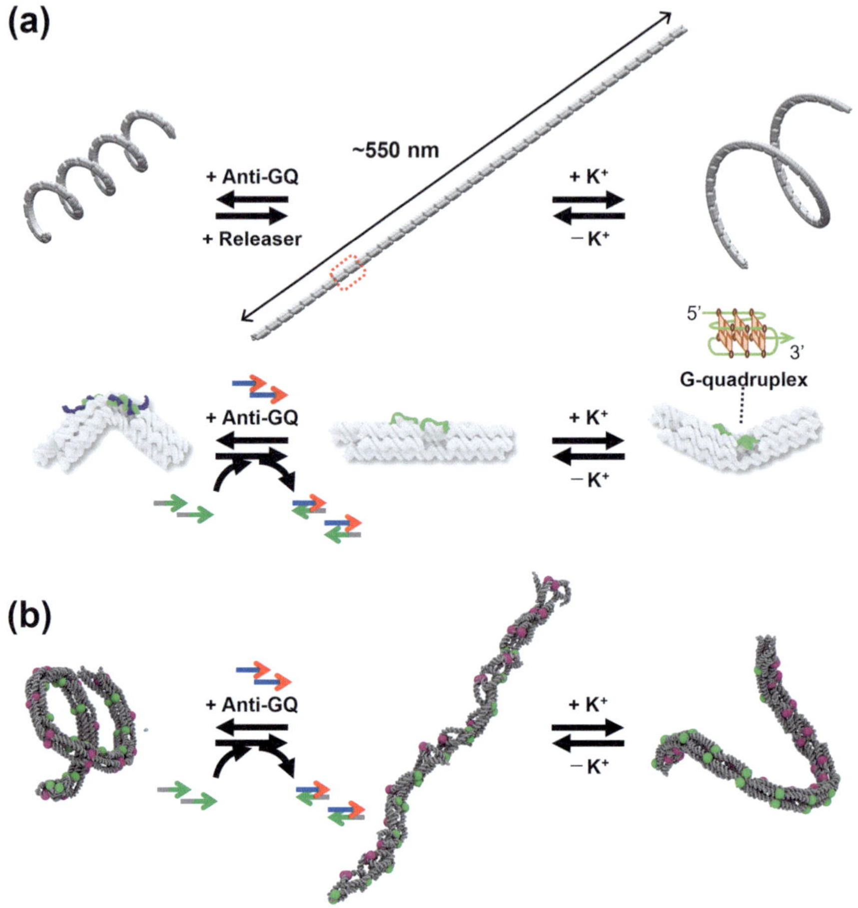

Fig. 5.4 DNA origami nanoactuator exhibiting coiling transformation **a** Model diagram of the reversible transformations of a DNA origami bundle into different spring shapes through the cumulative actuation of transformable modules (red dotted lines). **b** Model diagram of the bending of the module (dashed red box). **c** Molecular dynamics simulation image of the partial structure of the DNA device. Depending on the type of stimulus, coiled springs with specific shapes are formed; the spring shape induced by the K^+ and that induced by anti-strands differs from each other in terms of coil radius and helical pitch. Furthermore, a comparison of the distribution of particles highlighted in green and magenta shows that the inside and outside of the coils are inverted with respect to each other. Reproduced in part from [17] with permission

5.4 Summary and Outlook

In the fields of reconstitutive biology and molecular robotics, motor proteins and their associated filaments, such as kinesins and microtubules, have been extensively utilized as molecular actuators responsible for the deformation of liposome-based artificial cells [22–24]. These are primarily dependent on ATP and GTP; however,

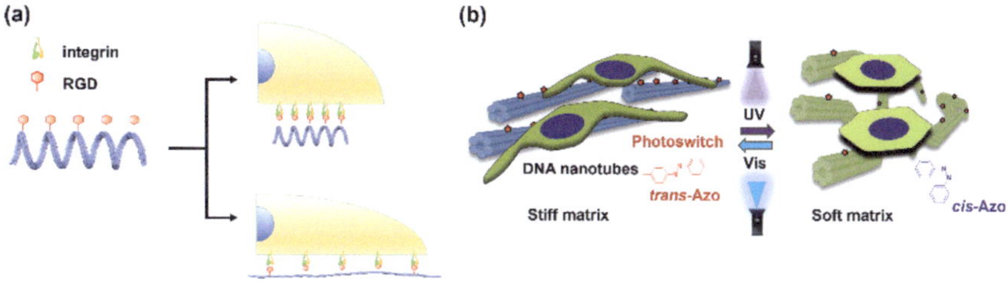

Fig. 5.5 Application of DNA origami nanoactuator for controlling cells **a** Schematics of the changes in the HeLa cell morphology induced by the RGD-modified DNA origami nanosprings carrying i-motif forming sequences. **b** Schematics of cell morphology change due to photocontrollable DNA origami nanotubes. A stiff nanoscaffold with fibrous DNA nanotubes and a soft nanoscaffold with flexible DNA nanotubes are reversibly switched using UV/VIS irradiation. Reproduced from [21] with permission

DNA-based actuators can be designed to respond to a variety of stimuli, including DNA strands, metal ions, pH, and combinations of different signals. This paves the way for the realization of "intelligent actuators," whose motions are directly coupled with a logic gate in response to input combinations. Notably, DNA origami actuators are not only distinct from conventional cytoskeletal actuators, but also potentially co-usable with them. The combined and cooperative operation of different types of molecular actuators may facilitate the development of artificial cells that exhibit more complex and sophisticated motions than those observed in living systems.

References

1. NC Seeman HF Sleiman 2018 DNA nanotechnology. Nat. Rev. Mater. 3 17068
2. PW Rothemund 2006 Folding DNA to create nanoscale shapes and patterns Nature 440 297 302
3. H Dietz SM Douglas WM Shih 2009 Folding DNA into twisted and curved nanoscale shapes Science 325 725 730
4. SM Douglas H Dietz T Liedl B Hogberg F Graf WM Shih 2009 Self-assembly of DNA into nanoscale three-dimensional shapes Nature 459 414 418
5. D Han S Pal J Nangreave Z Deng Y Liu H Yan 2011 DNA origami with complex curvatures in three-dimensional space Science 332 342 346
6. A Kuzuya Y Sakai T Yamazaki Y Xu M Komiyama 2011 Nanomechanical DNA origami 'single-molecule beacons' directly imaged by atomic force microscopy Nat. Commun. 2 449
7. T Gerling KF Wagenbauer AM Neuner H Dietz 2015 Dynamic DNA devices and assemblies formed by shape-complementary, non-base pairing 3D components Science 347 1446 1452
8. AE Marras L Zhou HJ Su CE Castro 2015 Programmable motion of DNA origami mechanisms Proc. Natl. Acad. Sci. U. S. A. 112 713 718
9. S Nummelin B Shen P Piskunen Q Liu MA Kostiainen V Linko 2020 Robotic DNA nanostructures ACS Synth. Biol. 9 1923 1940
10. ES Andersen M Dong MM Nielsen K Jahn R Subramani W Mamdouh MM Golas B Sander H Stark CL Oliveira JS Pedersen V Birkedal F Besenbacher KV Gothelf J Kjems 2009 Self-assembly of a nanoscale DNA box with a controllable lid Nature 459 73 76

11. T Tomaru Y Suzuki I Kawamata SM Nomura S Murata 2017 Stepping operation of a rotary DNA origami device Chem. Commun. 53 7716 7719
12. B Yurke AJ Turberfield AP Mills Jr FC Simmel JL Neumann 2000 A DNA-fuelled molecular machine made of DNA Nature 406 605 608
13. JL Mergny D Sen 2019 DNA quadruple helices in nanotechnology Chem. Rev. 119 6290 6325
14. K Watanabe I Kawamata S Murata Y Suzuki 2023 Multi-reconfigurable DNA origami nanolattice driven by the combination of orthogonal signals JACS Au 3 1435 1442
15. Y Suzuki I Kawamata K Mizuno S Murata 2020 Large deformation of a DNA-origami nanoarm induced by the cumulative actuation of tension-adjustable modules Angew. Chem. Int. Ed. 59 6230 6234
16. D Karna M Stilgenbauer S Jonchhe K Ankai I Kawamata Y Cui YR Zheng Y Suzuki H Mao 2021 Chemo-mechanical modulation of cell motions using DNA nanosprings Bioconjugate Chem. 32 311 317
17. D Karna E Mano J Ji I Kawamata Y Suzuki H Mao 2023 Chemo-mechanical forces modulate the topology dynamics of mesoscale DNA assemblies Nat. Commun. 14 6459
18. A Kuzyk Y Yang X Duan S Stoll AO Govorov H Sugiyama M Endo N Liu 2016 A light-driven three-dimensional plasmonic nanosystem that translates molecular motion into reversible chiroptical function Nat. Commun. 7 10591
19. P Zhan PK Dutta P Wang G Song M Dai SX Zhao ZG Wang P Yin W Zhang B Ding Y Ke 2017 Reconfigurable three-dimensional gold nanorod plasmonic nanostructures organized on DNA origami tripod ACS Nano 11 1172 1179
20. Y Ke T Meyer WM Shih G Bellot 2016 Regulation at a distance of biomolecular interactions using a DNA origami nanoactuator Nat. Commun. 7 10935
21. S Sethi T Emura K Hidaka H Sugiyama M Endo 2023 Photocontrolled DNA nanotubes as stiffness tunable matrices for controlling cellular behavior Nanoscale 15 2904 2910
22. T Kaneko TJ Itoh H Hotani 1998 Morphological transformation of liposomes caused by assembly of encapsulated tubulin and determination of shape by microtubule-associated proteins (MAPs) J. Mol. Biol. 284 1671 1681
23. Y Sato 2017 Micrometer-sized molecular robot changes its shape in response to signal molecules. Science Robot. 2 eaal3735
24. N Nakatani H Sakuta M Hayashi S Tanaka K Takiguchi K Tsumoto K Yoshikawa 2018 Specific spatial localization of actin and DNA in a water/water microdroplet: self-emergence of a cell-like structure ChemBioChem 19 1370 1374

Chapter 6
Expansion of Molecular Systems Function with the Use of Artificial Nucleic Acids

Ken Komiya

The introduction of artificial biomolecular analogues with specific properties not found in natural biomolecules can endow synthetic molecular systems with advanced functions. For DNA-based molecular devices to perform sensing, computing or actuation, molecular robotic systems integrating them, and molecular cybernetic systems capable of learning and memory, it is expected to implement various systems functions with the use of artificial nucleic acids. The construction of light-responsive molecular systems using light-responsive artificial nucleic acids, described in Chapter 3, is one of the representative examples.

On the other hand, this section describes the reaction systems in which novel systems functions have been achieved by taking advantage of the properties that have not necessarily been recognized useful, via careful consideration of the individual properties of artificial nucleic acids. In DNA-based reaction systems to carry out information transduction and computation, the reliability and usefulness of signal amplification were improved by deliberately exploiting incompatibilities with natural enzymes, and multi-step DNA extensions as successive state transitions were also achieved. These studies would have practical implications towards the construction of a molecular cybernetics world that goes beyond the living systems as paragons of functional molecular systems in nature.

6.1 Leak-Free Signal DNA Amplification Circuit

Various types of artificial nucleic acids, whose backbone-, sugar-, and/or base-moiety was modified, have been so far developed. In general, the sugar moiety of nucleotides can take two representative conformations, $C_{2'}$-*endo* and $C_{3'}$-*endo* conformations [1]

K. Komiya (✉)
Japan Agency for Marine-Earth Science and Technology (JAMSTEC), Yokosuka, Japan
e-mail: komiyak@jamstec.go.jp

S. Murata et al. (eds.), *Molecular Robotics II*,
https://doi.org/10.1007/978-981-95-2394-8_6

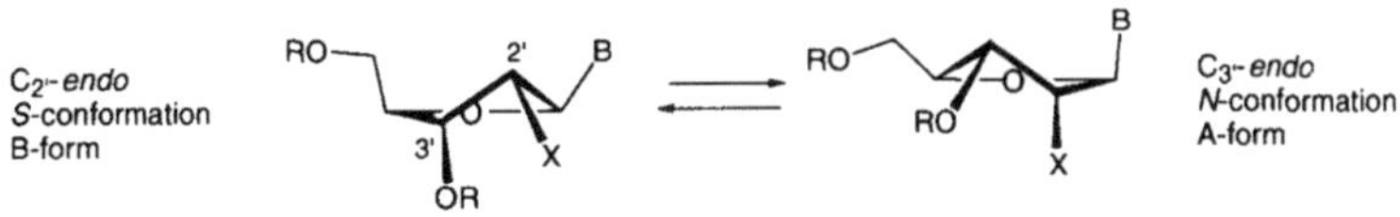

Fig. 6.1 Two representative conformations of the sugar moiety of nucleotides. Reprinted from [1], with permission from Elsevier

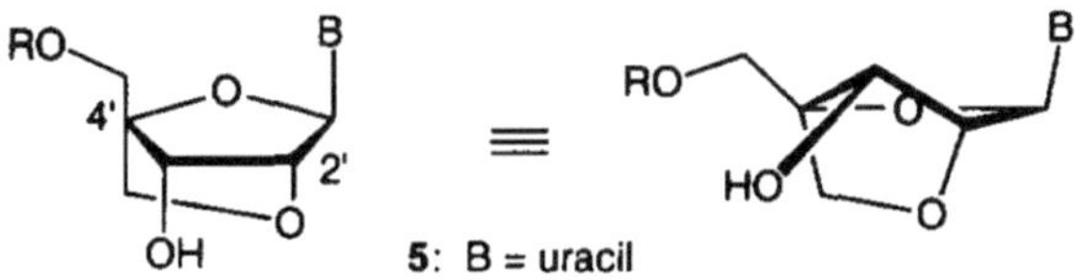

5: B = uracil
6: B = cytosine

Fig. 6.2 The molecular structure of LNA. The $C_{3'}$-*endo* sugar puckering is fixed with a methylene bridge over the 2′-*O* and 4′-*C* positions. Reproduced from [1], with permission from Elsevier

(Fig. 6.1). Natural nucleotides exist in an equilibrium mixture between these conformations. It is known that the B-form DNA duplex possesses $C_{2'}$-*endo* sugar puckering and the A-form RNA duplex has the $C_{3'}$-*endo* one [1]. Locked nucleic acid (LNA, also termed BNA), is an artificial bicyclic nucleic acid with a methylene bridge over the 2′-*O* and 4′-*C* positions showing the fixed $C_{3'}$-*endo* sugar puckering [1, 2] (Fig. 6.2). LNA, which was developed for the purpose of improving nucleic acid recognition, exhibits increased thermal stability and selectivity upon hybridization with DNA and RNA due to the increasing conformational inflexibility of the sugar moiety. In a more detailed investigation, multiple internal LNA residues increased the duplex thermal stability [3], while oligonucleotides with single terminal LNA residues conversely reduced the stability [4]. In addition, the property of LNA, showing nuclease resistance due to its unnatural structure incompatible with natural enzymatic activities, encouraged their application to antisense [5].

In molecular robotics and molecular cybernetics, development of reaction that can produce relatively short DNAs with prescribed lengths and sequences to allow information transduction and directing operation of systems components as signals is an essential demand. Furthermore, the function to rapidly and specifically amplify DNA signals is required when the number of input molecules and those required for the output response differ by orders of magnitude in a series of systems operations. Among many reported isothermal DNA amplification reactions that eliminate the need for temperature-changing cycles like PCR [6], synthetic reaction systems that combined DNA polymerization to transcribe a DNA template and nicking reaction to break a DNA duplex only at one strand, seems to meet the aforementioned demands. Such reactions include "EXPAR" for nucleic acid detection by fast single-stranded DNA amplification under relatively high-temperature conditions and "PEN DNA toolbox" achieving programming of non-linear dynamics, which have been intensively studied since 2000s and 2010s, respectively [7–10]. However, the combined use of DNA polymerase with restriction endonucleases or nicking endonucleases

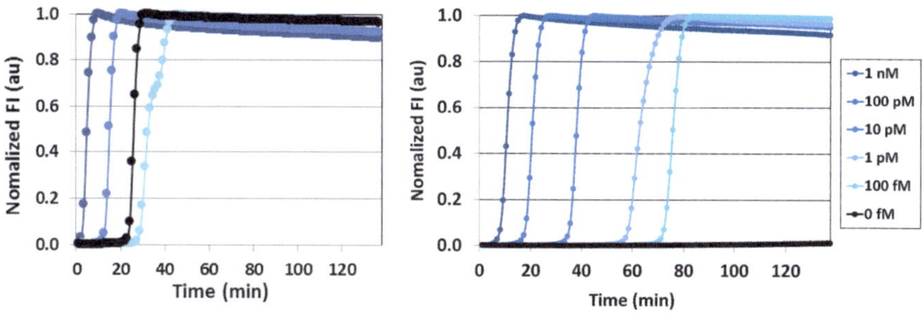

Fig. 6.3 Leak-free signal DNA amplification. (Left) Results of the amplification reaction using the two DNA template strands without LNA modification. (Right) Results of the amplification reaction using the templates with LNA modification. Reproduced from [15] with permission from the Royal Society of Chemistry

is known to enhance primer- and template-independent DNA synthesis, termed "ab initio DNA synthesis [11, 12]." It is a difficult challenge to achieve a specific response to input molecules down to very low concentrations at sub-pM or lower since non-specific amplification, which might be due to ab initio DNA synthesis, commonly occurs in the above synthetic reaction systems even in the absence of input DNA [13].

It is not possible to rationally devise a method of suppressing the ab initio DNA synthesis and non-specific amplification because their mechanisms have not yet been revealed, but if they are caused by non-typical interactions between the component DNAs and enzymes used in the reaction, the incompatibility between LNA and enzymes could be effectively utilized. It is known that the effect of LNA modification upon DNA degradation by nuclease and primer extension by DNA polymerase is dependent on the modified position [3, 14]. Komiya et al. investigated the position-dependent effects of LNA modifications introduced limitedly into the primer-binding region, but not the template region, of the two DNA strands used in the amplification reaction. Non-specific amplification observed in approximately 20 min without LNA modification was suppressed for longer than 120 min by the introduction of LNA modification, resulting in the detection of 100 fM input nucleic acid by amplifying DNA more than one million-fold under a physiological temperature condition [15] (Fig. 6.3).

6.2 Defined Hybridization and Polymerization Stop for DNA-Based Computing

Primer extension by DNA polymerase on a DNA template can be considered as a state transition of a computing nanomachine. From this perspective, the 3′-terminal sequence of a single-stranded DNA represents the current state as the nanomachine, and the process of hybridization for the priming DNA extension can be seen as a search for its complement on the DNA template. DNA extension by DNA polymerase then adds a new sequence, which represents the updated current state, at the 3′ end according to the DNA template encoding a transition rule, thereby implementing one round of state transition. Various types of computation can be performed by successive state transitions implemented via repeated DNA extension using a set of DNA templates, each encoding a transition rule. Therefore, it is required to develop a reaction mechanism whereby the newly added current state sequence autonomously becomes a single-stranded form after each DNA extension, followed by hybridization with its complement on the DNA template for the next round of state transition.

In "Whiplash PCR," successive state transitions are performed by autonomously repeating hairpin-structure formation of DNA, DNA extension by DNA polymerase, and denaturation of the hairpin structure under high temperature conditions [16, 17] (Fig. 6.4). Each set of transition rules is linearly assembled into a single DNA molecule, and a number of DNA molecules perform computation in parallel as state machines encoding their own transition rules. As a result, multiple-instruction, multiple-data (MIMD) computation is implemented uniquely within a single reaction vessel [17, 18]. For correctly proceeding the DNA extension to implement each state transition, the current state sequence should hybridize precisely with its complementary portion as designed. To prevent hybridization deviating from the binding frame, the use of artificial nucleic acids having unnatural bases, iso-dC and iso-dG, which form base pairs orthogonal to the natural nucleobases [19] (Fig. 6.5), was proposed to specify the binding frame [16].

In "primer exchange reaction (PER)" cascades, successive state transitions are isothermally performed and single-stranded DNAs with arbitrary user-prescribed sequences are synthesized as computational results in an autonomous and stepwise fashion using a set of catalytic DNA hairpin mediators with the aid of a strand-displacing polymerase [20]. Its responsive DNA synthesis could implement various molecular behaviours and enable programmable dynamic molecular circuits with a wide range of molecular sensing, monitoring, recording, signal-processing and actuation capabilities. In the PER mechanism, primer extension on the template DNA forming a hairpin structure halts at the stop sequence, followed by spontaneous dissociation of the extended primer via three-way branch migration (Fig. 6.6). Introduction of artificial nucleic acids, iso-dG and iso-dC, into PER hairpins as the stop sequences was demonstrated, with the intention of allowing the use of all four kinds of DNA bases in the synthesized sequences [20].

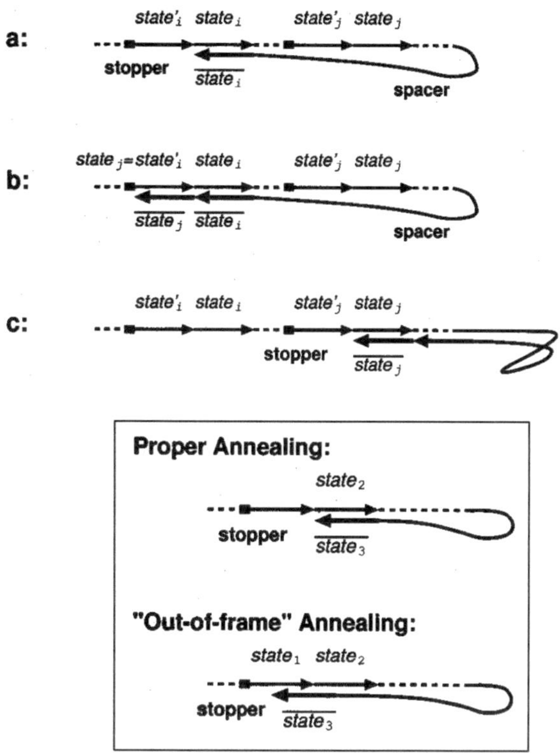

Fig. 6.4 Successive state transitions of Whiplash PCR. (Upper panel) Scheme of successive state transitions by self-templated DNA extension in the presence of DNA polymerase. (Lower panel) Correct and incorrect intramolecular hybridization by proper and "out-of-frame" annealing. Reproduced from [16], with permission from Elsevier

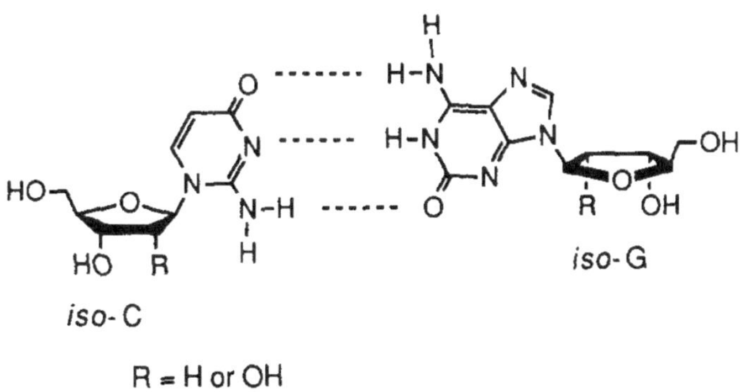

Fig. 6.5 Putative Watson–Crick base-paring between iso-C and iso-G. Reprinted with permission from [19]. Copyright 1989 American Chemical Society

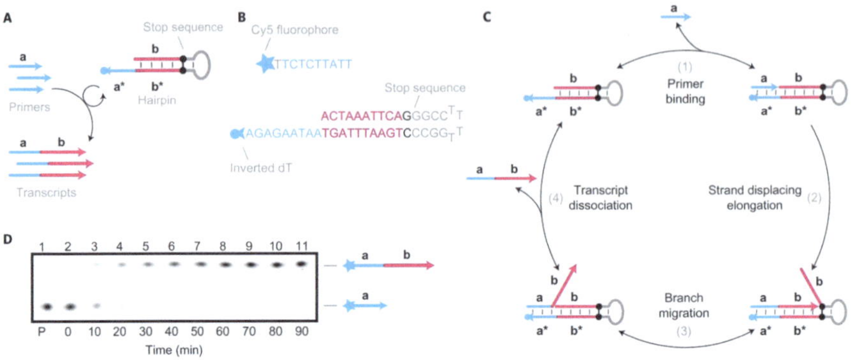

Fig. 6.6 PER mechanism. Catalytic hairpin-mediated primer extensions can be chained to form PER cascades. In this example, the stop sequence is implemented with a G–C pair by encoding the a (blue) and b (pink) domains with a three-letter code of As, Ts and Cs, and excluding dGTP from the reaction mixture. Reprinted from [20] with permission from Springer Nature

References

1. Obika, S., Nanbu, D., Hari, Y., Morio, K., In, Y., Ishida, T., Imanishi, T.: Synthesis of 2′-O,4′-C-Methyleneuridine and -cytidine. Novel bicyclic nucleosides having a fixed $C_{3'}$-endo sugar puckering. Tetrahedron Lett. **38**, 8735–8738 (1997)
2. Singh, S.K., Nielsen, P., Koshkin, A.A., Wengel, J.: LNA (locked nucleic acids): synthesis and high-affinity nucleic acid recognition. Chem. Commun. **4**, 455–456 (1998)
3. Di Giusto, D.A., King, G.C.: Strong positional preference in the interaction of LNA oligonucleotides with DNA polymerase and proofreading exonuclease activities: implications for genotyping assays. Nucleic Acids Res. **32**, e32 (2004)
4. Christensen, U., Christensen, N., Rajwanshi, V., Wengel, J., Koch, T.: Stopped-flow kinetics of locked nucleic acid (LNA)-oligonucleotide duplex formation: studies of LNA-DNA and DNA-DNA interactions. Biochem. J. **354**, 481–484 (2001)
5. Wahlestedt, C., Salmi, P., Good, L., Kela, J., Johnsson, T., Hokfelt, T., Broberger, C., Porreca, F., Lai, J., Ren, K., et al.: Potent and nontoxic antisense oligonucleotides containing locked nucleic acids. Proc. Natl. Acad. Sci. U.S.A. **97**, 5633–5638 (2000)
6. Zhao, Y., Chen, F., Li, Q., Wang, L., Fan, C.: Isothermal amplification of nucleic acids. Chem. Rev. **115**, 12491–12545 (2015)
7. Van Ness, J., Van Ness, L.K., Galas, D.J.: Isothermal reactions for the amplification of oligonucleotides. Proc. Natl. Acad. Sci. U.S.A. **100**, 4504–4509 (2003)
8. Carter, J.G., Iturbe, L.O., Duprey, J.-L.H.A., Carter, I.R., Southern, C.D., Rana, M., Whalley, C.M., Bosworth, A., Beggs, A.D., Hicks, M.R., Tucker, J.H.R., Dafforn, T.R.: Ultrarapid detection of SARS-CoV-2 RNA using a reverse transcription-free exponential amplification reaction. RTF-EXPAR. Proc. Natl. Acad. Sci. USA **118**, e2100347118 (2021)
9. Montagne, K., Plasson, R., Sakai, Y., Fujii, T., Rondelez, Y.: Programming an in vitro DNA oscillator using a molecular networking strategy. Mol. Syst. Biol. **7**, 466 (2011)
10. Baccouche, A., Montagne, K., Padirac, A., Fujii, T., Rondelez, Y.: Dynamic DNA-toolbox reaction circuits: a walkthrough. Methods **67**, 234–249 (2014)
11. Ogata, N., Miura, T.: Creation of genetic information by DNA polymerase of the thermophilic bacterium Thermus thermophilus. Nucleic Acids Res. **26**, 4657–4661 (1998)
12. Zyrina, N.V., Antipova, V.N., Zheleznaya, L.A.: *Ab initio* synthesis by DNA polymerases. FEMS Microbiol. Lett. **351**, 1–6 (2014)

13. Tan, E., Erwin, B., Dames, S., Ferguson, T., Buechel, M., Irvine, B., Voelkerding, K., Niemz, A.: Specific versus nonspecific isothermal DNA amplification through thermophilic polymerase and nicking enzyme activities. Biochemistry **47**, 9987–9999 (2008)
14. Crinelli, R., Bianchi, M., Gentilini, L., Magnani, M.: Design and characterization of decoy oligonucleotides containing locked nucleic acids. Nucleic Acids Res. **30**, 2435–2443 (2002)
15. Komiya, K., Komori, M., Noda, C., Kobayashi, S., Yoshimura, T., Yamamura, M.: Leak-free million-fold DNA amplification with locked nucleic acid and targeted hybridization in one pot. Org. Biomol. Chem. **17**, 5708–5713 (2019)
16. Sakamoto, K., Kiga, D., Komiya, K., Gouzu, H., Yokoyama, S., Ikeda, S., Sugiyama, H., Hagiya, M.: State transitions by molecules. Biosystems **52**, 81–91 (1999)
17. Komiya, K., Sakamoto, K., Kameda, A., Yamamoto, M., Ohuchi, A., Kiga, D., Yokoyama, S., Hagiya, M.: DNA polymerase programmed with a hairpin DNA incorporates a multiple-instruction architecture into molecular computing. BioSystems **83**, 18–25 (2006)
18. Hagiya, M., Arita, M., Kiga, D., Sakamoto, K., Yokoyama, S.: Towards parallel evaluation and learning of boolean μ-formulas with molecules. DIMACS Ser. Discrete Math. Theoret. Comput. Sci. **48**, 57–72 (1999)
19. Switzer, C., Moroney, S.E., Benner, S.A.: Enzymatic incorporation of a new base pair into DNA and RNA. J. Am. Chem. Soc. **111**, 8322–8323 (1989)
20. Kishi, J.Y., Schaus, T.E., Gopalkrishnan, N., Xuan, F., Yin, P.: Programmable autonomous synthesis of single-stranded DNA. Nature Chem. **10**, 155–164 (2018)

Part II
Importance of Artificial Membrane Devices in Molecular Robotics

Chapter 7
What Are Artificial Membrane Devices in Molecular Robotics?

Shin-ichiro M. Nomura

> The world is full of magic things, patiently waiting for our senses to grow sharper.
> —E. Phillpotts or W. B. Yeats

This section highlights that in order for an intelligent being to become aware of the wonders of the world, molecular robots must first possess efficient sensors. Refer to Part III for a discussion of the composition of the self, which is separated from the world, and Part IV for a discussion on "memory and learning," which serves as a basis for contemplating the wonders of the world.

7.1 Molecular Robot Membranes: Biological and Artificial Membranes

Molecular robots are artificial structures and have boundaries separating them from the outside world. It is desirable (in terms of time and effort) for molecules to be able to spontaneously divide their environment, rather than requiring another entity, a higher-order being, to divide the world for them. A wide variety of such molecular structures exist, including solids, gels, and liquid–liquid phase-separated structures. In particular, the structure used by living cells, the cell membrane, is separated from aqueous solutions by a lipid layer 5–7 nm thick, the main component of which is a lipid bilayer (Fig. 7.1). This two-dimensional plane is the focus of this chapter. On the surface of this evolutionarily selected and notably thin oily membrane are a variety of membrane proteins and glycolipids, the function of which is to respond to internal and external changes. Micrometer-sized molecular robots that mimic cells are often

S. M. Nomura (✉)
Department of Robotics, Graduate School of Engineering, Tohoku University, 6-6-01, Aza-Aoba, Aramaki, Aoba-ku, Sendai, Miyagi 980-8579, Japan
e-mail: shinichiro.nomura.b5@tohoku.ac.jp

S. Murata et al. (eds.), *Molecular Robotics II*,
https://doi.org/10.1007/978-981-95-2394-8_7

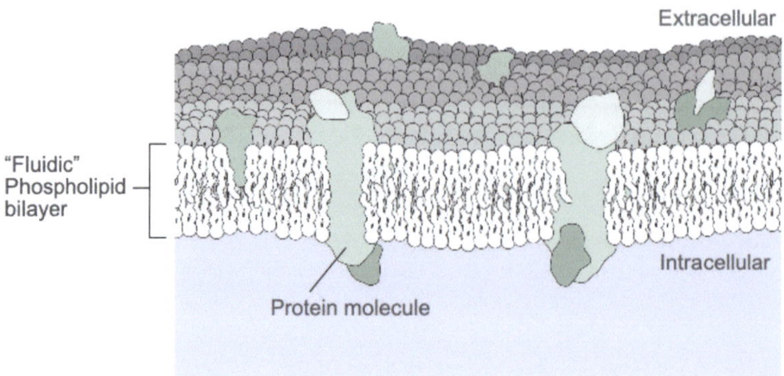

Fig. 7.1 Lipid bilayer structure

designed with lipid-enclosed vesicles (liposomes and GUVs), which spontaneously form in water [1–3].

We first review the structure of these lipid membranes. The discovery of lipid bilayers is believed to have started with a hypothesis published by Gorter and Grendel in the Netherlands in 1925. The area of lipid extracted from red blood cells and spread on the surface of water was twice the surface area of the red blood cells used, leading them to believe that the lipid comprised a double layer [4]. Although elaborate follow-up studies cast doubt on the "factor 2" hypothesis, according to a 1964 report by Lord Bangham, electron microscope images of liposomes created by dispersing extracted lipids in water showed that the lipids formed a bimolecular membrane [5].

In 1972, Singer and Nicholson [6] proposed the fluid mosaic model of lipid membranes, revealing that cell membranes are composed of highly fluid lipid bilayers and floating/sticking membrane proteins with hydrophobic moieties. The surface of the natural cell membrane is covered with glycans, provided by glycolipids and glycoproteins, and involves a complex structure with interactions between transmembrane proteins; further, it is lined with the cytoskeleton and compartmentalized in a flat, fenestrated shape. The discovery of the bilayer structure using synthetic amphiphilic molecules by Kunitake et al. in 1977 led to further developments [7], and it is now possible to fabricate artificial lipid bilayers using a variety of amphiphilic molecules.

The aggregates composed of these lipid molecules assume diverse structures in aqueous environments. In water, there are spherical micelles with an inner hydrophobic field; rod-shaped micelles; black films, flat membranes covering a hole in a substrate several millimeters in diameter in water; Langmuir–Blodgett membranes, lipid monolayers that develop at the air–liquid interface; lipid vesicles, which are closed membrane vesicles; and lipid nanoparticles, which interact with nucleic acids such as DNA and RNA, among other systems. Lipid nanoparticles and nanodiscs, in which the lipid bilayer structure is protected by amphiphilic polymers and the like in the form of disks, are among the various modalities. In particular, lipid vesicles, which are closed vesicles of lipid "bilayers" formed independently in water, are termed "liposomes". Furthermore, it has become simple in recent years to prepare them to a size equivalent to a cell of several micrometers to several tens of micrometers in diameter. Research on the installation and function of various molecules in these vesicles has been actively conducted worldwide in a wide range of basic and applied areas [3, 8, 9].

Molecular robots are artificial cell-type bodies that are created by combining artificially designed molecular devices with liposomes to perform a desired function. However, no matter how effective the system, it will not undergo changes if its internal space remains unreachable, enclosed in a flat membrane. The subsequent section considers molecular devices for functionalizing membranes, focusing on their similarities and differences with regard to natural membrane proteins.

7.2 Membrane Device Molecules Used in Molecular Robotics

Herein, we term molecules that localize and operate in lipid bilayers "membrane device molecules", of which there are various types. If they are amphiphilic or small hydrophobic molecules, they are spontaneously inserted into the membrane. Furthermore, when they comprise polymers, they also spontaneously attach to lipid membranes according to their configuration through hydrophobic groups on their main or side chains. However, hydrophobicity or amphiphilicity alone is not sufficient to ensure their insertion into the membrane and their desired functioning. Membrane device molecules in a lipid membrane environment search for an equilibrium point, a stable inter- and intramolecular conformation, with lipid molecules and other membrane device molecules. In biological membranes, the sequence of amino acids defines the optimal conformation in the lipid membrane environment, and in some cases, transmembrane or multiple transmembrane states are the most stable (12-transmembrane receptors exist). The modes of membrane devices that localize in lipid membranes can be classified as follows (Fig. 7.2):

(i). Anything that changes the physical properties of the membrane (modulation, increasing flowability, and increasing/reducing solubility)
(ii). Something that anchors another molecule to the membrane (linkers)
(iii). Those that impart substance permeability to the membrane (channels and transporters)
(iv). Those that transmit a change of state inside or outside the membrane (receptors and transducers)

These molecules do not function alone and can characteristically change the properties of the membrane by impinging on it; furthermore, they can affect the field of the membrane by penetrating matter or bonding with other molecules. Membrane molecular devices differ from electronic and software devices in that they are field-forming and inherently functional molecules.

Membrane devices mainly comprise of amphiphilic molecules. As the concentration of the monomer increases in aqueous solution, it reaches the critical micelle

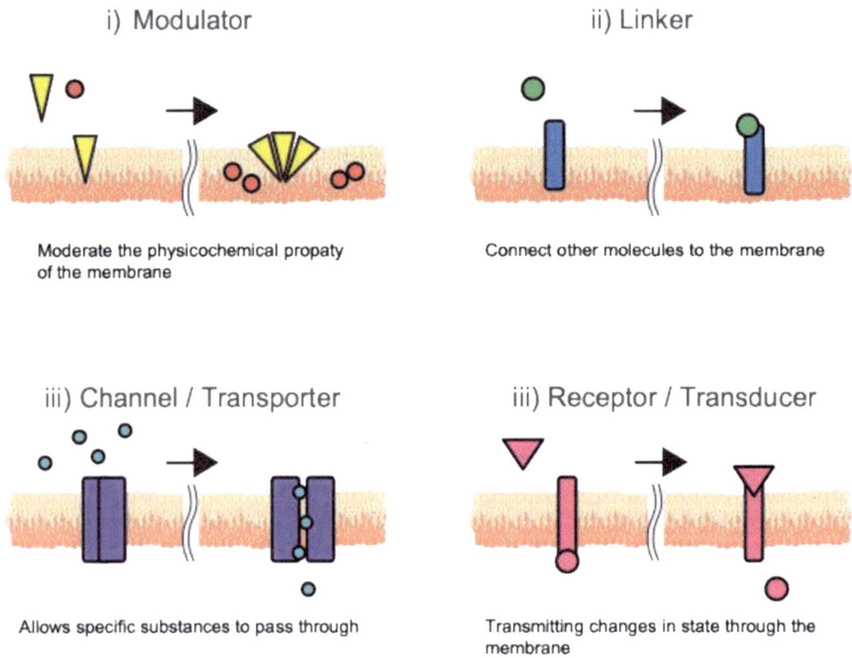

Fig. 7.2 Functioning of various membrane devices

concentration (CMC), leading to the formation of self-assemblies and micelle structures at concentrations above the CMC. It is then possible to predict the higher-order structure of an amphiphilic molecule from its geometric parameters (Fig. 7.3). Considering this, one can design amphiphilic molecules according to one's objectives.

A synergistic effect occurs between multiple types of membrane device molecules in the raft structure proposed by Simons [11]. This is a state in which a group of membrane proteins consisting of specific receptors form a planar complex with specific lipid components and drift across the lipid membrane like a raft. Studies

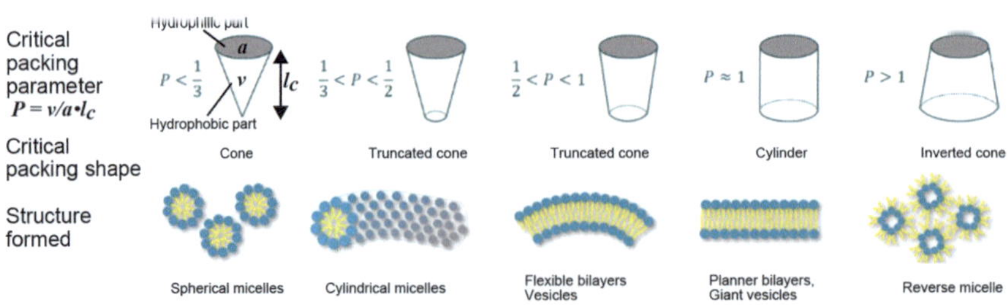

Fig. 7.3 Prediction of stable higher-order structures of amphiphilic molecules using geometric parameters [10]

have reported that similar structures can be obtained on the surface of artificial cell membranes such as GUVs—via phase separation between lipid molecules, for example [12]—and are the first choice of researchers seeking the localization of natural/artificial membrane devices.

In biological membranes, various amphiphilic molecules such as proteins, glycoproteins, lipids, and glycolipids exist in a mosaic-like structure to maintain diverse functions. Their function, in broad terms, involves being a transmission machine, allowing for changes in conditions to be conveyed or transmitted across a membrane. Created with this as a reference is an artificial molecular device that achieves the desired function.

The basic principle of engineering is to make use of what is available and build what is missing. Since there is much to be learned from natural cells, it is more beneficial to employ the molecular devices used by cells as they are, if possible. If this is not feasible, it is better to extract the essence of their function and design new molecules. As examples, we introduce "molecular clutches", "molecular nanopores", and "molecular sensors".

Membrane Device Example 1: Molecular Clutch

Amoeboid molecular robots are artificial cellular entities that use liposomes composed of lipid bilayers as "bodies" and kinesins and microtubules as "actuators" [13]. We designed the "molecular clutch" as a membrane device that controls the conformational changes of such molecular robots (Fig. 7.4). The molecular clutch functions to control the binding of kinesin to the membrane in response to specific DNA signals, switching the shape change on and off.

The molecular clutch consists of a "motor unit" and an "anchor unit"; kinesin is anchored to the membrane via DNA binding, which transfers the sliding motion of the microtubule to the membrane When the clutch is "engaged" by the DNA signal, the force of kinesin is transferred to the membrane, and the molecular robot undergoes a continuous shape change. In the "disengaged" state, on the other hand, the bound kinesin dissociates from the membrane, and the shape change ceases.

Furthermore, using photoresponsive DNA, it was possible to control the on/off functioning of the clutch by generating a DNA signal through UV irradiation. The shape change caused by switching the clutch on and off was confirmed by microscopic observations; the energy required for the shape change was calculated as approximately 10^{-18} J, suggesting that the clutch operates with the same amount of energy as the membrane deformation of the cell. Although the molecular robot remains levitated, there are still issues to be addressed before it can be attached to a substrate and realize locomotion with center-of-gravity movements. Since the interaction between the phospholipid membrane and the glass substrate is on the order of mJ/m^2, energy in excess of 10^{-14} J is required for a molecular robot with a ground area of approximately 10^{-11} m^2 to crawl and move. To overcome the large barrier between this energy and the 10^{-18} J required for the shape change of the amoeba-type molecular robot, it may be necessary to concentrate the force applied in the direction of movement by some means or to employ a "focal adhesion group" that limits the area of adhesion with the floor, as learned from cells. Furthermore, it is interesting to

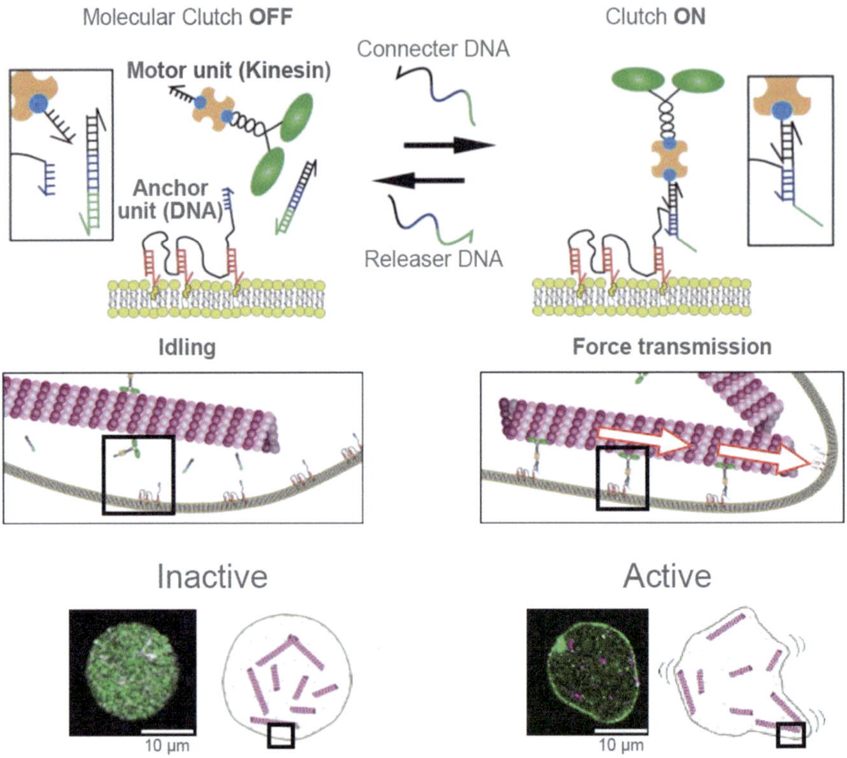

Fig. 7.4 Molecular clutch switching the motion state of an amoeba-type molecular robot on and off

note that a transmembrane clutch mechanism as a membrane device similar to this has been identified in cell motility using actin [14]. To control molecular-level movements by chemical–mechanical coupling, improving the efficiency and functionality of membrane devices is essential.

Membrane Device Example 2: DNA Nanopores

Lipid bilayers do not permit the permeation of DNA molecules. Therefore, if DNA is to be used as a control signal for cellular molecular robots, a device allowing its penetration into membranes is necessary. We have previously designed pore-size tunable molecular nanopores [15]. The nanopores, using DNA origami technology, form a square tube structure with a cross-sectional area of approximately 100 nm^2 and are inserted into giant liposomes (GUVs) that mimic artificial cell membranes, allowing for molecular penetration.

The nanopores are anchored to the membrane via hybridization with 47 cholesterol-modified DNA nanostructures (anchor-DNA), facilitating the permeation of hydrophilic molecules. The permeability was evaluated using an influx assay of the

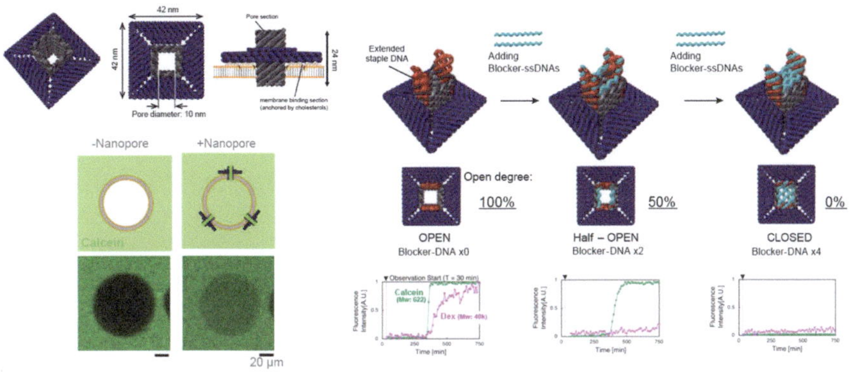

Fig. 7.5 Controlled permeation of GUV membranes using DNA origami nanopores

fluorescent dye calcein, and a clear increase in the fluorescence signal was observed in GUVs with nanopores incorporated into the membrane (Fig. 7.5).

In addition, single-stranded DNA (blocker-DNA), which acts as a "lid" for permeation control, was designed to control the pore size. By binding up to four blocker-DNAs, the size-selective transport of molecules becomes possible. When two blocker-DNAs are used, dextran, with a diameter of 4.8 nm, cannot pass through, while the use of four blocker-DNAs blocks almost all permeation.

This study demonstrates the design and operation of artificial nanopores with variable pore sizes using DNA nanostructures, which can control the permeation of DNA by DNA. This contributes to the development of molecular robots operated by DNA circuits and molecular information exchange systems in artificial cells.

However, they exhibit the disadvantage of low operating efficiency, and the coexistence of surfactants during folding and membrane insertion has been reported [16]. The significantly large pores that open in lipid membranes allow the permeation of small molecules and even short DNA chains, making it difficult to ensure the independence of internal and external environments.

Membrane Device Example 3: DNA Sequence-Dependent and Transmembrane Receptors

In building a circuit via DNA computing, one only aims to transmit information about a specific nucleic acid sequence across the membrane of the GUV; it is not desirable for other molecules to pass through the membrane. Therefore, we proposed a new membrane-permeable DNA signaling system named "Chabashira" [17]. This system uses DNA hybridization (binding of complementary DNA strands) to transmit single-stranded DNA sequences across the GUV membrane according to the following mechanism. First, a single-stranded DNA (driver) with multiple cholesterol modifications is fixed to the GUV membrane. When this binds to an externally introduced complementary single-stranded DNA (packet), the DNA sequence at the end is transmitted through and induces a fluorescent signal within the GUV. Experiments have also been performed using this system to activate a DNA amplification circuit, successfully inducing internal DNA amplification via an external DNA input. This research opens new possibilities for molecular communication in an environment that mimics the cell membrane; further, it is expected to serve as a fundamental technology for advanced information processing in molecular robots and artificial cells (Fig. 7.6).

Similarly, recent studies have indicated that single-stranded DNA can penetrate lipid membranes with a few hydrophobic groups and cholesterol modifications [18–20]. Further, the behavior of cholesterol-modified DNA on membranes is detailed in Dr. Diana Morzy's published PhD thesis [21]. Repositories have become important, and researchers seeking to design DNA membrane devices are advised to consult them.

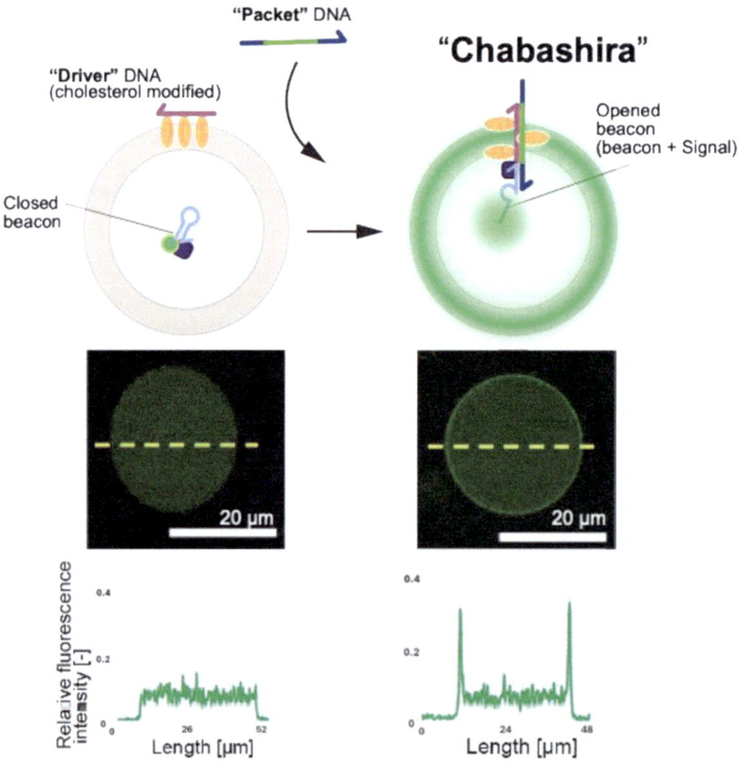

Fig. 7.6 Transmembrane DNA sequence information transfer by the Chabashira system

7.3 Technological Challenges and Future Prospects for Artificial Membrane Devices

Artificial membrane devices, as described in Sect. 7.2, are a new molecular communication technology for information transfer between molecular robots and artificial cells across lipid membranes. Their research and development provide the basis for more advanced DNA circuit design and molecular information processing using cellular membranes. In addition to improvements in sensitivity (accuracy) and operational speed, the following are some of the factors that need to be addressed regarding artificial membrane devices:

- Self-organization with high reproducibility
- Reusability (reset)
- Molecular communication between artificial cells (compartments)
- High durability

The following two factors relate to the fate of all nonliving molecules:

- Self-maintenance

Cellular molecular robots (unlike natural cells capable of self-maintenance and material production) should be discarded when they break down. It is not realistic to repair and reuse them on a large scale, as is the case with real robots. Performing self-maintenance, including in situ repair and metabolism, constitutes an important development requirement.

- Updating

Molecular robots that do not self-replicate are "vanishing objects". A meta-development and production system that can monitor the status of their use and update them for subsequent uses will be necessary.

It is expected that innovative molecular devices (including membrane structures) will be researched and developed for the next generation of molecular robots, leading to tougher and more efficient devices that are more tolerant of real-space applications while overcoming these challenges.

7.4 Conclusion

This section provides an overview of lipid bilayers as the structures that encase molecular robots, especially cellular-type ones; the molecular devices used in these membranes; and the associated challenges.

Designed molecules are often more directional in their performance than the products of biological evolution and require optimization, implying that the operating conditions are often more demanding (narrower operating window). In membrane devices in particular, in addition to the solution environment, the hydrophobic field environment on the lipid membrane side and even the conditions at the membrane interface must be considered.

On the other hand, molecular robotics aiming to operate in real space is expected to function in places where the survival of living organisms is usually difficult and in a wide range of activity locations. Therefore, robustness is required not only for molecular devices individually but also for molecular systems as a whole, including the membrane composition and the exchange of molecular information. In the future, it is important to conduct extensive and sophisticated large-scale operation tests with the shared recording of positive/negative data.

In this chapter, we describe recent research examples of various molecular devices that have been artificially designed and studied. Although some of the devices are not currently incorporated as "molecular robots", we hope that this chapter is informative and provides insights into various future applications.

References

1. Sato, Y.: Artificial molecular systems for complex functions based on DNA nanotechnology and cell-sized lipid vesicles. ChemSystemsChem. **6**, e202400021 (2024)
2. Peng, Z., Iwabuchi, S., Izumi, K., Takiguchi, S., Yamaji, M., Fujita, S., Suzuki, H., Kambara, F., Fukasawa, G., Cooney, A., Di Michele, L., Elani, Y., Matsuura, T., Kawano, R.: Lipid vesicle-based molecular robots. Lab Chip **24**, 996–1029 (2024)
3. Peng, H., Zhao, M., Liu, X., Tong, T., Zhang, W., Gong, C., Chowdhury, R., Wang, Q.: Biomimetic materials to fabricate artificial cells. Chem. Rev. (2024).
4. Gorter, E., Grendel, F.: On bimolecular layers of lipoids on the chromocytes of the blood. J. Exp. Med. **41**, 439–443 (1925)
5. Bangham, A.D., Horne, R.W.: Negative staining of phospholipids and their structural modification by surface-active agents as observed in the electron microscope. J. Mol. Biol. **8**, 660–668 (1964)
6. Singer, S.J., Nicolson, G.L.: The fluid mosaic model of the structure of cell membranes: cell membranes are viewed as two-dimensional solutions of oriented globular proteins and lipids. Science **175**, 720–731 (1972)
7. Kunitake, T., Okahata, Y.: A totally synthetic bilayer membrane. J. Am. Chem. Soc. **99**, 3860–3861 (1977)
8. Walde, P., Cosentino, K., Engel, H., Stano, P.: Giant vesicles: preparations and applications. ChemBioChem **11**, 848–865 (2010)
9. Toyota, T., Zhang, Y.: Identifying and manipulating giant vesicles; review of recent approaches Micromachines (Basel). **13** (2022).
10. Israelachvili, J.N.: Intermolecular and surface forces (2011)
11. Simons, K., Ikonen, E.: Functional rafts in cell membranes. Nature **387**, 569–572 (1997)
12. Baumgart, T., Hess, S.T., Webb, W.W.: Imaging coexisting fluid domains in biomembrane models coupling curvature and line tension. Nature **425**, 821–824 (2003)
13. Sato, Y., Hiratsuka, Y., Kawamata, I., Murata, S., Nomura, S.-I.M.: Micrometer-sized molecular robot changes its shape in response to signal molecules. Sci Robot. **2** (2017).
14. Elosegui-Artola, A., Oria, R., Chen, Y., Kosmalska, A., Pérez-González, C., Castro, N., Zhu, C., Trepat, X., Roca-Cusachs, P.: Mechanical regulation of a molecular clutch defines force transmission and transduction in response to matrix rigidity. Nat. Cell Biol. **18**, 540–548 (2016)
15. Iwabuchi, S., Kawamata, I., Murata, S., Nomura, S.-I.M.: A large, square-shaped, DNA origami nanopore with sealing function on a giant vesicle membrane. Chem. Commun. **57**, 2990–2993 (2021)

16. Iwabuchi, S., Nomura, S.-I.M., Sato, Y.: Surfactant-assisted purification of hydrophobic DNA nanostructures. Chembiochem. e202200568 (2022)
17. Yoshida, K., Abe, K., Sato, Y., Kawamata, I., Archer, R.J., Matsubayashi, H.T., Hamada, S., Murata, S., Nomura, S.-I.: Transmembrane DNA Sequence Signaling via Hybridization. ChemRxiv (2024).
18. Chen, H., Xu, W., Shi, H., Qiao, Y., He, X., Zheng, J., Zhou, S., Yang, X., Wang, K., Liu, J.: DNA-based artificial receptors as transmembrane signal transduction systems for protocellular communication. Angew. Chem. Int. Ed. Engl. **62**, e202301559 (2023)
19. Chen, H., Zhou, L., Li, C., He, X., Huang, J., Yang, X., Shi, H., Wang, K., Liu, J.: Controlled dimerization of artificial membrane receptors for transmembrane signal transduction. Chem. Sci. **12**, 8224–8230 (2021)
20. Liu, G., Huang, S., Liu, X., Chen, W., Ma, X., Cao, S., Wang, L., Chen, L., Yang, H.: DNA-based artificial signaling system mimicking the dimerization of receptors for signal transduction and amplification. Anal. Chem. **93**, 13807–13814 (2021)
21. Morzy, D.: DNA-Based Mimics of Membrane Proteins Lipid-DNA Interactions Determine Function [Apollo - University of Cambridge Repository] (2021)

Chapter 8
Synthetic DNA Nanopore

En Kawakubo, Towa Higashiho, and Yusuke Sato

Abstract DNA nanotechnology enables the bottom-up construction of membrane nanopores with programmable geometry and function. This chapter surveys design strategies for synthetic DNA nanopores, from DNA-origami pores with barrel-stem architectures to multi-helix bundles (e.g., 6HB) and minimal bundles (4HB) or membrane-spanning duplexes, and outlines how these structures are embedded in lipid bilayers despite DNA's hydrophilicity and negative charge. We highlight how hydrophobic anchoring (e.g., cholesterol), membrane curvature, and surfactants govern insertion yields and aggregation behavior, providing practical routes to robust reconstitution in vesicles. Functionally, DNA nanopores mediate size- and charge-selective transport of small dyes and large molecules (over several tens of kDa) depending on their design, while tunable "gates" respond to DNA inputs, temperature, or light to realize reversible open–close control. Combining pores with DNA reaction networks inside vesicles further enables reception-transduction-response cascades as an artificial molecular system. We conclude by outlining open challenges, such as ion selectivity, active transport against gradients, and cargo selectivity, and discuss the potential for advancement toward "Chemical AI" based on synthetic DNA nanopores.

Phospholipids consist of two hydrophobic fatty acid tails and a hydrophilic head group. Due to their amphipathic nature, phospholipids in water spontaneously form a bilayer membrane, with the hydrophobic tails buried inward and the hydrophilic heads exposed outward. This bilayer acts as a stable barrier, effectively separating aqueous environments. Vesicles, surrounded by lipid bilayer membranes, are often used as a compartment for artificial molecular systems [1], such as artificial cells [2] or molecular robots [3], where system components are encapsulated within the vesicle. However, due to the strong barrier function of lipid membranes, only small uncharged molecules, like water and glycerol, can pass through. To facilitate the

E. Kawakubo · T. Higashiho · Y. Sato (✉)
Kyushu Institute of Technology, Iizuka, Japan
e-mail: ysato@ics.kyutech.ac.jp

S. Murata et al. (eds.), *Molecular Robotics II*,
https://doi.org/10.1007/978-981-95-2394-8_8

transport of essential molecules, such as ions, ATP (as fuel), and nucleic acids (as signals), nanopores, nanoscale holes, must be embedded in the membrane.

Although synthetic chemistry offers a means to construct artificial nanopore molecules, DNA nanotechnology provides an alternative approach for creating artificial synthetic nanopores owing to its designability, which ranges from a few nanometers to several tens of nanometers in scale [4]. Researchers developed cylindrical structures made of DNA and demonstrated that they could be embedded in lipid membranes even though DNA is a negatively charged hydrophilic molecule that functions as nanopores. This chapter focuses on the history, design strategy, and function of synthetic DNA nanopores. In the last part of this chapter, we describe the perspective of DNA nanopores from the viewpoint of Chemical AI.

8.1 Development of DNA Nanopores

The design of DNA nanopores (often also called channels instead of nanopores) is mainly divided into DNA origami and six-helix bundle (6HB) types. The former is assembled based on the DNA origami technique [5, 6], whereas the latter is assembled from several short DNAs. A DNA origami nanopore was first developed by Langecker et al. in 2012 (Fig. 8.1) [7]. The origami nanopore comprised a transmembrane stem and a barrel-shaped cap (Fig. 8.1a). The origami nanopore adheres to lipid membranes, mediated by 26 cholesterol molecules facing the barrel surface. The stem is a 6-helix bundle with 42 nm in length and ~2 nm in inner diameter. The stem protrudes from the barrel, thereby puncturing the membrane. Transmission electron microscopy images showed the assembly and membrane attachment of the nanopores (Fig. 8.1b). Single-channel current measurements analyzed the electrical conductivity of the origami nanopore. The origami nanopore showed 0.87 ± 0.15 nS (Fig. 8.1c), which agrees with the predicted value (0.78 nS), suggesting the successful nanopore formation in lipid membranes.

The 6HB-type nanopore was first reported by Burns et al. in 2013 [8] (Fig. 8.1d). The 6HB has 15 nm in height, 5.5 nm in outer radius, and ~2 nm in inner radius, assembled from 14 DNA strands. The outside of the 6HB nanopore has a hydrophobic belt with 2.2 nm height, almost corresponding to the thickness of lipid membranes. The hydrophobic belt was realized by ethyl modifications to phosphates in the DNA backbones, which neutralized the negative charges and provided hydrophobic alkyl chains. Each duplex in the 6HB has 12 continuous modifications. The structural formation and purity were evaluated by size-exclusion chromatography (SEC), showing a 70% yield of the formation. The 6HB structures were visualized by an atomic force microscope (Fig. 8.1e), which showed 21.3 ± 0.3 nm in length, which agreed with the theoretical values (15 nm). Single-channel current measurements, similar to the DNA origami nanopore, analyzed the insertion of the 6HB nanopores in lipid membranes. In contrast to the origami nanopore, the conductance measured (395 $\pm$ 97 pS) did not agree with the theoretical value (1320 pS) that was calculated under the assumption where ions mobility in the nanopore is constant. Nevertheless,

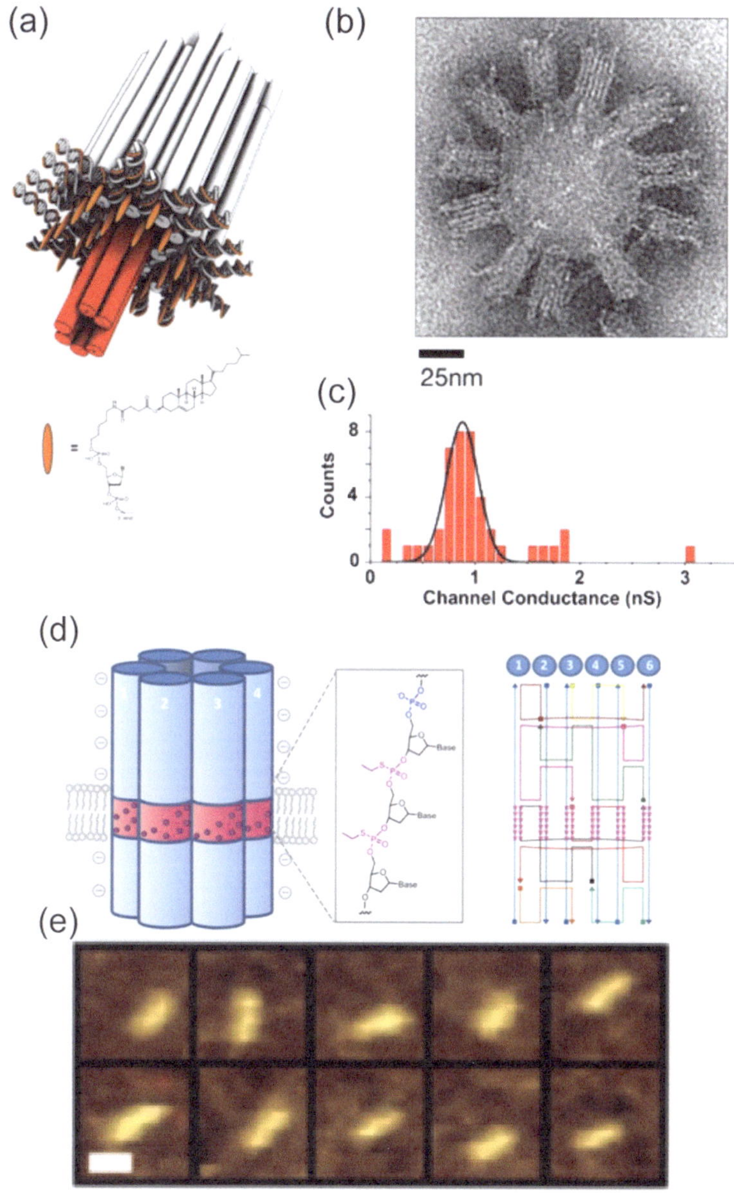

Fig. 8.1 DNA origami and six-helix bundle (6HB) type nanopores. **a** A schematic illustration of the origami nanopore. Cylinders indicate DNA duplexes. Orange ellipsoids indicate cholesterol-triethylene glycol modification. **b** A TEM image representing nanopore attachments onto a small unilamellar vesicle surface. **c** Histogram of conductance. **a**-**c** Reprinted with permission of [9]. **d** Schematic illustrations of 6HB DNA nanopores. The magenta-colored region on the left-hand side indicates a hydrophobic belt with alkyl modifications. Arrows on the right-hand side show pathways of assembled DNA strands. **e** AFM images of individual DNA nanopores. Scale bar: 100 nm. **d** and **e** Reprinted with permission of [8]

the other various experimental results suggested stable nanopore formation in lipid membranes.

A four-helix bundle (4HB) can be used as a smaller DNA nanopore [10] (Fig. 8.2a). The tiny gap at the center of the 4HB offers a space for an ion channel with 0.8 nm diameter as the nominal. The advantage of this design was simplicity, resulting in short preparation time and low cost. The electrical conductivity analysis showed the successful formation of nanopores in lipid membranes. On the other hand, although the conductance of this nanopore was expected to be 0.47 nS, experimental results showed that the conductance was sometimes greater than this value. It would be attributed to the combined effects of several factors: structural fluctuations due to the simplicity of the structure, increased diameter due to electrostatic repulsion, non-binding of DNA in parts of the structure due to voltage, and membrane deformation.

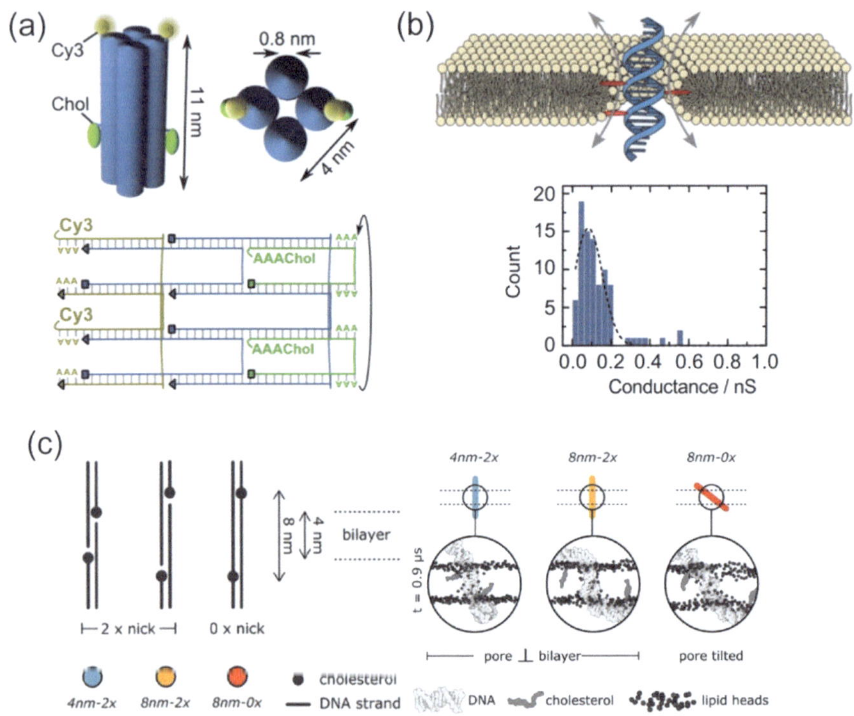

Fig. 8.2 Smaller nanopores in lipid bilayer membranes by DNA bundles or double-stranded DNA (dsDNA). **a** Schematic illustrations of a four-helix bundle DNA nanopore with 0.8 nm nominal diameter (top) and pathways of eight DNA strands assembled into the nanopore (bottom). Reprinted with permission of [10]. **b** A schematic illustration of membrane-spanning dsDNA modified with six porphyrin molecules (top) and histogram of conductance (bottom). Reprinted with permission of [11]. **c** Schematic illustrations of cholesterol-modified three different DNA with nicks (left) and representative snapshots of all-atom molecular dynamics simulations of the three dsDNA (right). Reprinted with permission of [12]

Transmembrane double-stranded DNA (dsDNA) also generates very small nanopores [11] (Fig. 8.2b). The dsDNA, modified with six porphyrin molecules, generated a toroidal nanopore at the interface between the dsDNA by insertion into the bilayer membrane. The nanopore conductance was 0.08 nS on average, which agreed with the simulated conductance (0.095 nS) by all-atom molecular dynamics simulation. This result suggested that all conductance in DNA nanopores is the sum of the conductance through the central pore in a structure and through the toroidal lipid pore surrounding it. The size of toroidal pores depends on the membrane-penetrating dsDNA design [12] (Fig. 8.2c). When the distance between the two hydrophobic molecules in dsDNA was larger than the thickness of the lipid bilayers (typically 4–5 nm), continuous dsDNA penetrated the membrane with a tilted orientation to solve the position mismatch between hydrophobic molecules and hydrophobic region in lipid membranes. This tilted orientation resulted in the formation of larger toroidal nanopores. On the other hand, dsDNA with discontinuities (nicks) did not tilt but deformed at the nicks to insert hydrophobic molecules into the lipid membrane. It provides insight into how the position of hydrophobic modifications in DNA induces deformation and affects the overall conductance in DNA nanopores.

Although the focus of this section so far was on membrane insertion of DNA nanopores, ion transport through DNA nanostructures was investigated and reported shortly before membrane insertion was reported (Fig. 8.3). Bell et al. have demonstrated DNA origami nanopore insertion into a solid-sate nanopore (SS nanopore) composed of silicone-nitride [13] (Fig. 8.3a). The authors designed a pyramidal-like DNA origami nanopore whose tip enters the SS nanopore. The origami nanopore has a double-stranded overhang at the tip to direct the nanopore insertion to the SS nanopore. Because the size of the SS nanopore was 13–18 nm and the origami nanopore geometry was 27.5×27.5 nm at the largest region, the origami nanopore could be trapped in the SS nanopore. This design enables repeated insertions of the origami nanopore in the SS nanopore by reversing the applied potential. The origami-SS hybrid nanopore showed 80% lower ionic current (Fig. 8.3a). In addition, the hybrid nanopore was used to detect λ-DNA strands. Wei et al. have also reported a hybrid nanopore, a DNA origami nanoplate on an SS nanopore [14]. The authors designed a double-layered DNA origami plate with 50 nm width and length and 6 nm thickness (Fig. 8.3b). The origami plate also has a central aperture of various sizes. The conductance of the hybrid nanopores decreased with the aperture size, suggesting that the origami plate covered the nanopore in a flat orientation. The nanoplate without the central aperture showed ~17% lower conductance than that without the nanoplate (Fig. 8.3b). It attributed that DNA origamis are mesh-like structures, and the double-layered design has a gap between the layers. DNA nanoplates with designated-sized apertures can not only function as size-selective molecular gates but can also enable chemical selectivity to SS nanopores. The authors demonstrated that extending a specific sequence from the aperture allows for sequence-specific detection of single-stranded DNA.

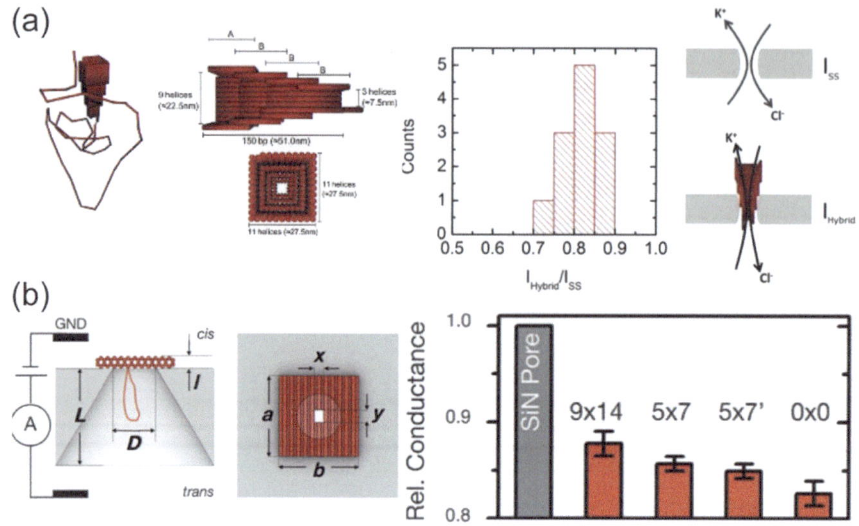

Fig. 8.3 Hybrid of DNA origami and solid-state (SS) nanopore. **a** Schematic illustrations of DNA origami nanopore and histogram of conductance. I_{Hybrid} and I_{SS} indicate current values obtained in the hybrid of DNA origami and SS nanopore and SS nanopore alone, respectively. Reprinted with permission of [13]. **b** Schematic of the experimental setup and relative conductance of a DNA origami nanoplate on a nanopore. Inset numbers in the graph indicate width × length in nm of the central aperture in the nanoplate. Prime in the number indicates ssDNA overhang elongation from the aperture. Reprinted with permission of [14]

8.2 Membrane Insertion of DNA Nanopores

The number of modifications of hydrophobic molecules is critically important to realize the insertion of DNA nanopores into lipid membranes. Birkholz et al. have reported a possible mechanism of the nanopore insertion into membranes [15]. 6HB DNA nanopores modified with one or three cholesterols (NP-1C or NP-3C) were used for the mechanism investigation. They found that NP-1C and -3C were initially bound to the membrane via a single cholesterol; then, NP-3C orients and inserts into the membrane (Fig. 8.4a). This transition was impossible for NP-1C due to the lack of multiple cholesterols required for complete membrane insertion. On the other hand, Luo et al. reported that 6HB modified with two cholesterols (6HB-2C) showed better insertion than 6HB-3C [16], although the structural design was slightly different from that used in [15]. The authors discussed that the aggregation of 6HB-3C in solution via hydrophobic interaction between cholesterols lowered the insertion rate.

Membrane curvature also impacts nanopore insertion. Burns et al. found that 6HB nanopores efficiently inserted into lipid vesicle membranes with high curvature (small unilamellar vesicles (SUVs) with 0.2 μm^{-1} curvature). In contrast, lower curvature (giant unilamellar vesicles (GUVs) with 20 μm^{-1} curvature) showed poor insertion [17] (Fig. 8.4b). The authors discussed that this preferential insertion into

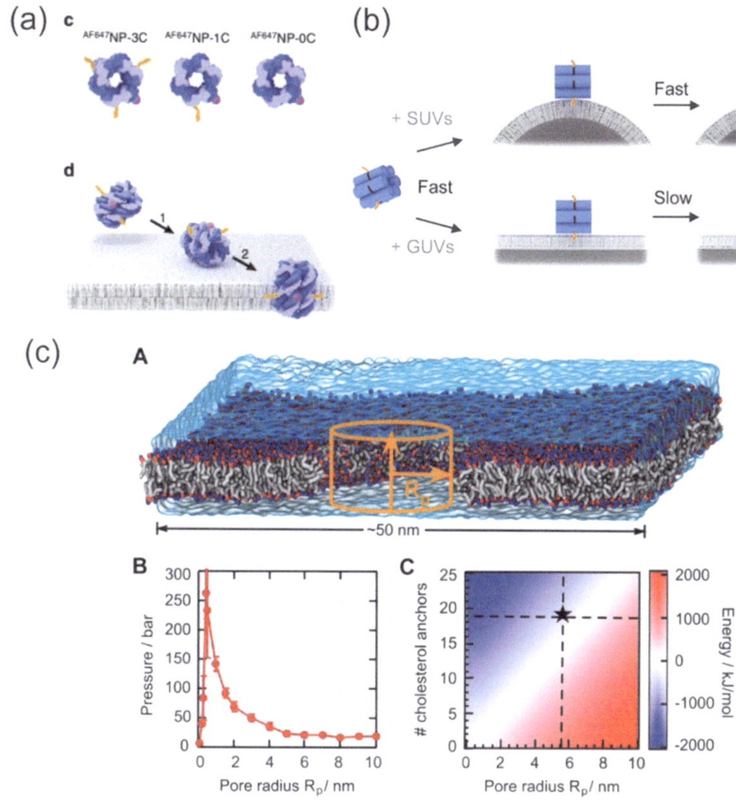

Fig. 8.4 Membrane insertion of DNA nanopores. **a** Schematics representing the possible process of 6HB nanopore insertion into membranes. 6HB is first bound to the membrane surface, followed by membrane insertion depending on the number of cholesterol modifications. Reprinted with permission of [15]. **b** Effects of membrane curvature on the nanopore insertion. High-curvature membranes enhance the insertion due to the membrane strain. Reprinted with permission of [17]. **c** Estimation of produced free energy by cholesterol insertion into lipid membranes and free energy penalty by the pore formation. Panel A: Set up the coarse-grained simulation. Panel B: Average pressure exerted by the lipid molecules on the potential wall to create the pore in the membrane as a function of the pore radius. Panel C: Free energy of the nanopore insertion as a function of pore radius and the number of cholesterols. Reprinted with permission of [18]

high-curvature membranes is most likely due to the membrane strain in strongly curved membranes. However, the number of cholesterol modifications is still dominant because the 6HB with one cholesterol did not show the insertion, yet membrane attachment. Göpfrich et al. theoretically estimated the number of cholesterol modifications required for nanopore formation (Fig. 8.4c) [18]. The authors focused on free energy produced by cholesterol insertion into a lipid bilayer membrane and free energy penalty by pore formation in the membrane. The calculation provided a free energy estimation as a function of pore size and the number of cholesterol modifications. It showed that the larger number of cholesterol modifications are suitable for nanopore insertion. However, as reported by Luo et al. [16], excess hydrophobic modifications can induce structural aggregation in solution mediated by hydrophobicity, which inhibits the membrane insertion of DNA nanopores.

One commonly used approach to facilitate membrane insertion is adding surfactants to the solution. Seifert et al. used n-octyloligooxyethylene (oPOE), a non-ionic surfactant, for the membrane insertion of porphyrin-modified 6HB nanopores [19]. Note that cholesterol has often been used for the hydrophobic anchors, but several studies used other hydrophobic molecules, including porphyrin, tocopherol, and peptide. oPOE was also used in other studies to achieve efficient insertion of DNA nanopores [20–23]. The mechanism of oPOE-mediated efficient insertion was investigated by Morzy et al. using 4HB [24]. They found that oPOE added below its critical micelle concentration can decrease the aggregation of 4HB with two cholesterols and lead to efficient insertion; however, oPOE did not decrease the aggregation of 4HB with four cholesterols, more strongly aggregating, and did not improve the insertion efficiency. The authors hypothesized that the main mechanism of the oPOE-mediated efficient insertion is likely due to the disordering of the membrane. The oPOE insertion into the membrane causes membrane strain, which is released as the formation of a small toroidal pore. Such membrane disruption might facilitate the insertion. This hypothesis is consistent with the discussion in [17] that strain in high-curvature membranes resulted in more efficient insertion.

Surfactants not only aid in membrane insertion but also in the challenging process of purification. Purifying DNA nanostructures modified with hydrophobic molecules is particularly difficult due to the tendency for aggregation. Iwabuchi et al. addressed this issue by developing a successful purification method for 6HB nanopores with four cholesterols [25]. They found that the addition of sodium cholate (SC) surfactant to the sample solution before structure folding effectively reduced aggregation. The 6HB nanopores prepared with SC were then purified using gel-extraction. The purified sample was successfully bound to the GUV membrane and functioned as nanopores.

8.3 Molecular Transport in DNA Nanopores

The molecular transport across the membrane is the most addressed property in DNA nanopores. It generally depends on the nanopore inner lumen size and cargo molecule. Krishnan et al. showed that a DNA origami nanopore with a rectangular aperture of $3.7 \times 4.2\ \mathrm{nm}^2$ can pass small dye molecules (Atto633) but cannot large dye-polymer conjugates (fluorescein-50 k dextran) [26]. Further details of the size dependency were reported by Thomsen et al. [27]. They constructed DNA origami nanopore with ~9 nm diameter and tested the size-selective molecular transport using GUVs and various dye molecules: a small sulforhodamine B (hydrodynamic radius; Rh ~ 0.5 nm; SRB), FITC-40 k dextran (Rh ~ 4.8 nm; dFITC-40 k), and FITC-500 k dextran (Rh ~15.9 nm; dFITC-500 k) (Fig. 8.5a). Experimental results showed that the smaller dye molecules (SRB and dFITC-40 k) passed through the nanopore inserted into

GUV membranes, whereas dFITC-40 k exhibited a slower flux than SRB. The large dFITC-500 k was unable to pass the pore. Additionally, by immobilizing molecular 'plug' (DNA-conjugated PEG polymer) within the nanopore, the flux of dFITC-40 k was blocked; however, unplugging by toehold mediated strand displacement reaction allowed for passing TAMRA-40 k dextran through the nanopore. Note that the largest DNA nanopore developed so far has ~30 nm of inner diameter, which can pass a molecule with 250 kDa [28].

The charge of the cargo molecule is also a key parameter for cargo transport. Diederichs et al. investigated the dye flux through a 6HB nanopore at the single molecular level [29]. They prepared an array of lipid membrane patches with micro-cavities containing dye molecules and compared the transport of Atto655 (0.6 kDa, hydrodynamic diameter; dh ~1.2 nm, and net charge –1), FAM (0.4 kDa, dh ~ 1.1 nm, and net charge –3), and Calcein (0.6 kDa, dh ~ 1.5 nm, and net charge –4) through 6HB nanopores with ~2 nm of inner diameter (Fig. 8.5b). Although the size of Atto655 and FAM is almost similar, FAM showed a lower transport rate and slower efflux rate (Atto655: 0.44 × 10-3 s-1, FAM 0.34 × 10–3 s-1) per single 6HB. Almost no transport was observed in Calcein, suggesting that electrostatic repulsion between the cargo and the lumen of negatively charged 6HB significantly reduced the transport rate. The authors also found that the electrostatic interaction between the 6HB and lipids alters the transport properties. FAM in GUVs containing 30 mol% of POPG, negatively charged lipids, exhibited efflux, but GUV without POPG did not result in FAM efflux. The authors discussed that electrostatic repulsion between POPG and 6HB induced pore dilation, enabling FAM transport.

8.4 Applications of DNA Nanopores

DNA molecules have a molecular sensing capability that enables gating function in DNA nanopores. A typical example is the recognition of DNA with specific base sequences. Many research groups have demonstrated the open/close function in DNA nanopores using DNA as a signal [30–33]. Dey et al. demonstrated a reversible gating function regulated by DNA hybridization and strand displacement reaction [32] (Fig. 8.6). They designed a DNA origami nanopore with a 20.4 × 20.4 nm square-shaped channel lumen at the center of a 70 × 70 nm plate shape. The gating function was realized by designing a square lid on the lumen (Fig. 8.6a). One side of the lid was connected to the base plate by flexible DNA hinges, whereas the other side was locked by hybridization with the plate, resulting in a close state of the nanopore. The unlock was induced by adding 'key' DNA strands that dissociate the DNA, hybridizing with the plate, by strand displacement reaction (Fig. 8.6b). The open lid can be switched back to the closed state by adding 'reverse key' DNA strands that release the key strands from the plate by strand displacement (Fig. 8.6b). The reversible gating function was confirmed by observing the transport of green fluorescence protein into GUVs (Fig. 8.6c). Other researchers demonstrated the gating function of DNA nanopores in response to temperature [34] and light [35]. Both were also based on

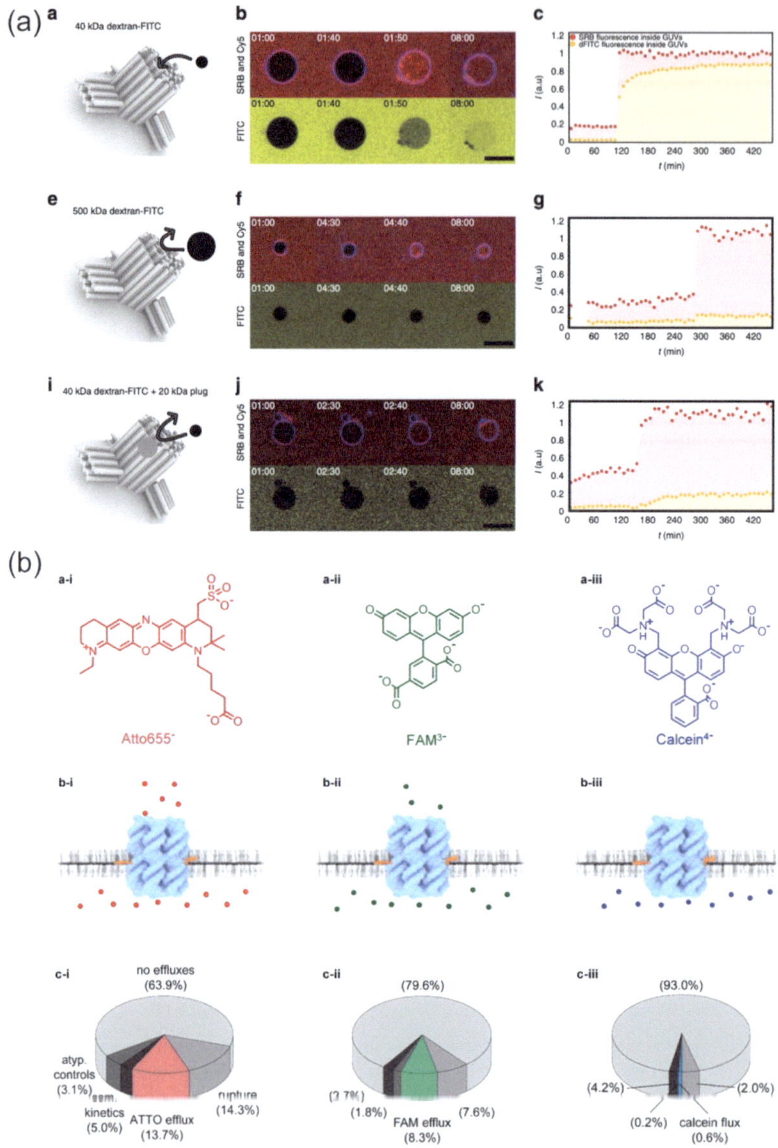

Fig. 8.5 Transport properties on DNA nanopores. **a** Size selectivity in cargo transport through DNA origami nanopore. Small molecules (sulforhodamine B; SRB) pass through the nanopore with ~9 nm of diameter (top). FITC-dextran 40 k (~4.8 nm in radius) exhibited slower transport (middle), and FITC-dextran500k (~15.9 nm in radius) was unable to pass through the nanopore (bottom). Scale bars: 10 μm. Reprinted with permission of [27]. **b** Size and charge-dependent transport properties. Atto655 (left), FAM (middle), and Calcein (right), with –1, –3, and –4 net charges were compared. Three times more negatively charged FAM than Atto655 showed a lower efflux rate, although they are similar in size. Calcein exhibited more than a 20-fold drop in efflux occurrence. Reprinted with permission of [28]

regulating DNA hybridization by external stimuli. As described here, the advantage of the use of DNA for nanopore construction is in the designability of the function, as well as the structural shapes. Another advantage of the DNA nanopore is the combination with a chemical reaction network composed of DNA.

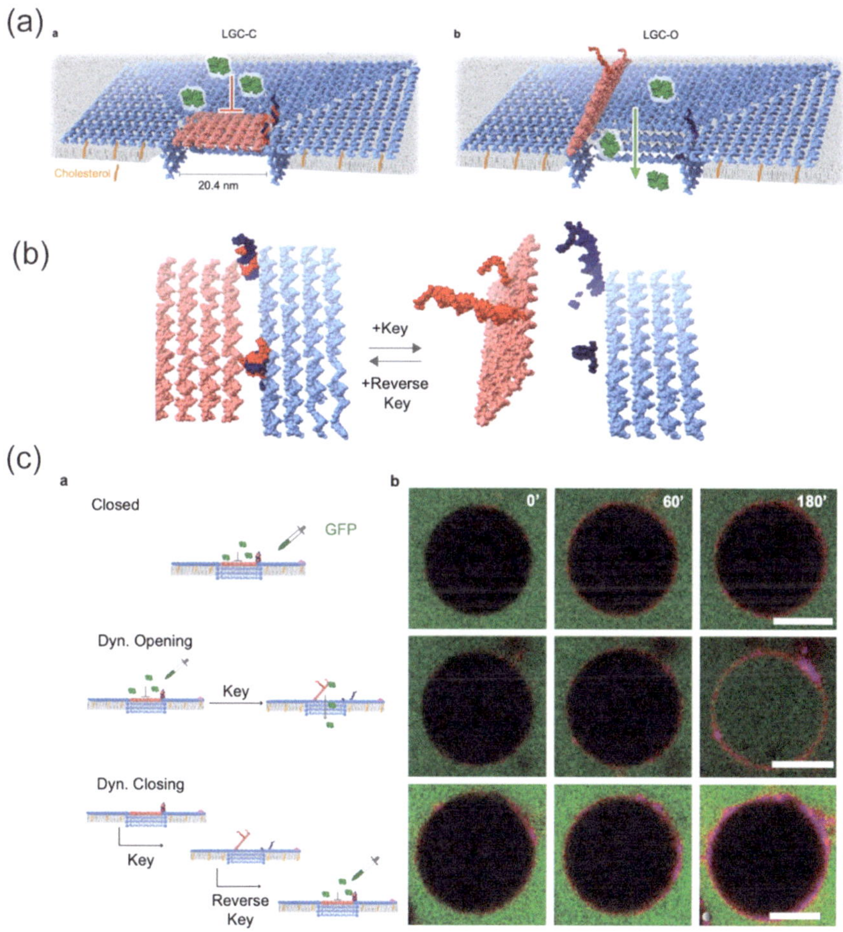

Fig. 8.6 Reversible gating function of DNA origami nanopore. **a** Schematics of the gating function of large gated channel (LGC) realized by regulating the lid (red colored) and the plate (blue colored) interaction. LGC-C and -O indicate LGC close and open, respectively. **b** Magnified schematics of the reversible open/close regulation of the lid. The key DNA hybridized with the DNA extended from the lid (red-colored strands). The reverse key DNA hybridized with the key DNA that removes the key from the plate, resulting in the hybridization of the lid and plate. **c** Schematics and microscope images representing the addition of green fluorescence protein (indicated as GFP) in the initial closed state (top), after the addition of the key DNA (middle), and after the addition of the key, followed by the reverse key DNA (bottom). Scale bars: 10 μm. Reprinted with permission of [32]

Peng et al. demonstrated a molecular signaling system encapsulated in GUVs equipping DNA nanopores (Fig. 8.7) [36]. The GUV inside contained DNA reaction networks consists of cascades of strand displacement reaction and enzymatic reactions (Fig. 8.7a). The initiation of the system was triggered by ATP recognition by the 6HB nanopore, which was initially closed by a 'locker' DNA strand. Because the locker is an ATP-binding aptamer sequence, ATP recognition resulted in opening the nanopore, leading to an influx of ions to GUV inside. The reaction network responded to specific metal ions, Zn^{2+} and Sr^{2+}. Zn^{2+}-dependent DNAzyme cleaved the substrate DNA, releasing signal DNA (indicated as S-I in Fig. 8.7b). The signal DNA was used for 'transduction' reactions, consisting of several strand displacement reactions, which eventually release a second signal DNA (indicated as S-II in Fig. 8.7b). The second signal was used for the enzymatic reaction to release DNA segments, which was finally detected by the rolling circular amplification reaction (Fig. 8.7b). Sr^{2+} was concurrently used to release a 'blocker' DNA strand, which switched back the nanopore to a closed state. The whole reaction successfully functioned in GUVs (Fig. 8.7c). This author concluded that this achievement may offer a strategy for engineering a smart molecular system.

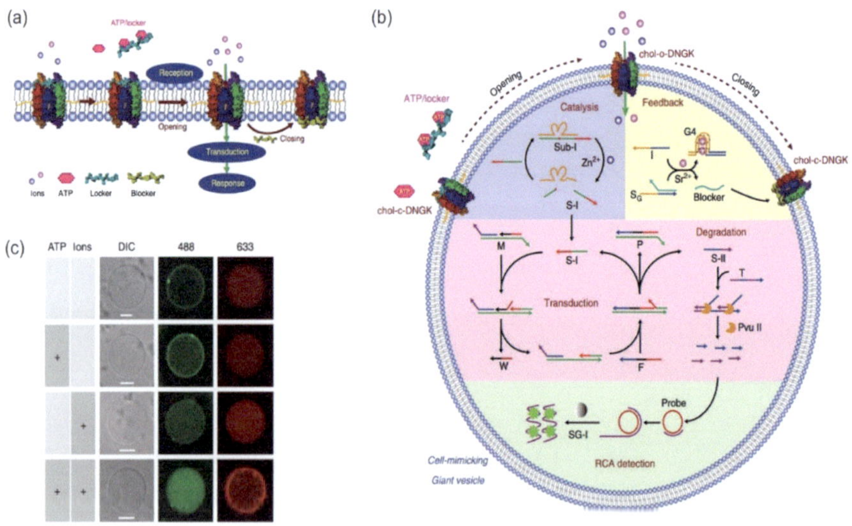

Fig. 8.7 An artificial signaling system in GUV realized by DNA nanopores and DNA reaction networks. **a** Schematics of whole system reactions. ATP opened the nanopore (reception). Then, ions pass through the nanopore, and reaction networks (transduction and response) are triggered. The nanopore switches to a closed state by releasing DNA through a reaction. **b** Whole reactions in this system. DNGK in this schematic indicates DNA nanogatekeeper. **c** Microscope images representing the system response under different conditions. Scale bars: 5 μm. Reprinted with permission of [36]

8.5 Conclusion and Perspective

The study of DNA nanopores has progressed rapidly since their first report only a decade and a few years ago. Initially focusing on ion transport through DNA nanopores in lipid membranes, research has now achieved the transport of molecules as large as 250 kDa. Moreover, autonomous opening and closing of nanopores in coordination with DNA reaction networks have been realized, promising to serve as critical components in complex molecular systems, such as chemical AI. However, current DNA nanopores have yet to achieve selective transport of ions and molecules and transport against concentration gradients (pumping function), which are critical functions in cellular systems. Developing ion-selective nanopores similar to sodium- or potassium-selective channels in neurons could enable electrical signaling between compartments (GUVs). Currently, DNA nanopores exhibit molecular size- and charge-selective transport. Incorporating molecular recognition mechanisms, such as aptamers, would allow for the selective transport of target molecules, including nucleic acids with specific sequences. This capability could facilitate information exchange between compartments via nucleic acid sequences, potentially enabling complex, large-scale information processing by combining with DNA computing. The DNA walker, which exhibits unidirectional motion on DNA nanostructures [37, 38], could offer insights into developing the pumping function in DNA nanopores. Achieving these functions may enable computation through DNA computing and various molecular reactions across multiple compartments, ultimately leading to the creation of an intelligent system composed of multiple cells as Chemical AI.

References

1. Sato, Y.: Artificial molecular systems for complex functions based on DNA nanotechnology and cell-sized lipid vesicles. ChemSystemsChem **6**, e20240021 (2024)
2. Herianto, S., et al.: Liposome-based artificial cells: from gene expression to reconstitution of cellular functions and phenotypes. Biomater. Adv. **142**, 213156 (2022)
3. Peng, Z., et al.: Lipid vesicle-based molecular robots. Lab Chip **5**, 996–1029 (2024)
4. Xing, Y., et al.: Functional nanopores enabled with DNA. Angew. Chem. Int. Ed. **62**, w20230313 (2023)
5. Rothemund, P.W., et al.: Folding DNA to create nanoscale shapes and patterns. Nature **440**, 297–302 (2006)
6. Douglas, S.M., et al.: Self-assembly of DNA into nanoscale three-dimensional shapes. Nature **459**, 141–418 (2009)
7. Langecker, M., et al.: Synthetic lipid membrane channels formed by designed DNA nanostructures. Science **338**, 932–936 (2012)
8. Burns, J.R., et al.: Self-assembled DNA nanopores that span lipid bilayers. Nano Lett. **13**, 2351–2356 (2013)
9. Bell, N.A.W., et al.: Nanopores formed by DNA origami: a review. FEBS Lett. **558**, 3564–3570 (2014)
10. Göpfrich, K., et al.: DNA-tile structures induce ionic currents through lipid membranes. Nano Lett. **15**, 3134–3138 (2015)

11. Göpfrich, K., et al.: Ion channels made from a single membrane-spanning DNA duplex. Nano Lett. **16**, 4665–4669 (2016)
12. Morzy, D., et al.: Membrane activity of a DNA-based ion channel depends on the stability of its double-stranded structure. Nano Lett. **21**, 9789–9796 (2021)
13. Bell, N.A.W., et al.: Ion DNA origami nanopores. Nano Lett. **12**, 512–517 (2012)
14. Wei, R., et al.: DNA origami gatekeepers for solid-state nanopores. Angew. Chem. Int. Ed. **51**, 4864–4867 (2012)
15. Birkholz, O., et al.: Multi-functional DNA nanostructures that puncture and remodel lipid membranes into hybrid materials. Nat. Commun. **9**, 1521 (2018)
16. Luo, L., et al.: DNA nanopores as artificial membrane channels for bioprotonics. Nat. Commun. **14**, 5364 (2023)
17. Burns, J.R., et al.: Defined bilayer interactions of DNA nanopores revealed with a nuclease-based nanoprobe strategy. ACS Nano **12**, 3263–3271 (2018)
18. Göpfrich, K., et al.: Large-conductance transmembrane porin made from DNA origami. ACS Nano **10**, 8207–8214 (2016)
19. Seifert, A., et al.: Bilayer spanning DNA nanopores with voltage-switching between open and close state. ACS Nano **9**, 1117–1126 (2015)
20. Ohman, A., et al.: A synthetic enzyme built from DNA flips 10^7 lipids per second in biological membranes. Nat. Commun. **9**, 2426 (2018)
21. Burns, J.R., et al.: Structural and functional stability of DNA nanopores in biological media. Nanomaterials **9**, 490 (2019)
22. Diederichs, T., et al.: Synthetic protein-conductive membrane nanopores built with DNA. Nat. Commun. **10**, 5018 (2019)
23. Morzy, D., et al.: Membrane activity of a DNA-based ion channel depends on the stability of its double-stranded structure. Nano Lett. **21**, 9789–9796 (2021)
24. Morzy, D., et al.: A surfactant enables efficient membrane spanning by non-aggregating DNA-based ion channels. Molecules **27**, 578 (2022)
25. Iwabuchi, S., et al.: Surfactant-assisted purification of hydrophobic DNA nanostructures. ChemBioChem **24**, e20220568 (2022)
26. Krishnan, S., et al.: Molecular transport through large-diameter DNA nanopores. Nat. Commun. **7**, 12787 (2016)
27. Thomsen, R.P., et al.: A large size-selective DNA nanopore with sensing applications. Nat. Commun. **10**, 5655 (2019)
28. Fragasso, A., et al.: Reconstitution of ultrawide DNA origami pores in liposomes for transmembrane transport of macromolecules. ACS Nano **15**, 12768–12779 (2021)
29. Diederichs, T., et al.: Principles of small-molecule transport through synthetic nanopores. ACS Nano **15**, 16194–16206 (2021)
30. Burns, J.R., et al.: A biomimetic DNA-based channel for the ligand-controlled transport of charged molecular cargo across a biological membrane. Nat. Nanotechnol. **11**, 152–156 (2016)
31. Iwabuchi, S., et al.: A large, square-shaped, DNA origami nanopore with sealing function on a giant vesicle membrane. Chem. Commun. **57**, 2990–2993 (2021)
32. Dey, S., et al.: A reversibly gated protein transporting membrane channel made of DNA. Nat. Commun. **13**, 2271 (2022)
33. Xing, Y., et al.: Multi-stimuli-responsive and mechano-actuated biomimetic membrane nanopores self-assembled from DNA. Adv. Matter. **35**, 2300589 (2023)
34. Arnott, P.M., et al.: A temperature-gated nanovalve self-assembled from DNA to control molecular transport across membranes. ACS Nano **13**, 3334–3340 (2019)
35. Offenbartl-Stiegert, D., et al.: A light-triggered synthetic nanopore for controlling molecular transport across biological membranes. Angew. Chem. Int. Ed. **61**, e202210886 (2022)

36. Peng, R., et al.: DNA-based artificial molecular signaling system that mimics basic elements of reception and response. Nat. Commun. **11**, 978 (2021)
37. Bath, J., et al.: Mechanism for a directional, processive, and reversible DNA motor. Small **5**, 1513–1516 (2006)
38. Ito, K., et al.: Design of artificial molecular motor inheriting directionality and scalability. Biophys. J. **123**, 858–866 (2024)

Chapter 9
Artificial Ion Channel Peptides

Junichi Taira

Abstract Peptides as short as 20 amino acid residues can form ion-conducting pores within lipid bilayers. Some of these peptides exhibit ion selectivity and can alter the pore size in response to changes in membrane potential. Artificial ion-channel peptides have been engineered by reverse-designing natural ion-channel proteins and peptides. Such artificial channels, capable of selective ion permeation across lipid membranes, are expected to serve as key functional components of molecular robots composed of liposomal units.

Living cells utilize transporter proteins for maintaining homeostasis and cellular signaling. Importing necessary molecules and exporting unnecessary ones are essential for maintaining cellular functions. Membrane transporters are also essential for intracellular signaling and intercellular communication. Among transporters, ion channels, which are responsible for the passive membrane permeation of ions, alter the intracellular charge and generate electrical signals. The physiological functions of ion-channel proteins are still studied, especially in neurophysiology. Their molecular functions have also been elucidated based on structural studies (e.g., ion filtering mechanisms and voltage sensors) [1, 2]. On the other hand, peptides with as few as 20 residues form ion-conducting pores in the lipid bilayer. Some of these peptides have ion selectivity and can change the size of the ion-conducting pore depending on the electric potential. Artificial ion-channel peptides have been developed based on the reverse engineering of natural ion-channel proteins and peptides. The artificial ion-channel peptide, which allows selective ion permeation through lipid membranes, is expected to be a key part of molecular robots distinguished by liposomal units.

J. Taira (✉)
Kyushu Institute of Technology, Fukuoka, Japan
e-mail: taira@bio.kyutech.ac.jp

S. Murata et al. (eds.), *Molecular Robotics II*,
https://doi.org/10.1007/978-981-95-2394-8_9

9.1 Natural Ion-Channel Peptides

Peptides that can form ion-conducting pores in the lipid bilayer are called ion-channel peptides or channel-forming ionophores and are mechanistically distinct from ion carriers like valinomycin. Natural ion-channel peptides have been isolated from various species, including fungi, insects, amphibians, and mammals. These peptides are believed to protect against bacterial invasion as antimicrobial agents. Interestingly, nonribosomal peptides (NRPs) [3] containing D-amino acids or unusual (non-protein-genic) amino acids have often been identified in fungal ion channel-forming peptides. This section overviews the structural property and pore-forming manners of alamethicin, magainin, and gramicidin as representative naturally occurring ion-channel peptides (Table 9.1). These peptides have been studied for many years, and their structures and ion-channel properties are intensively investigated [4].

Alamethicin is a 20-residue channel-forming peptide isolated from *Trichoderma virde* by Meyer and Reusser in 1967 [5]. It belongs to peptaibol (peptide with Aib and amino alcohol), a family of short peptides isolated from soil fungi. As its name indicates, peptaibol is an NRP characterized by a high content of 2-aminoisobutylic acid (Aib), an alpha-disubstituted amino acid, and C-terminal amino alcohol [6]. Alamethicin is a short peptide with many factors required to form an ion-conducting pore. Since Aib strongly induces a helical structure, most of the alamethicin structure is α-helix, with some 3_{10} helix structure. Alamethicin has no electric charge and is highly hydrophobic. The Gly-Leu-Aib-Pro sequence in its central portion is essential to provide flexibility and length to helices to form channel pores [7]. The barrel-stave model is an ion channel-forming mechanism in lipid bilayers, wherein parallel bundle formations of transmembrane helices surround a central pore. This model accounts for the unique multi-conductance of alamethicin-formed ion channels, and variations in the number of helices involved in the bundle formation can alter the pore size [8].

Magainin I and II are two closely related peptides isolated from *Xenopus laevis* with a broad spectrum of antimicrobial activity. Magainins form toroidal pores (toroidal model), supramolecular structures comprising lipid and peptide molecules [9]. The α-helix of magainin, which has an amphiphilic structure with distinct hydrophobic and cationic surfaces, allows peptide accumulation on bacterial

Table 9.1 Representative natural ion-channel peptides

Peptides	Sequence	Pore-forming manner
Alamethicin F-50	Ac-BPBABAQBVBGLBPVBBQQF*ol*	Barrel-stave model
Magainin I	GIGKFLHSAGKFGKAFVGEIMKS	Toroidal model
Gramicidin A	For-VGAlAvVvWlWlWlW-*EA*	Head-to-head dimer

The sequences are displayed using the standard single-letter code for amino acids. 2-Aminoisobutyric acid (Aib) is abbreviated as B. Lowercase letters indicate D-amino acids. The following abbreviations are used for terminal blocking groups: Ac, acetyl; *EA*, ethanolamine; *ol*, C-terminal amino alcohol; for, formyl.

membranes and the successive forming of the toroidal pore. Amphiphilicity and positive charge cannot be ignored in the de novo design of ion channel-forming peptides with α-helix structures.

Gramicidin is a mixture of structurally similar NRPs: gramicidin A, B, and C, reported by Dubos in 1939 [10]. It is the oldest antibiotic in clinical application. Gramicidin A, the main component of gramicidin, is a pentadecapeptide composed of seven L-amino acids alternating with six D-amino acids and glycine. The ion channel formed by gramicidin A is believed to comprise a β-helix with 6.3 residues per turn, which dimerizes head-to-head to form an ion-conducting pore of a specific size [11]. In this structure, the side chains of the amino acids orient toward the outside of the ion-conducting pore. This pore selectively permeates many monovalent cations while excluding divalent cations and anions.

9.2 De Novo Designed Ion-Channel Peptides

Various de novo designed artificial ion-channel peptides have been developed to obtain essential structural principles for ion-channel formation (Table 9.2). These peptides provide a straightforward model for studying the relationships between peptide structures and their ion channel-forming properties. Menestrina et al. reported the voltage-dependent channel formation of hydrophobic peptides composed of alanine and Aib repeated sequences [12]. However, most previously studied synthetic α-helical ion-channel peptides possess amphiphilic properties (Table 9.2) [13–19]. For instance, DeGrado et al. reported ion-channel formation by amphiphilic peptides comprising leucine and serine residues [13–15]. The channels are modestly cation-selective and exhibit asymmetric current–voltage curves, reflecting voltage-dependent conductance. Additionally, Lee et al. described ion-channel formation by amphiphilic peptides containing arginine residues [16, 17]. Kondo et al. designed an Aib-rich peptide sequence, Ac-(Aib-Xxx-Aib-Ala)$_5$-NH_2 (BXBA20), as a template of the amphiphilic helix, where the Xxx positions were used in a structure–activity relationship study [18, 19]. Ghadiri et al. reported ion-channel formation by cyclic peptides [20, 21]. An 8-residue cyclic peptide, formed by repeated D and L-amino acids, stacked to form a self-assembled cylindrical beta-sheet, resulting in nanotube formation. The conductance patterns suggest that ions passed through these nanotubes with a defined pore size. This ion-channel formation by cyclic peptides has many common features with the ion-conducting pores formed by gramicidin A. The transport rates of potassium and sodium ions through these channels were three times those of gramicidin A. A notable characteristic is the ability to alter the pore size by changing the number of amino acids constituting the cyclic peptide.

Table 9.2 Representative De Novo designed channel peptides

Peptides	Sequence	References
P5–P20	Boc-(ABABA)$_n$-OMe(n = 1–4)	[12]
LS-I	(LSSLLSL)$_3$-NH$_2$	[13, 14]
LS-IIb	(LSLBLSL)$_3$-NH$_2$	[15]
ARL$_8$	(Ala-Arg-Leu)$_8$	[16]
LRL$_8$	(Leu-Arg-Leu)$_8$	[16]
4$_6$	(LARL)$_3$-(LRAL)$_3$	[17]
BXBA20	Ac-(BXBA)$_5$-NH$_2$ (X = K, E, G, S)	[18, 19]
	cyclo[-(Trp-D-Leu)$_3$-Gln-D-Leu-]	[20]

9.3 Enhanced Pore Formation of Artificial Ion Channel Peptides Assisted by Interhelical Electrostatic Interaction

Ion-channel peptides have limited functions compared to ion-channel proteins. However, due to the difficulty in reconstituting ion-channel proteins, ion-channel peptides with a function to form or decay ion-permeable pores under specific conditions are preferable for applications in molecular robots. Among ion-channel peptides, amphiphilic α-helical peptides have the most promising structure. Many de novo designed ion channels comprise amphiphilic helices, as mentioned above. Extensive knowledge of helix-helix interactions, driven by hydrophobic and electrostatic interactions (e.g., coiled-coil), is also crucial in peptide-based supramolecular assembly [22–24]. The effective construction of stable helix bundles in lipid bilayers is one of the most critical issues in developing advanced artificial ion channels using helical peptides. Generally, the formation of helix bundles involves the binding and insertion of helices into lipid membranes, followed by the formation of higher-order structures through interactions between helices. Therefore, investigating helix-helix interactions in phospholipid bilayers would be an appropriate approach to address these processes. Electrostatic interactions between helices facilitate the insertion of charged regions of membrane-integral proteins into the lipid bilayer [25] (i.e., the polar amino acids are crucial in the interaction of transmembrane helices within the hydrophobic lipid bilayer) [26]. The author evaluated the effect of the helix-helix interaction facilitated by electrostatic interaction on ion-channel formation using a dimer of Aib-containing peptides with a simple structure [27]. The structure of the model peptide, referred to as K20E20, is illustrated in Fig. 9.1a. According to the helical net and wheel diagrams, clustered Lys or Glu residues are observed on the helix surface. The dimeric peptide is expected to orient hydrophobic regions toward the hydrophobic interior of the lipid bilayer and form intramolecular salt bridges.

Figure 9.1b represents the spectral changes in the circular dichroism (CD) measurement of K20E20 during titration with small unilamellar vesicles (SUVs) of dipalmitoylphosphatidylcholine (DPPC) under neutral (pH 7.4) and acidic (pH 3.0) conditions. Under neutral conditions, K20E20 exhibited spectral changes upon

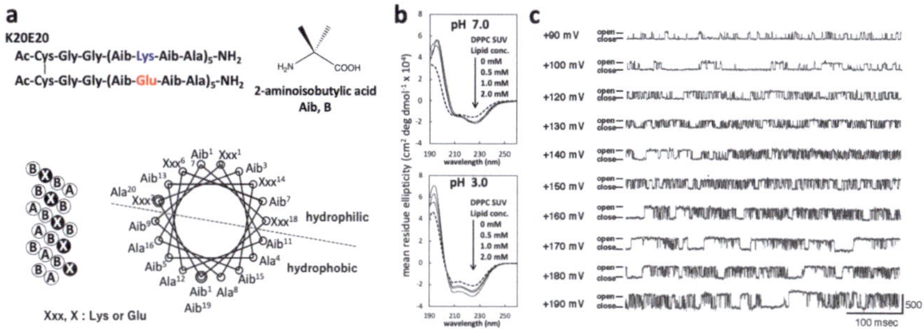

Fig. 9.1 Impact of salt bridge-assisted helix-helix interactions on ion-channel formation by amphiphilic α-helices. **a** Structure of the model peptide K20E20. Left panel, amino acid sequence of K20E20 and structure of 2-aminoisobutyric acid; right panel, helical net and wheel of BXBA20. **b** Changes in the circular dichroism spectra of K20E20 (**10 μM**) **pH 7.0** during titration with DPPC SUVs at pH 7.4 (upper panel) and pH 3.0 (lower panel). Dotted lines and solid lines represent spectra recorded in the absence or presence of DPPC SUVs, respectively. DPPC concentrations are 0.5, 1.0, and 2.0 mM. **c** The single-channel properties of K20E20 (1 nM) at pH 7.4 under variable potentials (+90 mV to + 190 mV). Reprinted with permission from Biochemistry 47, 3705–3714 (2008). Copyright 2008 American Chemical Society

titration with DPPC SUVs, though no lipid concentration dependence was observed. In contrast, lipid concentration-dependent spectral changes were observed under acidic conditions, where the charge on glutamic acid is lost. These results suggest that the intramolecular electrostatic interactions promoted peptide accumulation on hydrophobic membranes. Furthermore, the single-channel analysis demonstrated that K20E20 forms stable ion channels with a single pore size under neutral conditions, even with low (1 nM) peptide concentrations (Fig. 9.1c). Taken together, the electrostatic interactions within the K20E20 molecule are considered to expose the hydrophobic face of the dimeric helices, thereby enhancing their hydrophobicity, which facilitates their adsorption in the lipid bilayer and efficiently forms ion-permeable pores through the assistance of electrostatic interactions. On the other hand, under acidic conditions, where the charge of glutamic acid is lost, the affinity of K20E20 to the lipid membrane is diminished, suggesting that pH conditions affect the intramolecular electrostatic interaction in K20E20 and affects ion-conducting pore formation and disruption. E20E20 is expected to serve as a prototype of a functional molecule that regulates the rate of ion transport across lipid bilayer in response to pH.

The application of peptides in the development of molecular robots is expected to accelerate. Research findings on elucidating the functions of natural ion channel peptides, developing of artificial ion channel peptides, and construction of structures based on peptide–peptide interactions are beginning to lead to their application as components of molecular robots [28].

References

1. Doyle, D.A., et al.: The structure of the potassium channel: molecular basis of K+ conduction and selectivity. Science **280**, 69–77 (1998)
2. Payandeh, J., et al.: The crystal structure of a voltage-gated sodium channel. Nature **475**, 353–358 (2011)
3. Süssmuth, R.D., Mainz, A.: Nonribosomal peptide synthesis—principles and prospects. Angew. Chem. Int. Ed. Engl. **56**, 3770–3821 (2017)
4. Takada, Y., et al.: Discovery of gramicidin A analogues with altered activities by multidimensional screening of a one-bead-one-compound library. Nat. Commun. **11**, 4935 (2020)
5. Meyer, C.E., Reusser, F.: A polypeptide antibacterial agent isolated from *Trichoderma viride*. Experientia **2**, 85–86 (1967)
6. Rindfleisch, H., Kleinkauf, H.: Biosynthesis of alamethicin. FEBS Lett. **62**, 276–280 (1976)
7. Sansom, M.S.P., et al.: The mechanism of channel formation by alamethicin as viewed by molecular dynamics simulations. Novartis Found. Symp. **225**, 128–141 (1999)
8. Daniel, J.F.S., Filho, E.R.: Peptaibols of Trichoderma. Nat. Prod. Rep. **24**, 1128–1141 (2007)
9. Murzyn, K., Pasenkiewicz-Gierula, M.: Construction of a toroidal model for the Magainin pore. J. Mol. Model. **9**, 217–224 (2003)
10. Dubos, R.J.: Studies on a bactericidal agent extracted from a soil Bacillus. J. Exp. Med. **70**, 1–10 (1939)
11. Wallace, B.A.: Structure of gramicidin A. Biophys. J. **49**, 295–306 (1986)
12. Menestrina, G., et al.: Voltage-dependent channel formation by rods of helical peptides. J. Membr. Biol. **93**, 111–132 (1986)
13. Lear, J.D., et al.: Synthetic amphiphilic peptide models for protein ion channels. Science **240**, 1177–1181 (1988)
14. DeGrado, W.F., et al.: Protein design, a minimalist approach. Science **243**, 622–628 (1989)
15. DeGrado, W.F., Lear, J.D.: Conformationally-constrained a-helical peptide models for protein ion channels. Biopolymers **29**, 205–213 (1990)
16. Iwata, T., et al.: Design and synthesis of amphipathic 3_{10}-helical peptides and their interactions with phospholipid bilayers and ion channel formation. J. Biol. Chem. **269**, 4928–4933 (1994)
17. Agawa, Y., et al.: Interaction with phospholipids bilayers, ion channel formation, and antimicrobial activity of basic amphiphatic α-helical model peptides of various chain length. J. Biol. Chem. **266**, 20218–20222 (1991)
18. Higashimoto, Y., et al.: Structure-function relationship of model Aib-containing peptides as ion transfer intermembrane templates. J. Biochem. **125**, 705–712 (1999)
19. Hara, T., et al.: Side chain effect on ion channel characters of Aib rich peptides. J. Biochem. **130**, 749–755 (2001)
20. Ghadiri, M.R., et al.: Artificial transmembrane ion channels from self-assembling peptide nanotubes. Nature **369**, 301–304 (1994)
21. Bong, D.T., et al.: Self-assembling organic nanotubes. Angew. Chem. Int. Ed. Engl. **40**, 988–1011 (2001)
22. Marsden, H.R., Kros, A.: Self-assembly of coiled coils in synthetic biology: inspiration and progress. Ang. Chem. Int. Ed. **49**, 2988–3005 (2010)
23. Lapenta, F., et al.: Coiled coil protein origami: from modular design principles towards biotechnological applications. Chem. Soc. Rev. **47**, 3530–3542 (2018)
24. De Santis, E., et al.: Antimicrobial peptide capsids of de novo design. Nat. Commun. **8**, 2263 (2017)
25. Zhang, L., et al.: Contribution of hydrophobic and electrostatic interactions to the membrane integration of the Shaker K^+ channel voltage sensor domain. Proc. Nat. Acad. Sci. U.S.A. **104**, 8263–8268 (2007)
26. Gratkowski, H., et al.: Polar side chains drive the association of model transmembrane peptides. Proc. Nat. Acad. Sci. U.S.A. **98**, 880–885 (2001)

27. Taira, J., et al.: Ion channel formation assisted by electrostatic inter-helical interactions in covalently dimerized amphiphilic helical peptides. Biochemistry **47**, 3705–3714 (2008)
28. Chen, Z., et al.: Synthetic beta cells for fusion-mediated dynamic insulin secretion. Nat. Chem. Bio. **14**, 86–93 (2018)

Chapter 10
Probe-Type Artificial Cell Membrane System

Hiroki Koiwa, Hiromu Akai, Shun Okada, and Kan Shoji

Abstract *De novo-designed* artificial ion channels, including DNA, peptide, and organic compound, attract attentions as the sensor of molecular robots because they can emerge various stimulus responsiveness. To apply artificial ion channels as the sensor of molecular robots, it is necessary to understand the detailed characteristics of ion channels, such as channel structure, stability in the artificial cell membranes, molecular transportation, and molecular selectivity. Although the ion current measurement is one of the powerful methods to evaluate the characteristics of artificial ion channels, the low throughput of ion current measurements using conventional methods to form artificial cell membranes prevents the efficient evaluation of artificial ion channels. In this chapter, we introduce probe-type artificial cell membrane systems that can improve the throughput of ion current measurements.

10.1 Introduction

Liposome-based molecular robots that are constructed by integrating molecular-based sensors, intelligence, and actuators into giant liposomes attract much attention in the fields of medicine and drug delivery [1]. To develop molecular robots with stimulus-responsive properties analogous to living cells, functions that can recognize stimulations from external conditions and convert these stimulations to molecular

H. Koiwa · H. Akai · S. Okada · K. Shoji (✉)
Nagaoka University of Technology, Niigata, Japan
e-mail: kshoji@mech.nagaokaut.ac.jp

H. Koiwa
e-mail: s193043@stn.nagaokaut.ac.jp

H. Akai
e-mail: s215009@stn.nagaokaut.ac.jp

S. Okada
e-mail: s193022@stn.nagaokaut.ac.jp

S. Murata et al. (eds.), *Molecular Robotics II*,
https://doi.org/10.1007/978-981-95-2394-8_10

signals detectable by molecular robots, such as osmotic pressure and molecule transport through liposomal membranes, are required. To realize these functions, biological ion channels that form nanoscale pores across liposomal membranes have been utilized [2]. The biological ion channels can mediate molecular transport through the liposomal membrane, enabling communication between the inside and outside conditions of molecular robots. However, the biological ion channels that can detect specific stimulation or selectively discriminate molecules are limited.

Recently, de novo-designed artificial ion channels, including DNA, peptides, and organic compounds, have been developed and can emerge various functions that do not exist in nature [3]. For example, the artificial ion channels can reversibly open and close in the presence of specific molecules and external stimulations, including pH, temperature, and light. In addition, the artificial ion channels can selectively transport ions and molecules based on their size or charge across the liposomal membranes.

To apply artificial ion channels as the sensor of molecular robots, it is necessary to understand the detailed characteristics of ion channels, such as channel structure, stability in the artificial cell membranes, molecular transportation, and molecular selectivity. Currently, various evaluation methods of ion channels, such as cyro SEM, dye leakage assay, and ion current measurements were applied in the development of artificial ion channels [4]. Especially, ion current measurements potentially provide an easy and convenient evaluation for investigating the channel size and molecular transport of ion channels. Conventionally, numerous methods for ion current measurements that use artificial cell membranes prepared by the painting method and the Montal–Mueller method have been reported [5]. However, the low throughput of ion current measurements with these conventional methods prevents the efficient evaluation of artificial ion channels. Thus, we have developed probe-type artificial cell membrane systems to improve the throughput of ion current measurements [6, 7]. In these systems, the artificial cell membrane is formed at the tip of probes by simply inserting the electrode into a layered bath solution of an aqueous solution and an oil/lipid mixture. Then, ion currents through the artificial cell membrane can be monitored by applying the membrane potential between the probe and a reference electrode immersed in the bath solution. In this section, we introduce the mechanisms of membrane formation and the methods of ion current measurements with our recently developed probe systems. We further provide the DNA nanopore probe technology that can improve the insertion efficiency of DNA nanopores into artificial cell membranes [8].

10.2 Mechanism of Artificial Cell Membrane Formation with Probe Systems

First, we introduce the mechanism and procedures to form artificial cell membranes using probe systems. Although we have reported various probe-type artificial cell membrane systems [6, 7, 9, 10], in this section, we introduce two kinds of probe systems: a polyethylene glycol (PEG)-modified gold needle electrode [6] and a recessed-in-glass Ag/AgCl electrode [7, 11]. The artificial cell membranes are formed by the same mechanism in any probe systems and are formed based on the droplet contact method [12]. Specifically, the artificial cell membrane is formed by contacting the lipid monolayers around the tip of the microelectrode and at the interface between the aqueous solution and the oil/lipid layer by pushing down the microelectrode with a hydrophilic surface into the layered bath solution. The probe-type artificial cell membrane systems offer advantages such as highly stable bilayers and easily repeatable reformation of membranes because a supported membrane is formed, and membrane formation is controlled by pushing down and pulling up the microelectrode. Next, the fabrication methods of each microelectrode and the formation procedures of artificial cell membranes are described as follows.

Gold needle electrode

Gold needle electrodes are fabricated by electrochemical etching of gold wires (Fig. 10.1a). Then, thiol-PEG is modified onto the surface of gold needle electrodes to obtain a hydrophilic layer. Artificial cell membranes can be formed using the PEG-modified gold needle electrode in the following manner. First, the gold needle electrode is inserted into the oil/lipid mixture of the bath solution to form a lipid monolayer on the surface of the gold needle electrode. Then, the lipid bilayer is formed when contacting lipid monolayers by pushing down the gold needle electrode across the interface of the oil/lipid mixture and the aqueous solution (Fig. 10.1b).

Ag/AgCl electrode

Ag/AgCl electrodes are fabricated in the following procedures. First, a silver wire is sealed in a pulled glass capillary and connected to copper wires with solder. To expose the silver wire, the tip is polished with abrasive papers. Next, the glass-sealed silver wire is etched by soaking in the etchant to fabricate a microcavity at the tip [11]. Then, the AgCl surface is prepared by immersing the electrode in a bleach solution (Fig. 10.1c). Finally, the Ag/AgCl electrode is soaked in the aqueous solution to fill the microcavity at the tip with the aqueous solution. By using the Ag/AgCl electrode, artificial cell membranes can be formed in the same manner as the gold electrode (Fig. 10.1d).

10.3 Ion Current Measurements Using Probe Systems

Next, we introduce the method and advantages of the ion current measurement using the probe-type artificial cell membrane systems. The ion current measurement using the probe systems is performed in a Faraday cage with a simple setup that consists of the probe, a reference electrode, a micromanipulator, the bath chamber, and a patch-clamp amplifier (Fig. 10.2a). The probe is attached to the micromanipulator for controlling the vertical position of the probe, and the reference electrode, a Ag/AgCl electrode, is inserted into the bath electrolyte solution containing ion channels.

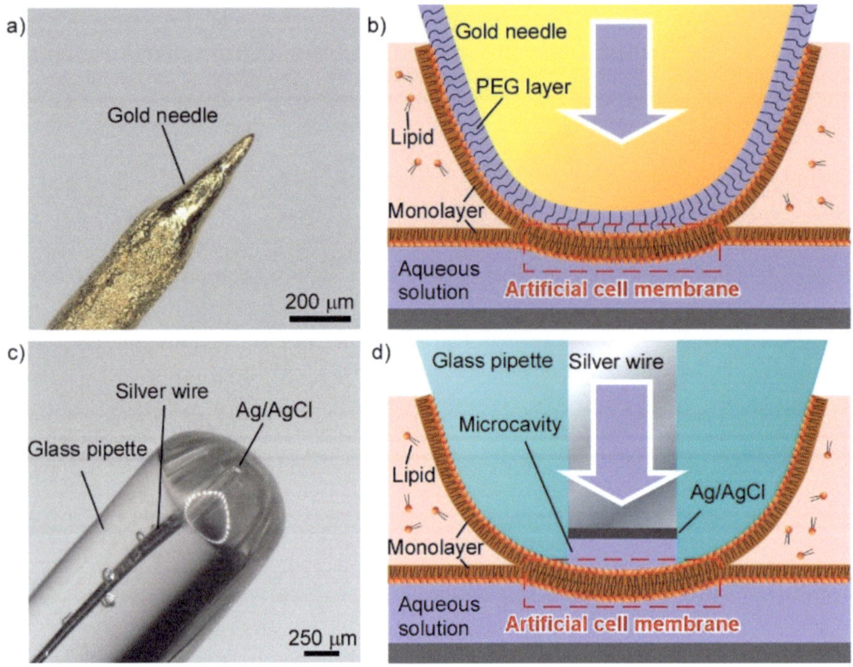

Fig. 10.1 Probe-type artificial cell membrane systems using a gold needle electrode and a Ag/AgCl microelectrode. **a** Microscopic image of a gold needle fabricated by etching a gold microwire. **b** Formation mechanism of an artificial cell membrane with a PEG-modified gold needle. **c** Microscopic image of a Ag/AgCl electrode fabricated by sealing a silver wire in a glass pipette. **d** Formation mechanism of an artificial cell membrane with a Ag/AgCl electrode

The patch clamp amplifier is connected to the electrodes and is used to apply the transmembrane potential and measure the ion current at the pico ampere level. The whole setup is constructed in a Faraday cage to cut off the environmental noises.

The ion current is measured by applying the transmembrane potential between the electrodes. Because ions, which are polar molecules, cannot pass through the artificial cell membrane, the ion current without the ion channels in the membrane is 0 A. On the other hand, because the ions can pass through the ion channels, constant ion currents depending on the channel size are observed when ion channels are reconstituted into the membrane. Thus, we can estimate the channel size by substituting the ion current, applied potential, channel length, and resistivity of the electrolyte solution into the Hille equation [13]. Moreover, we can evaluate the molecular transport through the ion channels by observing ion current changes when molecules pass through the ion channels.

The simple and highly reproducible membrane formation with the probe systems offers the advantage of high-throughput ion current measurements. To evaluate the channel characteristics, including the pore size and molecular transportation, based on ion current measurements, statistical analysis is required. In the probe systems,

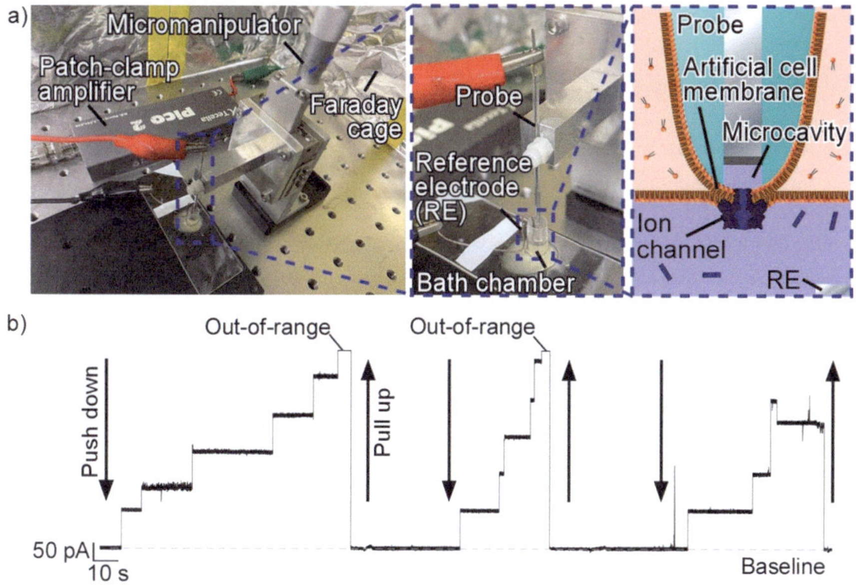

Fig. 10.2 **a** Setup for the ion current measurement using the probe-type artificial cell membrane system. **b** Ion current trace representing the reproducibility of the artificial cell membrane and reconstitution of ion channels with the probe system. Reproduced with permission from [7]

because we can easily collect current signals from multiple ion channels by repeatedly re-forming the artificial cell membranes (Fig. 10.2b) [7], statistical and accurate evaluations of the channel characteristics are achieved. For example, we previously reported high-throughput nanopore sensing using a 3D-printed chamber that has four microdroplets containing ion channels and different concentrations of single-stranded DNA (ssDNA) [7]. In this experiment, we successfully observed ssDNA translocation signals with different frequencies depending on the ssDNA concentration of each microdroplet within 10 min, indicating the highly efficient evaluation of molecular transport of the ion channels using the probe systems.

In this subsection, we introduced the detailed procedures and general advantages of the ion current measurement with the probe systems. From here, we explain own features of each probe system by using the ion current measurements of pore-forming proteins.

Gold needle electrode

The typical ion current trace of the α-hemolysin (αHL) obtained with the gold needle electrode is shown in Fig. 10.3a [6]. The step-like current increases indicate the reconstitutions of αHLs into the artificial cell membrane. Moreover, because the pore conductance calculated from the step-like current increase was 1.00 nS which was consistent with previous research [14], the successful ion current measurement of αHL using the gold needle electrode was confirmed. On the other hand, the current decay over time is typically observed in the ion current signals with the gold needle electrode (Fig. 10.3a). This phenomenon is caused by the formation of the electric double layer on the surface of the gold needle electrode. The experimental setup of the ion current measurement using the gold needle electrode (Fig. 10.3b) can be described as an equivalent circuit with the capacitance of the electric double layer (C_d) and the resistance of the αHL (R_p) (Fig. 10.3c). The capacitance is charged up by accumulating ions passing through the ion channel on the surface of the gold needle electrode. Consequently, the applied potential against the ion channel gradually decreases and the current decay is induced. This is the issue in the ion current measurement using the gold needle electrode because channel characterizations with a need for a high signal-to-noise (S/N) ratio, constant voltage, and long-time measurements is difficult. Nevertheless, we can resolve this issue by reducing the ion flow with the following methods: reduction of the salt concentration or applied potential. Furthermore, the solution using a Ag/AgCl needle, which is an equilibrium electrode, was reported [15].

On the other hand, by applying the ion current measurement using the gold needle electrode, we can classify two kinds of channel formation mechanisms: the additional channel insertion and expansion of existing channels [16]. Because the molecular transportation through ion channels is related to the pore diameter, changes in pore diameter induced by pore expansion potentially cause changes in the characteristics

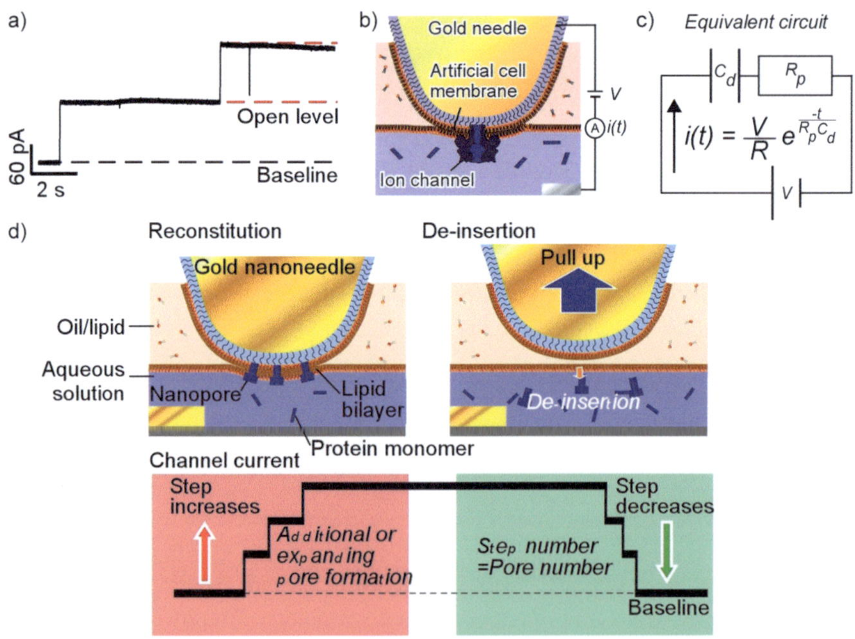

Fig. 10.3 **a** Representative current trace of αHL with a gold needle. Reproduced with permission from [6]. **b** Experimental setup with the gold needle and **c** equivalent circuit. **d** Concept of deinsertion current analysis for decerning pore formation mechanism. Reproduced with permission from [16])

of nanopore sensors in molecular robots. Therefore, evaluation of channel formation mechanisms is important in the development of molecular robots. Because both channel formation processes are expressed by very similar step current increases, it was very difficult to distinguish whether another channel is additionally inserted or the existing channel expands from the current signals. In the ion current measurement using the gold needle electrode, we can obtain current signals associated with deinsertions of ion channels by gently pulling up the electrode to gradually peel the lipid monolayers (Fig. 10.3d). Because the number of step decreases indicates the number of channels inserted into the artificial cell membrane, we can recognize the principal channel formation mechanism of the ion channel by comparing the step numbers and step currents of insertion and deinsertion.

From the above, the gold needle electrode allows not only the conventional channel evaluations, such as channel size and molecular transportation, with high throughput but also the analysis of channel formation mechanisms, although we need to take care of the current decay.

Ag/AgCl electrode

On the other hand, in the ion current measurement using the Ag/AgCl electrode which is the equilibrium electrode, we can measure ion currents without current decay caused by the formation of the electric double layer (Fig. 10.2b). Nevertheless, the current decay occurred by accumulating the ions passing through ion channels and changing the ion concentration in a microcavity on the tip of the Ag/AgCl electrode. Thus, we proposed a way of etching the silver to obtain the large microcavity [11]. By using the method, the ion current measurement almost without the current decay is enabled although a large number of αHL was reconstituted. Moreover, in the ion current measurement of another biological channel called streptolysin-O, which can construct the pore with an inner diameter of tens of nm, the current signal without current decay until 20 s after the reconstitution of the channel was obtained. This result indicated the applicability of the Ag/AgCl electrode for the evaluation of the ion channels with a large inner diameter.

From the above, the Ag/AgCl electrode with a large microcavity offers a valuable method for the characterization of ion channels with a large inner diameter and the long-term measurement of ion channels.

In summary, by using the probe systems, we can perform highly efficient evaluations of channel size and molecular transport. Moreover, the gold-needle-electrode-based system has the issue of current decay but offers the evaluation of channel formation mechanisms by measuring deinsertion currents of ion channels. On the other hand, because the Ag/AgCl electrode is characterized by almost no current decay, they enable the evaluation of ion channels with a large inner diameter and the long-term measurement of ion channels.

10.4 DNA Nanopore Probe

In the preceding subsection, we described the efficient evaluation of biological ion channels by ion current measurements using probe-type artificial cell membrane systems. However, because DNA nanopores, which are expected to be applied as sensors for molecular robots, offer lower insertion efficiency into cell membranes compared with biological nanopores, it was challenging to efficiently evaluate the characteristics of DNA nanopores. Thus, we introduce the DNA nanopore probe which can efficiently insert DNA nanopores by applying the probe systems. The DNA nanopores, artificial ion channels constructed by the structural DNA nanotechnology, are potentially utilized as sensors of molecular robots because of their high designability of size and shape, and functionality by chemical modification [4]. On the other hand, although hydrophobic modifications for DNA nanopores are required to insert hydrophilic DNA nanopores into artificial cell membranes, which have hydrophobic inside, the insertion processes of DNA nanopores into artificial cell membranes take time even with the hydrophobic modifications [17]. Additionally, hydrophobic molecules also induce the aggregation of DNA nanopores due to hydrophobic interactions, causing lower yields of DNA nanopores [18]. Hence, we developed the DNA nanopore probe technology which can efficiently insert DNA nanopores without hydrophobic modifications into the membrane. DNA nanopore probes are prepared by immobilizing DNA nanopores on gold needle electrodes, and DNA nanopores can be physically inserted into the artificial cell membrane formed on the surface of the probe by immersing the probe into the layered bath chamber of the aqueous solution and oil/lipid mixture [8, 19].

In this subsection, we describe the design of DNA nanopores to immobilize on the gold needle electrode and the insertion mechanism of DNA nanopores into artificial cell membranes using the DNA nanopore probe. Additionally, we explain the performance of the DNA nanopore probes based on the results of ion current measurements using the DNA nanopore probes.

Design of DNA nanopore and preparation of DNA nanopore probe

The DNA nanopore probe was constructed by hybridizing a sticky end extended from a six-helix bundle (6HB) DNA nanopore to ssDNAs premodified on gold electrodes (Fig. 10.4a). To physically insert the DNA nanopores into the artificial cell membrane using the DNA nanopore probe technology, the DNA nanopores immobilized on the gold needle electrode need to protrude from the PEG layer. Therefore, the thickness of the PEG layer and the length of the sticky end were optimized as follows. The theoretical and measured lengths of the 6HB DNA nanopore have been reported as 14 and 20.4 ± 4.5 nm, respectively [20]. Conversely, the thicknesses of the PEG (molecular weight (MW) = 3000) layer and the artificial cell membrane on the gold needle electrode were approximately 24 nm [6] and 6 nm, respectively. Thus, assuming that the maximum length of DNA nanopores is 24.9 nm, when the distance between the gold needle electrode and the DNA nanopore is in the range of 5.1–24 nm, DNA nanopores tethered to the electrode can be inserted into the membrane.

From the aforementioned estimation, we determined the length of double-stranded DNA, which connects the gold electrode and the DNA nanopores, as 30 bp (length = 9.86 nm) and designed the 6HB DNA nanopore with the sticky end of 30 nucleotides.

Next, we described the method to modify DNA nanopores onto the gold needle electrodes. First, the thiolated ssDNA, which can hybridize to the sticky end, was immobilized on the PEG-modified gold needle electrode by dipping the electrode into a ssDNA solution. Then, the DNA nanopores were hybridized to the premodified ssDNA by immersing the electrode in a DNA nanopore solution.

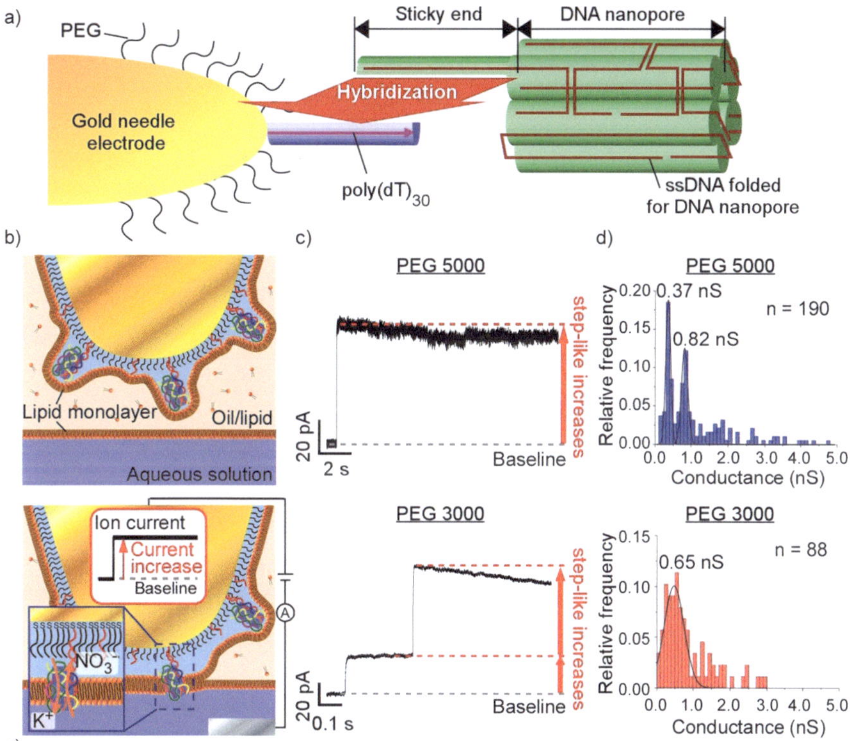

e)

	DNA nanopore probe	Conventional method
Number of membrane formations	63	52
Number of pore insertions	23	6
Insertion probability (%)	36.5	11.5

Fig. 10.4 Schematic diagram of DNA nanopore probe technology. **a** Method for immobilization of DNA nanopores on gold needle electrodes and **b** Physical membrane insertion model of DNA nanopores using DNA nanopore probes. **c** Representative ion current traces of DNA nanopores which were measured by DNA nanopore probes **d** Histograms of pore conductance. **e** The insertion probabilities of tethered and untethered DNA nanopores. Reproduced with permission from [8]

Ion current measurements with DNA nanopore probe

Next, we introduce the method and the result of the ion current measurements using the DNA nanopore probe. In the same way as the gold needle electrode, artificial cell membranes can be formed on the surface of the DNA nanopore probe by pushing down the probe into the layered bath solution (Fig. 10.4b, top). Then, DNA nanopores are physically inserted into the artificial cell membrane (Fig. 10.4b, bottom).

First, we experimentally confirmed that the thickness of the PEG layer on the electrode is a significant parameter for the insertion of DNA nanopores. We prepared thiol-PEGs with MWs of 3000 and 5000 (PEG 3000 and PEG 5000) and performed channel current measurements of the electrode-tethered DNA nanopores. As a result, although step-like current increases were observed under both conditions (Fig. 10.4c), the histograms of pore conductance showed clear differences between each condition (Fig. 10.4d). With PEG 5000, two conductance peaks of 0.37 and 0.82 nS were observed. Because the theoretical pore conductance of the single DNA nanopore was calculated as approximately 0.46 nS, the conductance peaks obtained with PEG 5000 indicated the insertions of single and double DNA nanopores (Fig. 10.4d). On the other hand, with PEG 3000, a single conductance peak of 0.65 nS with higher variability was obtained. Since this result suggested insertions of both premodified ssDNAs and DNA nanopores into the membranes, we can conclude that the PEG layer must be sufficiently thicker than the length of the sticky end to correctly measure the ion currents derived from the electrode-tethered DNA nanopores.

Finally, we evaluated and compared the insertion efficiency of DNA nanopores into the artificial cell membranes between tethered and untethered nanopores (Fig. 10.4e). The insertion probabilities of the DNA nanopores into the membrane were calculated using the number of artificial cell membrane formations and the event number of DNA nanopore insertion within 5 min after the membrane formation. With untethered nanopores, as a result, DNA nanopore insertion events were observed 6 times in 52 membrane formations. On the other hand, DNA nanopore insertion events were observed 23 times in 63 membrane formations with the DNA nanopore probe. The insertion probabilities of the DNA nanopores obtained from both experiments were 36.5 and 11.5%, respectively. From these results, we showed that the DNA nanopore probe can insert DNA nanopores into the membrane three times more efficiently than the use of untethered DNA nanopores. In conclusion, the DNA nanopore probe can be used as a fundamental tool for investigating the function and characteristics of DNA nanopores because it can provide efficient membrane insertion of DNA nanopores.

10.5 Conclusion and Future Outlook

In this section, we described easy and convenient methods to form artificial cell membranes and perform ion current measurements using two kinds of probe systems: a gold needle and a Ag/AgCl electrode. In the both probe systems, the artificial cell membranes are formed by immersing the microelectrodes into the bath solution consisting of the oil/lipid mixture and aqueous solution. In addition, the ion current measurement can be performed by applying a transmembrane potential between the microelectrodes and the reference electrode. By using probe systems, we can perform the high-throughput evaluation of channel size and molecular transportation through membrane channels. Moreover, the DNA nanopore probe that DNA nanopores are tethered on the surface of the gold needle electrode can provide a high insertion efficiency of DNA nanopores into artificial cell membranes by forming artificial cell membranes on the surface of the probes.

In the future, we expect that further studies of probe-type artificial cell membrane systems will offer various advantages to the development of molecular robots. For example, our systems will be potentially applied for the ion current measurements of other artificial ion channels as well as DNA nanopores. Furthermore, a bioinspired scanning ion conductance microscopy [9], constructed by the probe-type artificial cell membrane system, is expected to image activities of membrane channels on molecular robots with high spatial resolution. We believe that advancing probe-type artificial cell membrane systems will contribute to the realization of molecular robots.

References

1. Shoji, K., Kawano, R.: Recent advances in liposome-based molecular robots. Micromachines **11**, 788 (2020)
2. Cao, C., Long, Y.T.: Biological nanopores: confined spaces for electrochemical single-molecule analysis. Acc. Chem. Res. **51**, 331–341 (2018)
3. Sakai, N., Matile, S.: Synthetic ion channels. Langmuir **29**, 9031–9040 (2013)
4. Lanphere, C., et al.: Design, assembly, and characterization of membrane-spanning DNA nanopores. Nat. Protoc. **16**, 86–130 (2021)
5. Komiya, M., et al.: Advances in artificial cell membrane systems as a platform for reconstituting ion channels. Chem. Rec. **20**, 730–742 (2020)
6. Shoji, K., Kawano, R., White, R.J.: Spatially resolved chemical detection with a nanoneedle-probe-supported biological nanopore. ACS Nano **13**, 2606–2614 (2019)
7. Shoji, K., Kawano, R., White, R.J.: Recessed Ag/AgCl microelectrode-supported lipid bilayer for nanopore sensing. Anal. Chem. **92**, 10856–10862 (2020)
8. Ikarashi, S., et al.: DNA nanopore-tethered gold needle electrodes for channel current recording. ACS Nano **17**, 10598–10607 (2023)
9. Yoshihara, R., Nomi, S., Shoji, K.: Nanopore probes using hydrogel-filled nanopipettes as sensors for chemical imaging. ACS Appl. Nano Mater. **5**, 15808–15816 (2022)
10. Yanyao, Q., Shun, O., Shoji, K.: Nanopore analysis of microfluidically prepared samples using a probe-type lipid bilayer system. Microchem. J. **217**, 115084 (2025)
11. Hasegawa, N., Shoji, K.: Microcavity volume control on a tip of Ag/AgCl electrodes for stable channel current measurements of biological nanopores. Analyst **147**, 1191–1198 (2022)
12. Funakoshi, K., Suzuki, H., Takeuchi, S.: Lipid bilayer formation by contacting monolayers in a microfluidic device for membrane protein analysis. Anal. Chem. **78**, 8169–8174 (2006)
13. Hille, B.: Ion Channels of Excitable Membranes, 3rd edn. Sinauer Associate Inc., Sunderland (2001)
14. Tsuji, Y., et al.: Droplet split-and-contact method for high-throughput ttransmembrane electrical recording. Anal. Chem. **85**, 10913–10919 (2013)

15. Hussein, E.A., White, R.J.: Silver nanoneedle probes enable sustained DC current, single-channel resistive pulse nanopore sensing. Anal. Chem. **93**, 11568–11575 (2021)
16. Shoji, K., Kawano, R., White, R.J.: Analysis of membrane protein deinsertion-associated currents with nanoneedle-supported bilayers to discover pore formation mechanisms. Langmuir **36**, 10012–10021 (2020)
17. Burns, J.R., Howorka, S.: Defined bilayer interactions of DNA nanopores revealed with a nuclease-based nanoprobe strategy. ACS Nano **12**, 3263–3271 (2018)
18. Langecker, M., et al.: Synthetic lipid membrane channels formed by designed DNA nanostructures. Science **338**, 932–936 (2012)
19. Hiromu, A., Hirano, T., Mabuchi, T., Shoji, K.: Specific ATP detection using molecule-responsive DNA nanopores. Abstr. Small **21**(29) (2025)
20. Burns, J.R., et al.: Lipid-bilayer-spanning DNA nanopores with a bifunctional porphyrin anchor. Angew. Chem. Int. Ed. **52**, 12069–12072 (2013)

Chapter 11
Fully Synthetic Transmembrane Channels

Kohei Sato

Abstract Inspired by the structure and function of natural transmembrane channel proteins, various fully synthetic transmembrane channels have been developed. In this chapter, the author introduces the anomalous material permeation properties of fluorinated supramolecular channels and explains how fluorination imparts emergent functions and selectivity beyond those of conventional synthetic channels.

11.1 Introduction

Channel proteins, which possess a transmembrane pore with an inner diameter of several angstroms, are responsible for the selective permeation of ions and small molecules to play important roles in a variety of biological events such as transmembrane signaling and maintenance of homeostasis [1]. Inspired by the beautiful structure and sophisticated function of such channel proteins, a number of fully synthetic channel-forming molecules have been developed with the expectation of application in the treatment of ion channel-related diseases (channelopathies) [2] and in industrial water desalination processes [3]. Representative molecular designs of synthetic channels include unimolecular nanotubes such as cucurbituril [4], pillararene [5], and carbon nanotubes [6]. Other types of synthetic channels that have been extensively studied are supramolecular channels, which are formed by self-assembly of small building blocks into one-dimensional nanotubes via non-covalent interactions such as hydrogen bonding, electrostatic interactions, aromatic interactions, and hydrophobic effects [7]. Although supramolecular channels tend to be less stable than unimolecular channels due to their intrinsically dynamic nature, they allow for diverse structural modifications and excellent responsiveness to external stimuli. In this section, the author presents some recent examples of supramolecular synthetic transmembrane channels developed in the author's laboratory, with emphasis on their structural features and their unique functions. Artificial transmembrane channels formed

K. Sato (✉)
Kwansei Gakuin University, Hyogo, Japan
e-mail: ksato@kwansei.ac.jp

S. Murata et al. (eds.), *Molecular Robotics II*,
https://doi.org/10.1007/978-981-95-2394-8_11

by biomacromolecules such as peptides and nucleic acids are beyond the scope of this section.

11.2 Supramolecular Ion Channels Formed by Multiblock Amphiphiles

A variety of natural channel proteins can be included in a type of multipass transmembrane proteins, which are composed of repeating rigid hydrophobic domains and flexible hydrophilic domains (Fig. 11.1, left) [8]. Inspired by this structural feature, Muraoka, Kinbara, and co-workers developed a synthetic multiblock amphiphile (MA) containing rigid hydrophobic oligo(phenylene–ethynylene) units and flexible hydrophilic oligo(ethylene glycol) chains (Fig. 11.1, right) [9]. They reported that MA adopts a folded conformation within the hydrophobic layer of the lipid bilayer membranes formed by 1,2-dioleoyl-*sn*-glycero-3-phosphocholine (DOPC), with its rod-shaped aromatic units oriented perpendicular to the membrane plane. In addition, MA self-assembled within the membranes to form a tetrameric supramolecular transmembrane ion channel with a pore diameter of 0.53 nm. Since this seminal report, they have developed several types of MAs with ligand-gated ion permeation properties [10] and mechanosensitivity [11].

To further explore the potential of MAs as synthetic ion channels, we have focused on the unique properties of the fluorine atom, which has the largest electronegativity with the second smallest atomic size among all elements [12], and designed fluorinated MAs. In 2021, we developed a V-shaped fluorinated MA that possesses a

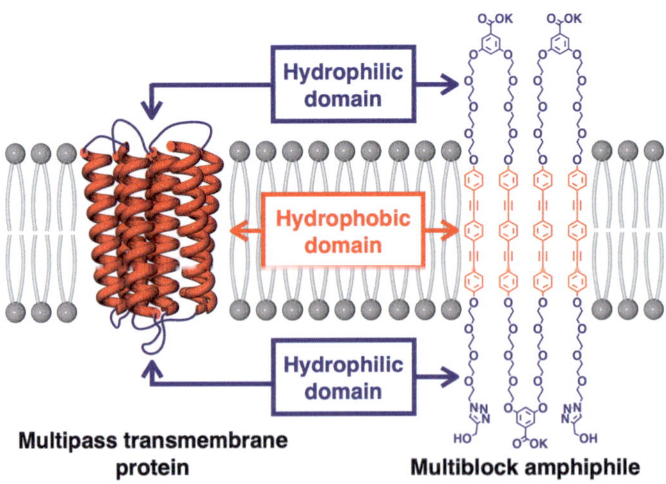

Fig. 11.1 Schematic representation of the multipass transmembrane protein (left) and the multiblock amphiphile (right). Both molecules are composed of repeating hydrophobic and hydrophilic domains

permanent dipole moment along the aromatic units due to multiple asymmetrically functionalized fluorine atoms (Fig. 11.2a) [13]. We also incorporated highly polar phosphate ester groups between fluorinated aromatic units and oligo(ethylene glycol) chains to control the molecular orientation across the membrane plane, as they are unable to penetrate the hydrophobic layer of the membranes [10].

To investigate the transmembrane ion transport property of the V-shaped fluorinated MA, we performed current recording measurements of a planar black lipid membrane (BLM) containing the V-shaped fluorinated MA. Interestingly, the V-shaped fluorinated MA exhibited a distinct voltage response; it showed ion currents only when the direction of an applied voltage was antiparallel to the dipole moment of the aromatic units. Under these conditions, the dipole–dipole repulsion between the aromatic units was suppressed and the formation of the supramolecular ion channel was likely enhanced. In addition, the V-shaped fluorinated MA showed the ligand response. In the presence of (*R*)-propranolol, a sodium ion channel antagonist, the transmembrane ion transport property of the V-shaped fluorinated MA was blocked due to the electrostatic and aromatic interaction between (*R*)-propranolol

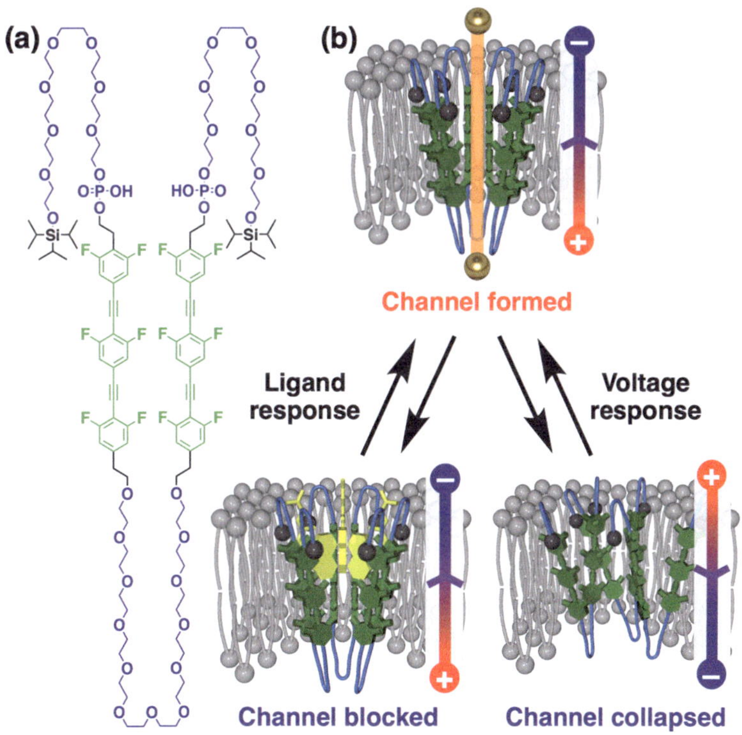

Fig. 11.2 **a** Chemical structure of a V-shaped fluorinated MA. **b** Schematic representation of ligand and voltage-responsive transmembrane ion permeation through the supramolecular channel formed by fluorinated MA

and MA. Successive addition of *β*-cyclodextrin as a host molecule to remove (*R*)-propranolol successfully restored the original ion transport property, demonstrating the reversible ligand response. These results provided the first example of anisotropic dual-stimuli-responsive transmembrane ion transport by a fully synthetic channel (Fig. 11.2b).

In 2022, we developed another type of fluorinated MA in which the fluorinated aromatic units were linked by two oligo(ethylene glycol) chains to form a macrocyclic structure [14] (Fig. 11.3a). We envisioned that this molecular design would suppress the rigorous molecular motion of the fluorinated MA within the membranes and enable the formation of a stable supramolecular transmembrane channel. Based on the current recording measurements of a planar BLM, we confirmed that the newly designed macrocyclic fluorinated MA formed a supramolecular ion channel with an improved pore stability compared to the previously developed V-shaped MA (Fig. 11.2a). Based on the current values observed under the application of voltage, the pore diameter was calculated to be 0.44 nm. We also performed the transmembrane ion transport assay using DOPC vesicles encapsulating 8-hydroxypyrene-1,3,6-trisulfonic acid (HPTS), a pH-sensitive fluorescent dye, in the presence of a pH gradient across the intravesicular and extravesicular media. We then added the macrocyclic fluorinated MA to the preformed DOPC vesicles and monitored the increase in fluorescence intensity caused by transmembrane ion transport under different conditions.

Interestingly, the macrocyclic fluorinated MA showed enhanced ion transport property for potassium ion compared to other alkali metal cations ($k_K / k_{Li} = 4.0$, $k_K / k_{Na} = 3.6$, $k_K / k_{Rb} = 1.7$, and $k_K / k_{Cs} = 2.0$, *k*: rate constant of a pseudo-first order reaction). To understand the mechanism of the potassium selective ion transport, we performed the quantum mechanics-molecular mechanics (QM/MM) simulations of the macrocyclic fluorinated MA in the membrane-water system. As shown in Fig. 11.3b, the QM/MM simulation revealed that a potassium ion was surrounded by the fluorinated aromatic units in which eight fluorine atoms interacted with the potassium ion through $C - F{\cdot}M^+$ interactions, resulting in the potassium selectivity of the supramolecular channel. In addition, we found that the macrocyclic fluorinated MA possesses mechanosensitivity. Ion permeation assay using HPTS and current recording measurements using BLM revealed that the transmembrane ion transport property of the macrocyclic fluorinated MA is enhanced under membrane-contracting conditions, whereas it is diminished under membrane-expanding conditions because of the dissociation of MA molecules. These results demonstrated the first example of a fully synthetic channel that exhibited both selective ion transport properties and mechanosensitivity. It is noteworthy that these two features are somewhat mutually exclusive; the selective ion transport requires "static" properties with a defined pore structure, whereas stimuli-responsiveness requires "dynamic" properties.

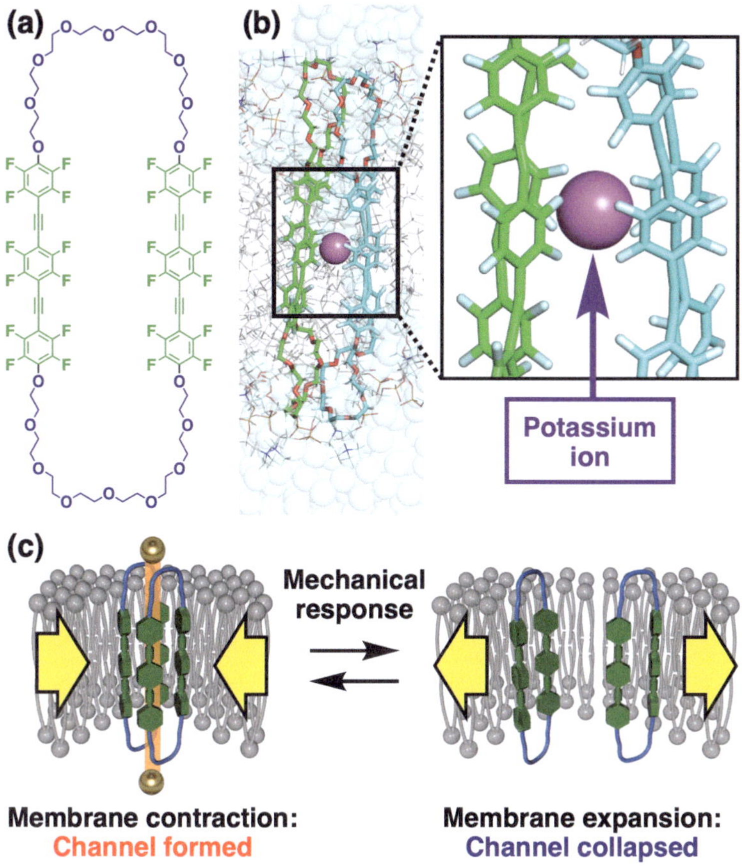

Fig. 11.3 **a** Chemical structure of a macrocyclic fluorinated MA. **b** Schematic representation of the membrane-water system from the QM/MM simulation at 500 ps. A potassium ion was located at the center of two molecules of macrocyclic fluorinated MAs. **c** Schematic representation of mechanosensitive transmembrane ion permeation through the supramolecular channel formed by macrocyclic fluorinated MAs

11.3 Supramolecular Water Channel Formed by Fluorinated Macrocycles

In 2022, our research group developed fully synthetic transmembrane channels for water permeation [15]. In this study, we first synthesized four types of nanorings with different inner diameters and different numbers of fluorine atoms (Fig. 11.4). Using transmission electron microscopy, we confirmed that all nanorings self-assembled on top of each other to form supramolecular nanotubes under hydrophobic environment.

We then added these supramolecular nanotubes to BLMs formed by 1,2-dipalmitoyl-*sn*-glycero-3-phosphocholine (DPPC) and recorded currents generated

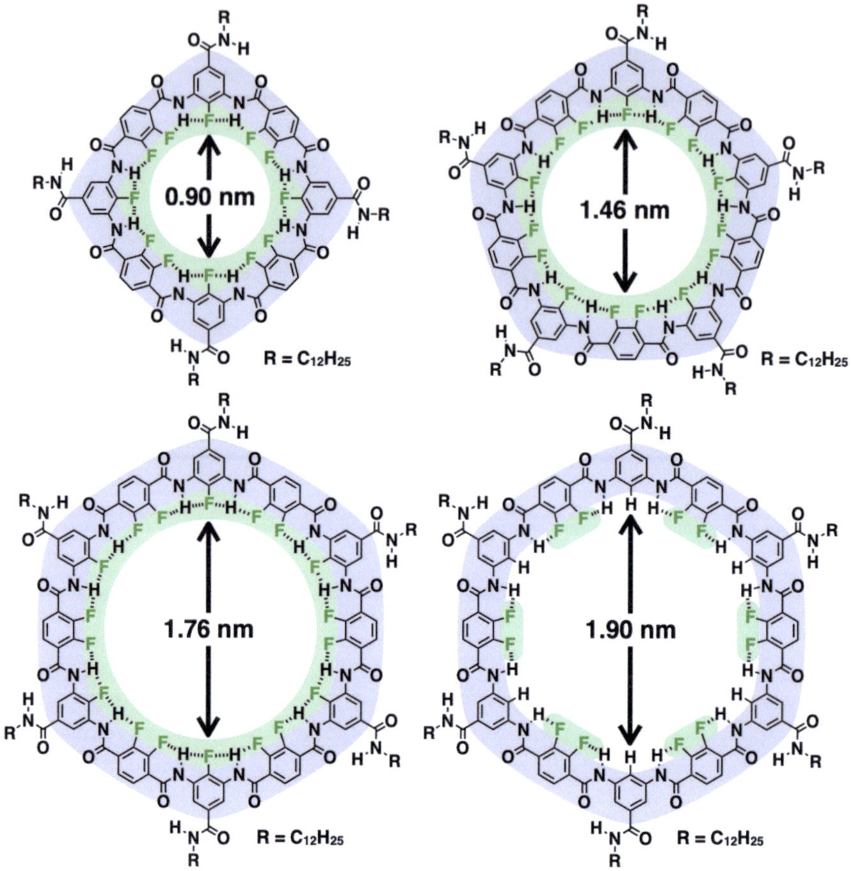

Fig. 11.4 Chemical structures of nanorings with different inner diameters and different numbers of fluorine atoms

by transmembrane ion permeation and found that the nanoring with an inner diameter of 0.90 nm showed the lowest ion permeability among other nanorings. However, through the water permeation assay, we found that the smallest nanoring possesses outstanding functions. We prepared DPPC vesicles encapsulating carboxyfluorescein (CF), a dye that exhibits fluorescence self-quenching at high concentrations, in the presence of supramolecular nanotubes formed by nanorings and exposed the vesicles to hypertonic buffer solution using stopped flow systems (Fig. 11.5, left). We then monitored the efflux of water through the nanotubes caused by the osmotic pressure based on the decrease in fluorescence intensity of CF.

Surprisingly, the supramolecular nanotube formed by the nanoring with an inner diameter of 0.90 nm exhibited 2 orders of magnitude higher water permeability compared to aquaporin-1, a natural water channel protein. Supramolecular nanotubes formed by larger nanorings and fewer numbers of fluorine atoms showed lower water

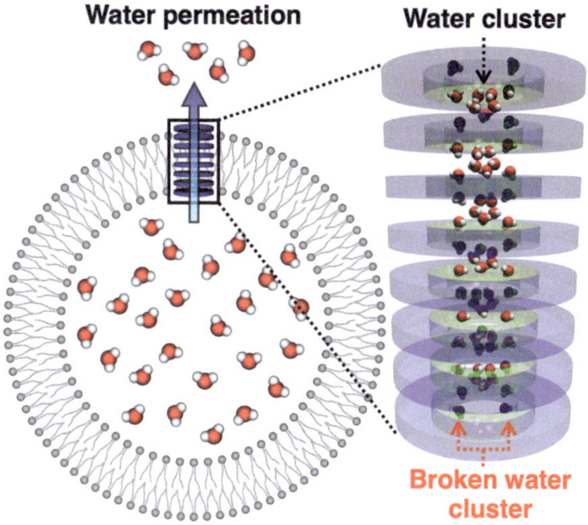

Fig. 11.5 Schematic representation of the flow of water clusters and broken water clusters through the supramolecular channel formed by fluorinated nanorings

permeability than that of the smallest nanoring, indicating that the number of fluorine atoms per channel cross-sectional area is a critical factor for the ultrafast water permeation. Using molecular dynamics simulations, we investigated the effect of wall hydrophobicity and found that water clusters can be broken inside the channel, increasing the rate of water permeation (Fig. 11.5, right). Moreover, we found that the supramolecular nanotube exhibits negligible permeability of chloride ion (Cl^-) due to the electrostatic barrier originating from fluorine atoms, demonstrating the promising application in water purification technologies.

11.4 Conclusions and Perspectives

In this section, the author has given a brief introduction to the recently developed fully synthetic transmembrane channels. One of the great advantages of synthetic channels is their structural feasibility, which allows the introduction of "unnatural" elements such as fluorine atoms to drastically modify the physicochemical properties of channel-forming molecules. Indeed, fluorinated water channel exhibited outstanding water permeability. However, the mechanism of the anomalous functions of fluorinated transmembrane channels described in this section remains largely unknown. Further studies from a theoretical point of view are necessary to fully understand the permeation phenomena through the synthetic channels. Moreover, since most of the synthetic channels require extensive efforts in their synthesis, their substantial improvement is also inevitable for their practical applications in medical and industrial uses.

References

1. Alberts, B., Heald, R., Johnson, A., Morgan, D., Raff, M., Roberts, K., Walter, P.: Molecular Biology of the Cell, 7th edn. W. W. Norton & Co., New York (2022)
2. Graves, T.D., Hanna, M.G.: Postgrad. Med. J. **81**, 20–32 (2005)
3. Huang, L.B., Di Vincenzo, M., Li, Y., Barboiu, M.: Chem. Eur. J. **27**, 2224–2239 (2021)
4. Jeon, Y.J., Kim, H., Jon, S., Selvapalam, N., Oh, D.H., Seo, I., Park, C.S., Jung, S.R., Koh, D.S., Kim, K.: J. Am. Chem. Soc. **126**, 15944–15945 (2004)
5. Hu, X.B., Chen, Z., Tang, G., Hou, J.L., Li, Z.T.: J. Am. Chem. Soc. **134**, 8384–8387 (2012)
6. Tunuguntla, R., Henley, R.Y., Yao, Y.C., Pham, T.A., Wanunu, M., Noy, A.: Science **357**, 792–796 (2017)
7. Zheng, S.P., Huang, L.B., Sun, Z., Barboiu, M.: Angew. Chem. Int. Ed. **59**, 2–34 (2020)
8. Sato, K., Muraoka, T., Kinbara, K.: Acc. Chem. Res. **54**, 3700–3709 (2021)
9. Muraoka, T., Shima, T., Hamada, T., Morita, M., Takagi, M., Tabata, K.V., Noji, H., Kinbara, K.: J. Am. Chem. Soc. **134**, 19788–19794 (2012)
10. Muraoka, T., Noguchi, D., Kasai, R.S., Sato, K., Sasaki, R., Tabata, K.V., Ekimoto, T., Ikeguchi, M., Kamagata, K., Hoshino, N., Noji, H., Akutagawa, T., Ichimura, I., Kinbara, K.: A synthetic ion channel with anisotropic ligand response. Nat. Commun. **11**, 2924 (2020)
11. Muraoka, T., Umetsu, K., Tabata, K.V., Hamada, T., Noji, H., Yamashita, T., Kinbara, K.: J. Am. Chem. Soc. **139**, 18016–18023 (2017)
12. Sato, K.: Langmuir **40**, 2809–2814 (2024)
13. Sasaki, R., Sato, K., Tabata, K.V., Noji, H., Kinbara, K.: Synthetic ion channel formed by Multiblock Amphiphile with anisotropic dual-stimuli-responsiveness. J. Am. Chem. Soc. **143**, 1348–1355 (2021)
14. Sato, K., Sasaki, R., Matsuda, R., Nakagawa, M., Ekimoto, T., Yamane, T., Ikeguchi, M., Tabata, K.V., Noji, H., Kinbara, K.: Supramolecular mechanosensitive potassium channel formed by fluorinated Amphiphilic Cyclophane. J. Am. Chem. Soc. **144**, 11802–11809 (2022)
15. Itoh, Y., Chen, S., Hirahara, R., Konda, T., Aoki, T., Ueda, T., Shimada, I., Cannon, J.J., Shao, C., Shiomi, J., Tabata, K.V., Noji, H., Sato, K., Aida, T.: Ultrafast water permeation through nanochannels with a densely fluorous interior surface. Science **376**, 738–743 (2022)

Chapter 12
Protein and Peptide Nanopores

Zugui Peng, Jiajue Ji, Tao Ren, Shoko Fujita, Shoji Iwabuchi, and Ryuji Kawano

12.1 Introduction

In living organisms, membrane receptors, transporters, and ion channels function as sensors, enabling cells to detect and react to various chemical and physical signals. These biological mechanisms have been adapted to endow molecular robots with sensing abilities. Nanopores, which create nanometer-scale openings in lipid membranes, have proven effective in facilitating the transport of large molecules, thus aiding molecular robots in their detection tasks [1]. The gating of these nanopores controlled by factors such as pH, light, temperature, and osmotic pressure further enhances the robots' sensing precision. Various materials have been utilized to assemble nanopores, including proteins, peptides, DNA, and other nanostructures. This chapter explores the properties, distinctions, and recent advancements in the assembly of nanopores using peptides and proteins (For more details on the introduction of nanopores assembled by DNA and other nanostructures, we direct readers

Z. Peng · J. Ji · T. Ren · S. Fujita · S. Iwabuchi · R. Kawano (✉)
Tokyo University of Agriculture and Technology, Tokyo, Japan
e-mail: rjkawano@cc.tuat.ac.jp

Z. Peng
e-mail: skhou@go.tuat.ac.jp

J. Ji
e-mail: ji-jj@st.go.tuat.ac.jp

T. Ren
e-mail: tao-ren@st.go.tuat.ac.jp

S. Fujita
e-mail: shoko-fujita@st.go.tuat.ac.jp

S. Iwabuchi
e-mail: s-iwabuchi@go.tuat.ac.jp

S. Murata et al. (eds.), *Molecular Robotics II*,
https://doi.org/10.1007/978-981-95-2394-8_12

to other chapters). Additionally, we will discuss methods for integrating nanopores into molecular robots.

12.2 Nanopore Measurement

Protein nanopores are most extensively used in planar lipid membrane systems for rapid, simple, and label-free single-molecule detection, known as nanopore sensing. In nanopore sensing, ionic current flowing through nanopores changes as molecules translocate the nanopores under a steady voltage. The sensitivity of the detection is largely determined by the size and structure of nanopores, prompting extensive research into various pore-forming proteins for detecting diverse target molecules [2]. In a landmark study in 1996, Kasianowicz et al. utilized α-hemolysin (αHL) from *Staphylococcus aureus*, with a pore size of 1.4 nm for the translocation of single-stranded DNA (ssDNA), providing the basis for the use of nanopores in molecular detection [3]. In 2010, Derrington et al. identified *Mycobacterium smegmatis* porin A (MspA) as capable of distinguishing between individual nucleotide bases [4]. These foundational studies eventually led to the development and commercial release of the first nanopore DNA sequencer, MinION by Oxford Nanopore Technologies in 2015, which uses the *E. coli* curli transport channel CsgG [5] (Fig. 12.1). This device has proven effective for large-scale DNA sequencing, drastically reducing the cost of human genome sequencing to $1,000. Advances in bioinformatics and improvements in nanopores have improved the sequencing accuracy from 66 to 97% [6].

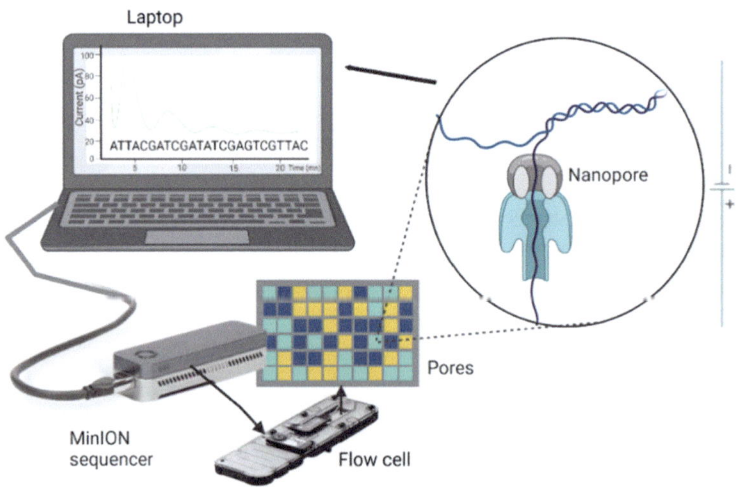

Fig. 12.1 MinION flow cell contains multiple channels for a total of 2,048 nanopores used to sequence DNA or RNA. (Reprinted from *Current Clinical Microbiology Reports* [7])

Following the achievement of nanopore DNA sequencing, single-molecule analysis of folded proteins [8, 9] and amino acid sequencing in proteins [10] are currently underway. Although discrimination of amino acids is particularly challenging due to their smaller size and varied charge compared to nucleotides, the detection of single amino acid mutations in peptides using aerolysin from *Aeromonas hydrophila* [11], sequential readout of peptides using MspA [12, 13], and identification of digested protein fragments using fragaceatoxin C (FraC) from *Actinia fragacea* [14] have been reported in recent years. However, further technological advances are required to achieve full-length protein sequencing in single amino acid-level resolution.

The power of nanopore is not only limited to molecule sequencing. To further exploit their potential, the Kawano group have applied nanopores for decoding nucleotide sequence information from DNA computing [15, 16]. In a practical demonstration, Hiratani et al. recognized the expression patterns of two microRNAs (miRNAs) using a DNA logic AND gate [17]. This method translates structural changes in nucleic acid molecules into electrical signals, measured as variations in current blocking duration. Most recently, Takeuchi et al. developed a system to recognize patterns of five miRNAs overexpressed in bile duct cancer (BDC) using αHL, diagnostic DNA, and a planar lipid membrane [18] (Fig. 12.2). This approach enabled label-free detection of miRNA expression patterns in the plasma of BDC patients at sub-femtomolar concentrations, showcasing the nanopore's potential for highly sensitive diagnostic applications.

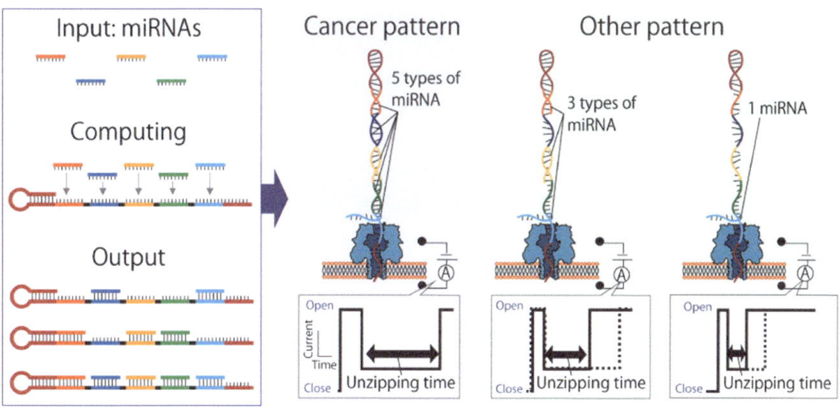

Fig. 12.2 Recognizing patterns of five miRNAs overexpressed in bile duct cancer (BDC) using αHL, diagnostic DNA, and a planar lipid membrane. (Reprinted from *JACS Au* [18])

12.3 Exploration of New Protein Nanopores

Despite the various protein nanopores reported, with diameters ranging from 1.2 to 42 nm [2], exploration of novel protein nanopores remains fascinating. Here, we will introduce protein nanopores that have been reported in recent years, with the hope of their future application in molecular robots. In 2023, the Maglia group identified the detection capabilities of an α-helical pore-forming toxin YaxAB from *Yersinia enterocolitica* [19]. The characteristic feature of YaxAB is its large conical shape, featuring openings of 15 nm on the *cis* side and 3.5 nm on the *trans* side (Fig. 12.3). YaxAB is notable as the largest protein nanopore utilized for the analysis of folded proteins in planar membrane systems, capable of capturing proteins ranging in size from 35 to 125 kDa. Furthermore, YaxAB demonstrated its practical diagnostic potential by detecting CRP at clinically relevant concentrations (>2.5 mg/L) in human serum diluted 160-fold. This highlights the capability of nanopores for sensitive protein detection in complex biological samples.

Detecting molecules larger than the nanopore or selectively distinguishing molecules of similar size and hydrodynamic radius are difficult challenges. To address this, researchers have used target-specific binding moieties such as partial antibody domains [20] and aptamers [21, 22]. However, the configuration of detection systems must be tailored to each recognition element used. Recently, Movileanu and colleagues developed a versatile system for the sensitive detection of target proteins [23]. Their study utilized the ferric hydroxamate uptake component A (FhuA) of *E. coli* covalently attached to a monobody, a recombinant protein based on the fibronectin type III (FN3) domain. Monobodies offer several advantages: they can be selected from a wide range of variants against numerous target proteins, and their relatively small hydrodynamic radius prevents interference with the pore opening, unlike larger antibody-derived proteins. The monobody-conjugated tFhuA detected human small ubiquitin-related modifier 1 (hSUMO1) and WD40 repeat protein 5 (WDR5) and achieved the identification of the 20 nM-epidermal growth factor receptor (EGFR), which is a 180 kDa prognostic protein biomarker associated

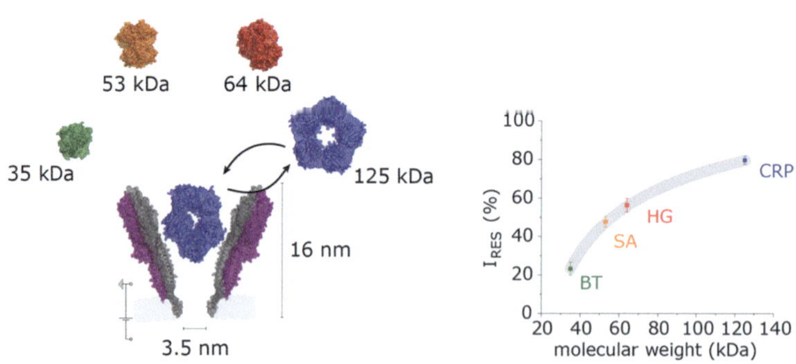

Fig. 12.3 Differentiation of proteins with different molecular weights by YaxAB nanopores with openings of 15 nm. (Reprinted from *ACS Nano* [19])

with lung, colon, and breast cancer, in 5% (v/v) fetal bovine serum (FBS), suggesting potential applications for molecular robots as nanomedicine.

Recently, the exploration of nanopores has extended to include de novo design approaches, rather than relying solely on natural membrane proteins (Fig. 12.4). De novo design harnesses computer-aided protein design to engineer proteins with precisely customized structures and functionalities [24], offering the exciting prospect of creating nanopores with unique structures not found in nature and expanding the range of detectable molecules. Conventionally, de novo design of transmembrane proteins posed a significant challenge due to the intricate relationship between membrane insertion, folding mechanisms, and the corresponding amino acid sequences. However, in 2021, Baker and colleagues achieved the de novo design of an eight-strand transmembrane β-barrel from scratch through (i) the refinement of the loop structure and adjustment of hydrophobicity, (ii) backbone design and sequence optimization using Rosetta, (iii) additional design guidelines appropriate for the membrane environment, and (iv) exhaustive expression trials [25]. The expressed structure was confirmed to be highly consistent with the designed structure (1.1 Å backbone RMSD over all residues). Its application in single-molecule detection has not been reported, likely due to its smaller pore size compared to natural ones. However, with the increasing availability and accessibility of de novo protein design tools, the design of protein nanopores with molecular selectivity, transport capacity, stability, and durability suitable for molecular robotics applications is expected.

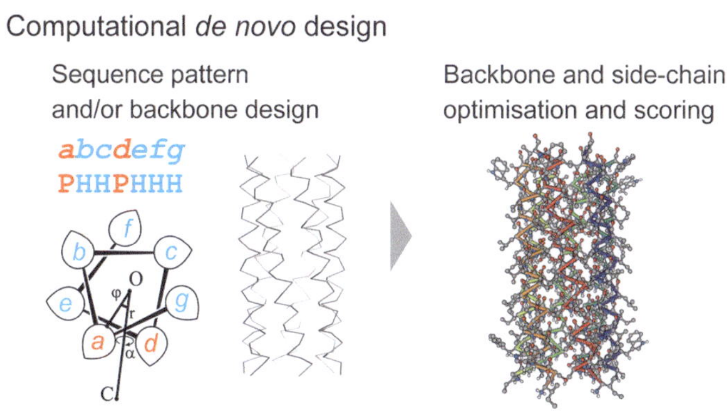

Fig. 12.4 Computational de novo design of proteins. The process includes designing sequence patterns and backbones, and then repeatedly refining the backbone and sidechain structures. (Reprinted from *Physical Chemistry Chemical Physics* [26])

12.4 Development of Peptide Nanopores

Peptides that assemble into transmembrane pores offer intriguing sensors for molecular robotics applications. Peptide nanopores have advantages over protein ones, including enhanced tractability and scalability. Originally, peptides such as gramicidin and alamethicin demonstrated the ability to form pores in bacterial membranes, exhibiting inherent antimicrobial activity [27] (Fig. 12.5). The peptide nanopores have been reported to enable the permeation of small molecules such as fluorophores to vesicles [28], but it has been reported as difficult to transport molecules with larger molecular weights [29]. Recent efforts have focused on designing more stable and uniform peptide nanopores.

One strategy for achieving stable peptide nanopores involves the redesign of transmembrane regions found in natural membrane proteins [30]. The advantage of this approach is that complex design principles are not required, and a certain degree of stability is expected. The Bayley group reported cWza peptide, which is a consensus sequence derived from the D4 domains of the 340 kDa Wza polysaccharide transporters of Gram-negative bacteria [31] (Fig. 12.6). cWza showed size and large conductance, facilitating the translocation of cyclodextrins under applied voltage in a planar membrane system.

Similar to protein nanopores, de novo design is another approach to obtain stable nanopores. Several α-helical barrel nanopores have been reported in recent years by the precise design of inter-helix packing interfaces. In 2021, Woolfson and collaborators reported an α-helix peptide designed to form stable, monodisperse pores [32]. This began with the creation of a water-soluble α-helical barrel, achieved by strategically introducing polar residues within the lumen of an α-helical bundle, a parallel hexamer water-soluble pore. By incorporating hydrophobic leucine and membrane

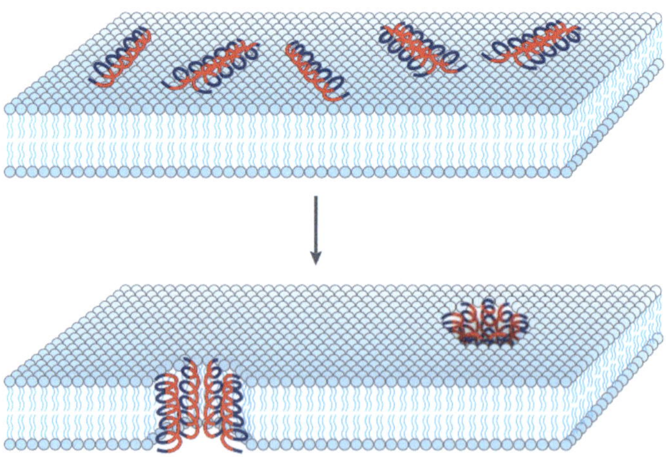

Fig. 12.5 Formation of nanopores in lipid membranes by antimicrobial peptides. (Reprinted from *BioMed Research International* [27])

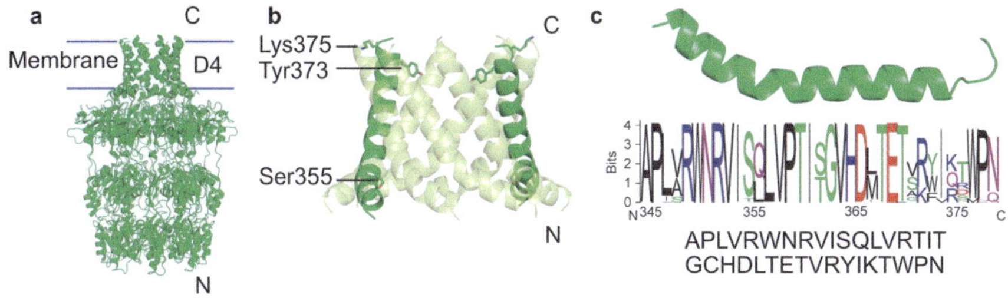

Fig. 12.6 Structure of *E. coli* Wza and peptide design of the cWza pore. **a**, Structure of native *E. coli* Wza. **b**, Structure of the Wza D4 domain. **c**, BLAST search for the Wza D4 domain (top) shows the consensus Wza peptide (cWza) sequence (bottom). (Adapted from *Nature Chemistry* [30])

interface-preferring tryptophan residues on the outer surface of the water-soluble pore, peptides capable of forming monodisperse hexameric pore with longer lifetimes and fewer gating events in the membrane environment were engineered. While single-molecule detection remains unexplored, likely due to the small pore size, they previously reported the design of water-soluble helical barrels ranging from 5 to 9-mers by modulating residue sizes at specific positions on the helical wheel [33], leading to the possibility of pore size expansion and translocation of molecules.

In 2022, Shimizu et al. reported the de novo design of β-hairpin peptides SV28 and SVG28, which form transmembrane β-barrel nanopores following four simple design principles: (i) alternating hydrophilic and hydrophobic transmembrane residues, (ii) introduction of membrane interface-anchoring Tyr, (iii) introduction of charged residues for membrane insertion by electrophoresis, and (iv) introduction of a glycine kink to reduce the strain of the β-barrel (only for SVG28) [34] (Fig. 12.7). Under optimized conditions, SVG28 formed almost monodisperse heptameric pores and exhibited discrimination between L-PLL (molecular weight 30,000–70,000) and S-PLL (molecular weight 10,000). Additionally, SVG28 surpasses the commonly used protein nanopore αHL in efficiently detecting cationic peptides, owing to its enhanced terminal charge. Furthermore, SV28, without the glycine-kink of SVG28, can form a wide range of pore sizes from 1.7 to 6.3 nm and can translocate the G-quadruplex structure of DNA when the pore size is sufficiently large. This finding suggests the feasibility of identifying SVG28 variant sequences capable of forming nanopores with tailored sizes. Recently, Fujita et al. reported that hydrophilic variants of SVG28 can be synthesized using cell-free protein synthesis [35], offering easy access to SVG28 for molecular robotics.

12.5 Nanopores in Molecular Robots

So far, we have discussed about nanopores as sensors in planar lipid membrane systems. In lipid vesicle-based molecular robots, the ability of nanopores to transfer substances and molecules through lipid membranes enables lipid vesicles to achieve

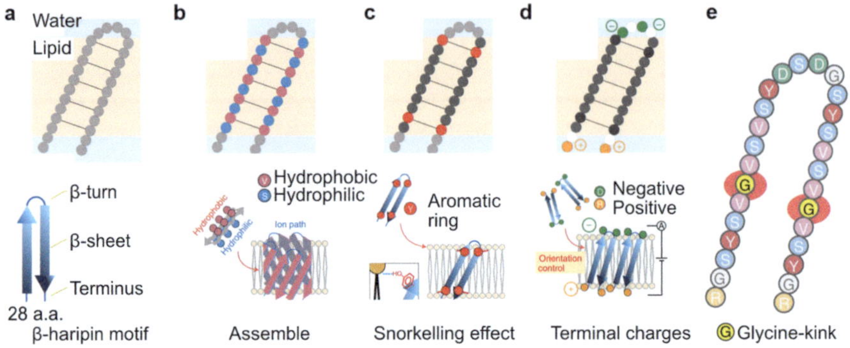

Fig. 12.7 Design strategy of SVG28. **a**, The design of β-hairpin peptide with 28 amino acids (a.a.) divided into three sections: β-turn, β-sheet transmembrane, and the terminals. **b**, Hydrophilic and hydrophobic amino acids are arranged in an alternating fashion. **c**, Interaction of aromatic rings stabilizes the β-barrel pore. **d**, Designing specific charges at the terminus allows control of peptide orientation on an applied voltage. **e**, Introducing G-kink into the transmembrane region. (Adapted from *Nature Nanotechnology* [34])

molecular exchange with their external environment. The power of nanopores was initially demonstrated by using nanopores to transport small fluorophore molecules from/into lipid vesicles [36]. As the field processes, focus has shifted to transport substances with information to achieve lipid vesicle-based applications. For example, Deshpande et al. utilized αHL to transport ATP into lipid vesicles to trigger pLL/ATP coacervate formation, building a platform to understand intracellular phase behavior [37] (Fig. 12.8a). Takagi et al. realized the transmission of external environmental pH information to the lipid vesicle through the yeast voltage-dependent anion channel (VDAC, 2.5–3 nm) and created a lipid vesicle-based pH sensor [38]. These approaches indicate the ability of nanopores to sophisticate the molecular system in lipid vesicles.

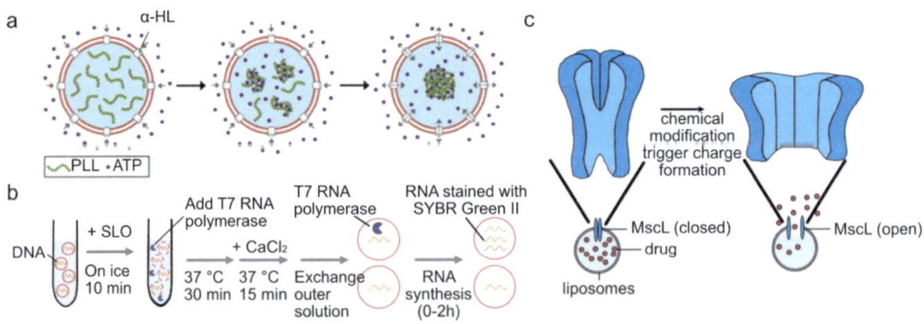

Fig. 12.8 Nanopores in lipid vesicle-based molecular robots. **a**, pLL/ATP coacervate formation within liposomes through influx of ATP by αHL. (Adapted from *Nature Communications* [37]) **b**, Process of triggered liposomal release through MscL. **c**, Transport of T7 RNA polymerase by the SLO

Nanopores with controllable opening and closing functions can further unlock the potential of lipid vesicles. Koçer et al. incorporated the mechanosensitive channel MscL, into the lipid vesicles [39]. This channel could be controlled to open and close in response to ambient pH, light, or a combination of both, thereby enabling controlled release of content inside lipid vesicles (Fig. 12.8b). Tsuji et al. used streptolysin O nanopores (SLO) to exchange fluorescent proteins and introduce T7 RNA polymerase (98 kDa) into liposomes by forming reversible pores that can be closed with calcium ions [40] (Fig. 12.8c). The development of new opening-controllable nanopores will represent a critical step towards the deployment of lipid vesicle-based molecular robots for practical application.

Soap bubbles, consisting of two surfactant layers and one water layer, are potential candidates as molecular robots in the atmosphere. By introducing functional molecules into the water layer, previous studies have demonstrated the soap bubbles to detect methamphetamine in aerosols [41] or to pollinate flowers [42]. Similar to the lipid vesicle-based systems, nanopores have great potential to sophisticate the soap bubble-based systems. Because of the opposite structure of the surfactant and the lipid membrane, most pore-forming proteins or peptides are unlikely to form pores on surfactant layers. Some antifungal antibiotics, such as Amphotericin B [43, 44], have shown ability to form pores through lipid monolayers (Fig. 12.9), but their ability to transform molecules from air to water remains unknown. We expect that functional nanopores in the soap bubble-based system can be realized in the near future.

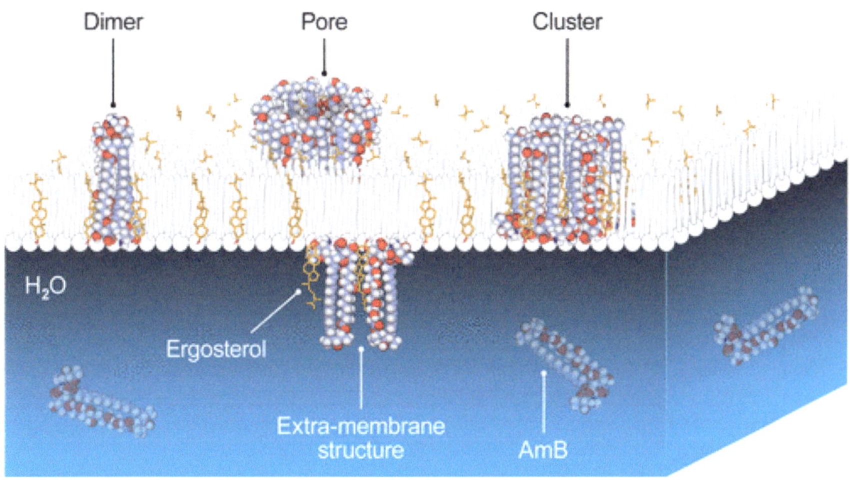

Fig. 12.9 Amphotericin B forms various nanopores and other structures on DOPC-Ergo monolayer. (Reprinted from *The Journal of Physical Chemistry Letters* [44])

12.6 Conclusions

In this chapter, we have discussed recent progress in protein/peptide nanopores and their applications in molecular robots. These progresses are the result of significant advancements in biophysics, biochemistry, and systems engineering. Meanwhile, nanopores formed by DNA and other nanostructures, which are discussed in other chapters, also have unique advantages. Accordingly, we anticipate that combining the proteins/peptides with other materials to form nanopores will be a strategy to achieve new nanopores with homogeneity and selectivity, thereby accelerating the development of molecular robots.

References

1. Shoji, K., Kawano, R.: Recent advances in liposome-based molecular robots. Micromachines **11** (2020)
2. Watanabe, H., et al.: Analysis of pore formation and protein translocation using large biological nanopores. Anal. Chem. **89**, 11269–11277 (2017)
3. Kasianowicz, J.J., Brandin, E., Branton, D., Deamer, D.W.: Characterization of individual polynucleotide molecules using a membrane channel. Proc. Natl. Acad. Sci. U.S.A. **93**, 13770–13773 (1996)
4. Derrington, I.M., et al.: Nanopore DNA sequencing with MspA. Proc Natl Acad Sci U S A **107**, 16060–16065 (2010)
5. Jain, M., Olsen, H.E., Paten, B., Akeson, M.: The Oxford Nanopore MinION: delivery of nanopore sequencing to the genomics community. Genome Biol. **17** (2016)
6. Lu, H.Y., Giordano, F., Ning, Z.M.: Oxford nanopore MinION sequencing and genome assembly. Genomics Proteomics Bioinform. **14**, 265–279 (2016)
7. Wasswa, F.B., Kassaza, K., Nielsen, K., Bazira, J.: MinION whole-genome sequencing in resource-limited settings: challenges and opportunities. Curr Clin Microbiol Rep **9**, 52–59 (2022)
8. Huang, G., et al.: PlyAB nanopores detect single amino acid differences in folded Haemoglobin from blood. Angewandte Chemie-International Edition **61** (2022)
9. Zhang, S.L., et al.: Bottom-up fabrication of a proteasome-nanopore that unravels and processes single proteins. Nat. Chem. **13**, 1192-+ (2021)
10. Motone, K., Cardozo, N., Nivala, J.: Herding cats: label-based approaches in protein translocation through nanopore sensors for single-molecule protein sequence analysis. Iscience **24** (2021)
11. Ouldali, H., et al.: Electrical recognition of the twenty proteinogenic amino acids using an aerolysin nanopore. Nat. Biotechnol. **38**, 176–181 (2020)
12. Yan, S.H., et al.: Single molecule ratcheting motion of peptides in a mycobacterium smegmatis Porin A (MspA) Nanopore. Nano Lett. **21**, 6703–6710 (2021)
13. Brinkerhoff, H., Kang, A.S.W., Liu, J.Q., Aksimentiev, A., Dekker, C.: Multiple rereads of single proteins at single-amino acid resolution using nanopores. Science **374**, 1509-+ (2021)
14. Lucas, F.L.R., Versloot, R.C.A., Yakovlieva, L., Walvoort, M.T.C., Maglia, G.: Protein identification by nanopore peptide profiling. Nat. Commun. **12** (2021)
15. Ding, T., et al.: Nucleic acids research vol. 48, pp. 2791–2806 (2020)
16. Takiguchi, S., Takeuchi, N., Shenshin, V., Gines, G., Genot A.J., Nivala, J., Rondelez, Y., Kawano, R.: Harnessing DNA computing and nanopore decoding for practical applications: from informatics to microRNA-targeting diagnostics. Chemical Society Reviews Under review (2024)

17. Hiratani, M., Kawano, R.: DNA logic operation with nanopore decoding to recognize microRNA patterns in small cell lung cancer. Anal. Chem. **90**, 8531–8537 (2018)
18. Takeuchi, N., Hiratani, M., Kawano, R.: Pattern recognition of microRNA expression in body fluids using nanopore decoding at subfemtomolar concentrations. JACS Au **2**, 1829–1838 (2022)
19. Straathof, S., et al.: Protein sizing with 15 nm conical biological Nanopore YaxAB. ACS Nano **17**, 13685–13699 (2023)
20. Zhang, X.L., et al.: Specific detection of proteins by a nanobody- functionalized nanopore sensor. ACS Nano **17**, 9167–9177 (2023)
21. Rotem, D., Jayasinghe, L., Salichou, M., Bayley, H.: Protein detection by nanopores equipped with aptamers. J. Am. Chem. Soc. **134**, 2781–2787 (2012)
22. Soskine, M., et al.: An engineered ClyA nanopore detects folded target proteins by selective external association and pore entry. Nano Lett. **12**, 4895–4900 (2012)
23. Ahmad, M., et al.: A generalizable nanopore sensor for highly specific protein detection at single-molecule precision. Nat. Commun. **14** (2023)
24. Joh, N.H., et al.: De novo design of a transmembrane Zn2+-transporting four-helix bundle **346**, 1520–1524 (2014)
25. Vorobieva, A.A., et al.: De novo design of transmembrane beta barrels. Science **371**, 801–810 (2021)
26. Niitsu, A., Sugita, Y.: Towards de novo design of transmembrane α-helical assemblies using structural modelling and molecular dynamics simulation. Phys. Chem. Chem. Phys. **25**, 3595–3606 (2023)
27. Ebenhan, T., Gheysens, O., Kruger, H.G., Zeevaart, J.R., Sathekge, M.M.: Antimicrobial peptides: their role as infection-selective tracers for molecular imaging. Biomed. Res. Int. **2014**, 867381 (2014)
28. Schmidtchen, A., et al.: Boosting antimicrobial peptides by hydrophobic oligopeptide end tags. J. Biol. Chem. **284**, 17584–17594 (2009)
29. Ladokhin, A.S., Selsted, M.E., White, S.H.: Sizing membrane pores in lipid vesicles by leakage of co-encapsulated markers: pore formation by melittin. Biophys. J. **72**, 1762–1766 (1997)
30. Mahendran, K.R., et al.: A monodisperse transmembrane α-helical peptide barrel. Nat. Chem. **9**, 411–419 (2017)
31. Krishnan, R.S., et al.: Assembly of transmembrane pores from mirror-image peptides. Nat. Commun. **13** (2022)
32. Scott, A.J., et al.: Constructing ion channels from water-soluble α-helical barrels. Nat. Chem. **13**, 643-+ (2021)
33. Dawson, W.M., et al.: Coiled coils 9-to-5: rational *de novo* design of α-helical barrels with tunable oligomeric states. Chem. Sci. **12**, 6923–6928 (2021)
34. Shimizu, K., et al.: De novo design of a nanopore for single-molecule detection that incorporates a β-hairpin peptide. Nat. Nanotechnol. **17**, 67-+ (2022)
35. Fujita, S., Kawamura, I., Kawano, R.: Cell-free expression of de novo designed peptides that form fl-barrel nanopores. Acs Nano (2023)
36. Ota, S., Yoshizawa, S., Takeuchi, S.: Microfluidic formation of monodisperse, cell-sized, and unilamellar vesicles. Angewandte Chemie-International Edition **48**, 6533–6537 (2009)
37. Deshpande, S., et al.: Spatiotemporal control of coacervate formation within liposomes. Nat. Commun. **10** (2019)
38. Takagi, K., Ohgita, T., Yamamoto, T., Shinohara, Y., Kogure, K.: Transmission of external environmental pH Information to the inside of liposomes *via* pore-forming proteins embedded within the liposomal membrane. Chem. Pharm. Bull. **64**, 432–438 (2016)
39. Kocer, A.: A remote controlled valve in Liposomes for triggered liposomal release. J. Liposome Res. **17**, 219–225 (2007)
40. Tsuji, G., Sunami, T., Oki, M., Ichihashi, N.: Exchange of proteins in liposomes through streptolysin O pores. ChemBioChem **22**, 1966–1973 (2021)
41. Vannoy, K.J., Tarolla, N.E., Kauffmann, P.J., Clark, R.B., Dick, J.E.: Detecting methamphetamine in aerosols by electroanalysis in a soap bubble wall. Anal. Chem. **94**, 6311–6317 (2022)

42. Yang, X., Miyako, E.: Soap bubble pollination. Iscience **23** (2020)
43. Wang, J., Wang, X.Z., Wang, Z.: Effect of amphotericin B on the thermodynamic properties and surface morphology of the pulmonary surfactant model monolayer during respiration. Molecules **28** (2023)
44. Janik, S., Luchowski, R., Grela, E., Grudzinski, W., Gruszecki, W.I.: How does the antibiotic amphotericin B enter membranes and what does it do there? J. Phys. Chem. Lett. **15**, 4823–4827 (2024)

Part III
Cellular Compartments for Cell-Mimetic Molecular Robots: Giant Vesicles

Chapter 13
Nonequilibrium State of Giant Vesicles in a Microfluidic Environment

Taro Toyota, Haruto Obuchi, Yiting Zhang, and Hironori Sugiyama

Giant vesicles (GVs), which are closed lipid bilayer membranes with diameters greater than 1 μm, have attracted attention as capsules for constructing cell-mimetic molecular robots. GVs can encapsulate not only molecules and macromolecules but also micrometer-sized molecular complexes and particles. The lipid bilayer membrane of GVs provides a reaction field for functional amphiphiles, including membrane proteins. These intramembrane molecules and macromolecules enable the influx and efflux of specific chemical species and information through molecular recognition at both the inner and outer leaflets of the GVs. Due to their size, GVs can be transported and finely aligned to construct network structure within a microfluidic device. These morphological and chemical reaction networks can be used to develop highly functionalized information circuits. In this section, we review and discuss the morphological dynamics and membrane permeability of GVs in nonequilibrium states achieved using microfluidic devices.

T. Toyota (✉) · H. Obuchi
The University of Tokyo, Tokyo, Japan
e-mail: cttoyota@mail.ecc.u-tokyo.ac.jp

H. Obuchi
e-mail: haruto-obuchi342@g.ecc.u-tokyo.ac.jp

Y. Zhang
Rikkyo University, Tokyo, Japan
e-mail: i_show@rikkyo.ac.jp

H. Sugiyama
Institute of Science Tokyo, Tokyo, Japan
e-mail: h.sugiyama@elsi.jp

S. Murata et al. (eds.), *Molecular Robotics II*,
https://doi.org/10.1007/978-981-95-2394-8_13

13.1 Spontaneous Morphological Changes of GVs in a Microfluidic Channel

GVs have attracted considerable attention as cell membrane models. The semipermeable nature of the vesicular membrane allows its shape to change in response to osmotic pressure and other stimuli, resulting in deformed shapes that resemble those of living cells. In equilibrium state, the experimentally observed shapes of GVs include spherical, oblate, prolate, budded, invaginated, and starfish types. These shapes can be explained using the spontaneous-curvature model [1], the bilayer-coupling model [2], or the area difference elasticity model [3]. The spontaneous-curvature model estimates the bending energy of the vesicle based on structural parameters of the lipid components. An additional constraint for the area difference between the inner and outer leaflets is imposed in the bilayer-coupling model. The area difference elasticity model includes the effect of the relative areal stretching of the two leaflets. These models have been used to rationalize the observed shapes and transformations of the GVs. For example, sequential deformation and division of GVs was realized using two-component GVs that respond to heating and cooling cycles [4] (Fig. 13.1). When the GV membrane is extended by internal forces, such as the elongation of encapsulated fibrous proteins [5] or external aspiration forces [6], the GV protrudes and forms tubes until membrane tension is primarily balanced by the applied forces.

When GVs are transported by continuous flow in a narrow channel, representing a nonequilibrium state, they deform from spherical to oblate and develop a wrinkled shape due to the shear force around them [7]. Interestingly, the inner and outer leaflets are forced differently by the shear force field; each leaflet responds separately to the

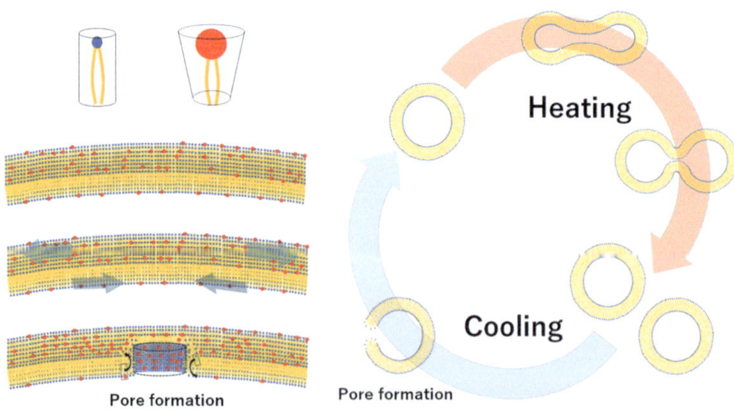

Fig. 13.1 Schematic illustration of the deformation and division of a two-component GV caused by heating and cooling cycles [4]. On the left panel, a sequence shows a membrane, with two different lipid molecules, undergoing structural changes leading to pore formation. On the right panel, a circular flow chart represents the cycle of heating and cooling. The heating phase shows membrane deformation and division, while the cooling phase shows pore formation

force [8]. These aspects of the vesicular membrane surrounded by flow indicate that even unilamellar GVs (known as giant unilamellar vesicles, GUVs) are not simple single-membrane compartments, but can be separately functionalized based on the design of the inner and outer leaflets of the vesicular membrane.

13.2 Manipulation and Arrangement of GVs Using a Microfluidic Device

The vesicular membrane of GVs possesses a surface electric charge, dielectric constant, and refractive index different from those of water (or buffered solutions), enabling manipulation using electric/magnetic fields, acoustic pressure fields, and optical tweezers [9]. Microfluidic devices allow for simultaneous manipulation of a large amount (10^2–10^3), facilitating the statistical analysis of changes in GVs [10]. The advantage of using microfluidic device design allows for precise channel configurations, which can be used to manipulate GV dispersions. For example, GV purification, which separates GVs from a mixture of GVs containing additives (including nanometer-sized particles), has been performed using a pinched-flow fractionation mechanism.

In construction of cell-mimetic molecular robots, the manipulation and arrangement (and immobilization) of GVs are important for continuous observation under an optical microscope and for the geometrical positioning of functionalized GVs to facilitate signal transfer. In particular, the techniques that strategically position functionalized GVs enhance the hierarchy of information processing among GVs, aiding in the development of molecular robots that function similarly to multicellular organisms. For example, the deterministic lateral displacement (DLD) [11] is a promising mechanism for GV size sorting; integrating DLD with the regions containing trapping microstructures enables the arrangement of monodisperse GVs with a variation coefficient of approximately 10% [12]. Improvement in trapping microstructures can lead to the formation of GV dyads (Fig. 13.2).

In an equilibrium state, multiple lipid components for forming GVs, such as a mixture of 1,2-dioleoyl-*sn*-glycero-3-phosphocholin (DOPC), sphingomyelin (SM), and cholesterol, cause intramembrane phase separation under specific conditions determined by dispersion temperature and composition ratios [13]. This is because the DOPC-rich domains exist in the liquid-disordered phase, whereas SM-rich domains are in the liquid-ordered phase. This phenomenon is valuable because it allows for the localization of functional molecules such as membrane proteins and amphiphilic macromolecules within a GV, resembling the differentiation of membrane protein functions within local domains on cell membranes, known as the "raft" model. Recent studies on the phase separation of GVs have shown that the pattern of these domains changes from polka dots to stripes and waves under steady flow when GVs are immobilized in microfluidic devices due to the shear force [14] (Fig. 13.3). This indicates that the flow itself can play a role in shifting and tuning GV functions.

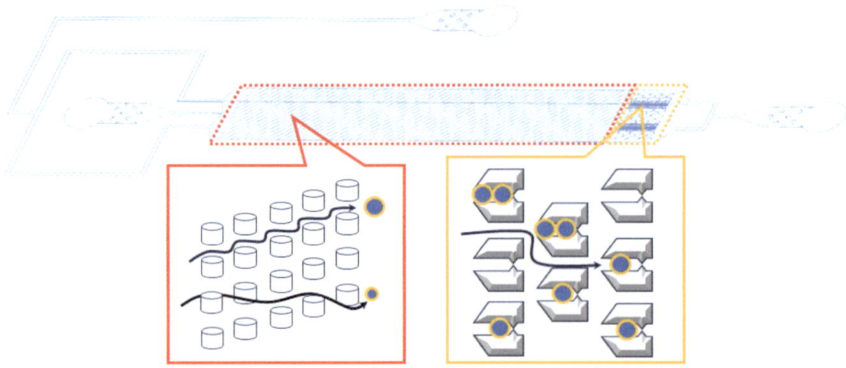

Fig. 13.2 Schematic illustration of GV sorting and trapping for dyad formation using a microfluidic device. The device contains two highlighted regions of GV size sorting (outlined in red) and GV trapping (outlined by yellow)

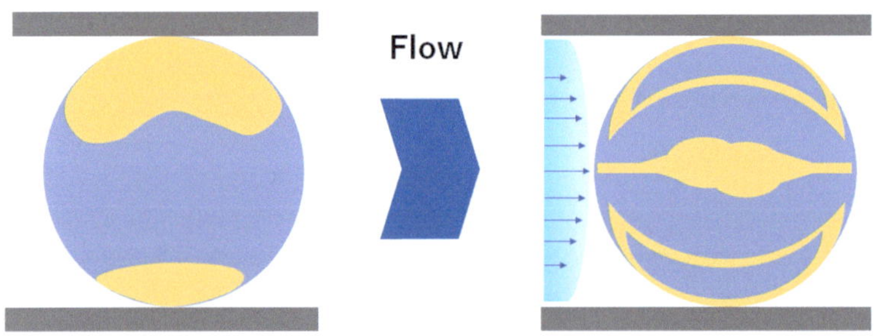

Fig. 13.3 Schematic illustration of phase separation in GVs under a steady flow condition. The phase separation of GV at the initial state is shown on the left panel. The pattern of the domains changes to stripes and waves under a steady flow, represented on the right panel

13.3 Membrane Permeability of GVs Under Steady Flow

The membrane permeability of GVs is regulated by their lipid components at equilibrium. In addition to neutral water-soluble molecules, ionic molecules can pass through the GV membrane and accumulate when there is an imbalance in cation/anion concentration across the membrane [15]. This accumulation can facilitate the replenishment of substrates after their depletion in GVs containing enzymes. Moreover, GVs trapped in microstructures have shown atypical membrane permeability under flow conditions, even without a concentration bias of solutes across the GV membrane [12]. When GVs composed of a mixture of 1-palmitoyl-2-oleoyl-*sn*-glycero-3-phosphocholine (POPC), a zwitterionic phospholipid, and 1-oleoyl-2-palmitoyl-*sn*-glycero-3-phosphoglycerol (POPG), a negatively charged phospholipid, were immobilized in the microstructures, the permeation of fluorescent ionic molecules, including a fluorescent polyethylene glycol analogue, across the lipid membrane, resulting in the accumulation of these molecules against the concentration gradient (termed hydrodynamic accumulation; HDA) [16]. In addition to POPG, other negatively charged phospholipids, such as 1,2-dioleoyl-*sn*-glycero-3-phospho-(1'-rac-glycerol) (DOPG), 1,2-dioleoyl-*sn*-glycero-3-phospho-$_L$-serine (DOPS), and 1,2-dioleoyl-*sn*-glycero-3-phosphoric acid (DOPA), contributed to HDA, and no significant differences among the types of negatively charged phospholipids were observed. This is plausibly because the external flow induces a compositional asymmetry of negatively charged phospholipids between the inner and outer leaflets, which allows for the accumulation of these molecules against the concentration gradient. This is another example where flow itself can influence GV functions based on membrane permeability.

13.4 Summary

In this review, microfluidic devices for handling GVs are highlighted as tools for the simultaneous manipulation and arrangement of GVs in large quantities. However, flow-induced shear forces affect both the transport and shape of trapped GVs, phase separation, and membrane permeability due to differential forces acting on the outer and inner leaflets of the GVs. This implies that when designing functionalized GVs for construction of cell-mimetic molecular robots, the flow of input signals and fuels solutions should be considered as a key experimental condition to ensure robust operation under nonequilibrium states. In the future, a GV-based molecular robot functioning as a power engine [17] could lead to advanced molecular robots capable of autonomously utilizing input signals and fuel.

References

1. Helfrich, W.: Elastic properties of lipid bilayers: theory and possible experiments. Z. Naturforsch, Teil C **28**, 693 (1973)
2. Seifert, U., Berndl, K., Lipowsky, R.: Shape transformations of vesicles: phase diagram for spontaneous-curvature and bilayer-coupling models. Phys. Rev. A **44**, 1182 (1991)
3. Miao, L., et al.: Budding transitions of fluid-bilayer vesicles: the effect of area-difference elasticity. Phys. Rev. E **49**, 5389 (1994)
4. Jimbo, T., et al.: Role of inverse-cone-shape lipids in temperature-controlled self-reproduction of binary vesicles. Biophys. J. **110**, 1551–1562 (2016)
5. Hayashi, M., et al.: Reversible morphological control of tubulin-encapsulating giant liposomes by hydrostatic pressure. Langmuir **32**, 3794–3802 (2016)
6. Kwok, R., Evans, E.: Thermoelasticity of large lecithin bilayer vesicles. Biophys. J. **35**, 637–652 (1981)
7. Abreu, D., et al.: Fluid vesicles in flow. Adv. Colloid Interf. Sci. **208**, 129–141 (2014)
8. Sebastian, B., Favero, T., Dittrich, P.S.: The effects of shear force transmission across vesicle membranes. J. Phys. Chem. Lett. **8**, 6128–6134 (2017)
9. Toyota, T., Zhang, Y.: Identifying and manipulating giant vesicles: review of recent approaches. Micromachines. **13**, 644 (2022)

10. Robinson, T.: Microfluidic handling and analysis of giant vesicles for use as artificial cells: a review. Adv. Biosys. **3**, 1800318 (2019)
11. Huang, L.R., et al.: Continuous particle separation through deterministic lateral displacement. Science **304**, 987–990 (2004)
12. Sugiyama, H., et al.: Hydrodynamic accumulation of small molecules and ions into cell-sized liposomes against a concentration gradient. Commun. Chem. **3**, 32 (2020)
13. Dimova, R.: Giant vesicles and their use in assays for assessing membrane phase state, curvature, mechanics, and electrical properties. Annu. Rev. Biophys. **48**, 93–119 (2019)
14. Hamada, T., Mizuno, S., Kitahata, H.: Shear-induced nonequilibrium patterns in lipid bilayer membranes exhibiting phase separation. Langmuir **40**, 8843–8850 (2024)
15. Ishida, N., et al.: Uphill accumulation of ionic species into a lipid vesicle by the concentration gradient of counter ions. Langmuir **38**, 14208–14216 (2022)
16. Sugiyama, H., et al.: Role of negatively charged lipids achieving rapid accumulation of water-soluble molecules and macromolecules into cell-sized liposomes against a concentration gradient. Langmuir **38**, 112–121 (2021)
17. Shoji, K., Kawano, R.: Osmotic-engine-driven liposomes in microfluidic channels. Lab Chip **19**, 3472–3480 (2019)

Chapter 14
Preparation of Monodisperse GUVs Using a Microfluidic Device

Mamiko Tsugane and Hiroaki Suzuki

Giant unilamellar vesicles (GUVs), which mimic cell membranes, are primarily used as exteriors of cell-type molecular robots. A method for producing GUVs that are uniform in size and structure and have high encapsulation efficiency is required to ensure the designed performance of robots. In this section, we describe a system for producing GUVs using microfluidic devices. This system, utilizing the interfacial transfer of water-in-oil (W/O) droplets, can produce monodisperse GUVs with diameters of 25–45 μm with a transfer yield of ~100%. To apply this system to constructing molecular robotics, we also present the results of encapsulating an in vitro transcription/translation system in GUVs and examining the conditions under which protein synthesis and reconstitution of membrane proteins are processed.

14.1 From Classical GUV Preparation Methods to Microfluidics-Based GUV Production Systems

Lipid vesicles, or liposomes, are vesicles with a lipid bilayer membrane artificially synthesized using amphiphilic phospholipids. These vesicles possess structures and functions that mimic those of cells, making them practical platforms for constructing cell membrane models and artificial cells [1–3]. Molecular robots are often developed interchangeably with artificial cells by integrating molecular sensors, processors, and actuators in the lumen and on the lipid membrane [4]. Because artificial cells and molecular robots often contain large and complex molecular systems, vesicles

M. Tsugane · H. Suzuki (✉)
Department of Precision Mechanics, Faculty of Science and Engineering, Chuo University, Tokyo, Japan
e-mail: suzuki@mech.chuo-u.ac.jp

M. Tsugane
e-mail: tsugane@nano.mech.chuo-u.ac.jp

S. Murata et al. (eds.), *Molecular Robotics II*,
https://doi.org/10.1007/978-981-95-2394-8_14

should have a diameter larger than 1 μm [5]. Moreover, the lipid membrane must be unilamellar for membrane proteins to function. Lipid vesicles with these structural characteristics are called giant unilamellar vesicles (GUVs).

Various methods for preparing cell-sized GUVs, including bulk techniques and newly developed microfluidic systems, have been developed. Bulk methods can be categorized into two types: (1) swelling-based methods, which involve the rehydration of dried lipid films, and (2) emulsion-based methods, in which lipid-stabilized water-in-oil (W/O) emulsions containing cytosolic components are transferred across an oil-water (O-W) interface to form a bilayer [6].

Classical swelling-based methods for vesicle production include natural swelling [7, 8] and electroformation [9, 10]. These methods produce a large number of polydisperse and often multilamellar vesicles. In addition, it is difficult to prepare vesicles in physiological (ion-rich) solutions, and their encapsulation efficiencies are low.

In contrast, emulsion-based methods improve the encapsulation efficiency of water-soluble macromolecules in vesicles and can efficiently prepare GUVs composed of a single bilayer membrane [11, 12]. In the emulsion-transfer method, when a water droplet with a lipid monolayer in an oil phase is transferred across another O-W interface covered with a lipid monolayer by external forces, such as centrifugal force, the hydrophobic faces of the two monolayers meet to form a bilayer membrane. This method has been predominantly used in the artificial-cell society, because it allow excellent encapsulation efficiency of micromolecules, complex systems, and even particulate materials [29, 30, 34, 35]. A possible disadvantage of this method is that these GUVs may contain solvent remaining in the hydrophobic layer of the bilayer. In addition, because pipetting or stirring was used to prepare the W/O emulsion, the resultant GUVs were polydisperse [5].

When GUVs are used for artificial cells or molecular robots, the uniformity of the structure and size is important to ensure the reproducibility of their functionalities. Several GUV fabrication methods, including size uniformity, high encapsulation efficiency, and high throughput, have been developed using microfluidic devices. Microfluidic systems can precisely control the flow rate of small amounts of solution and generate monodisperse droplets at high speeds [13–15]. Thus, GUV production methods that apply droplet generation have become mainstream. GUV preparation methods using microfluidic devices are classified as (1) pulsed jet flow, (2) double-emulsion template, and (3) emulsion transfer.

The pulsed jet flow method produces GUVs by ejecting a pulsed water jet from a micro nozzle against a planar phospholipid bilayer membrane formed by the droplet contact method [16]. In this method, a micronozzle is manually positioned, which hinders reproducibility and throughput and makes it difficult to continuously produce large quantities of GUVs.

In the double-emulsion template method, water-in-oil-in-water (W/O/W) double-emulsion droplets were first prepared by the flow-focusing method using a glass microcapillary or microfluidic device, which were transformed into GUVs [17–19]. Recently, Deshpande et al. developed an octanol-assisted liposome assembly (OLA) [20]. The double-emulsion droplets with a thin oil layer, generated by a flow-focusing type PDMS microfluidic device, spontaneously transformed into GUVs after the 1-octanol layer accumulated as a droplet and separated. Deng et al. used a coaxial glass microcapillary device to prepare W/O/W droplets as templates for GUV [21]. The separation of the oil layer was promoted by adding Pluronic F-68, a triblock copolymer surfactant, to the outer phase.

In addition to the aforementioned flow-focusing approaches, an emulsion-transfer method was implemented using a microfluidic device. Hu et al. produced monodisperse W/O emulsions using microchannels via flow focusing [22]. The collected emulsion droplets were transferred across the O-W interface via centrifugation, similar to the bulk emulsion-transfer method. In addition, Matosevic and Paegel developed a device that combined the production of W/O emulsions and their transfer across the O-W interface [23] in a single microchannel device. Lipid-stabilized droplets produced by flow-focusing were transferred across the O-W interface with the aid of a triangular post, forming lipid bilayer vesicles. However, the yield of GUVs obtained using this method was only 5%.

Although the GUV production method using microfluidic devices can produce GUVs with high throughput and with a coefficient of variation (CV) of 5% or less, the production technologies still have limitations. In glass microcapillary devices, the coaxial alignment of capillaries, which has a significant effect on the generation of W/O/W droplets, must be performed manually, which is cumbersome and depends on the researcher's skill. In contrast, dimethylpolysiloxane (PDMS) microfluidic devices can be mass-produced with good reproducibility; however, controlling the wetting of oil and water inside the intrinsically hydrophobic flow channel is necessary [24]. Practically, it is necessary to partially hydrophilize the channel section where the W/O/W droplets are generated and conveyed, which requires skilled techniques. Precise flow control is essential for the formation of double emulsions with thin oil layers using flow-focusing-based methods.

14.2 Preparation of Monodisperse GUV Production Device Using Droplet Transfer at the W/O Interface

Our group developed a PDMS microfluidic device for the production of monodisperse GUVs using an emulsion-transfer method that does not require skilled fabrication techniques, partial surface treatment of channels, or precise flow-rate tuning [25].

Microfluidic Device: Design and Concept

The design of our microfluidic channel (Fig. 14.1) has three inlets for the inner aqueous phase, the oil phase with lipids, and the outer aqueous phase. A PDMS channel with this design was fabricated and assembled to introduce these solutions using a three-channel pressure-controlled pump. A schematic diagram of the GUV generation process is shown in Fig. 14.2a. First, W/O droplets comprising a lipid monolayer were generated using a flow-focusing method. Next, the outer aqueous phase was introduced to narrow the oil flow, and a W/O interface with the lipids was formed in the channel. In the curved part of the downstream channel, the width of the oil layer became smaller than the diameter of the droplet, and the W/O droplet was pressed against the W/O interface and passed through it to become a W/O/W droplet (Fig. 14.2b, c). After collection in a tube, the oil layer of the W/O/W droplets accumulates due to the dewetting phenomenon, and the droplets become GUVs with a lipid bilayer. In the flow-focusing part of this device, W/O droplets were readily formed because PDMS is hydrophobic. Downstream, the wettability between PDMS and oil was adjusted such that the oil layer of the W/O/W droplets did not wet the channel walls without partial hydrophilization.

Success Rate Improvement in W/O Droplets Transfer Across W/O Interface

After screening various oils, including squalene, squalane, hexadecane, 1-octanol, paraffin oil, soybean oil, and mineral oil, we found that squalene, with its low viscosity, was the best choice for maintaining a stable laminar flow in microchannels at a moderate pressure range. Although W/O droplets can be readily generated using

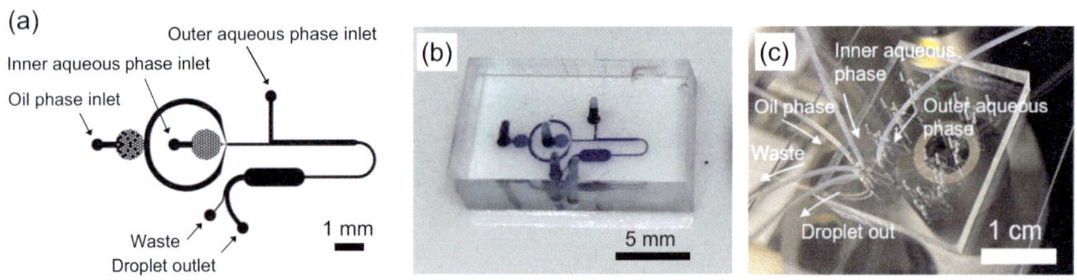

Fig. 14.1 **a** Design of the microfluidic channel **b** image of the PDMS microfluidic device **c** image of the microchannel with connected tubes placed on a microscope stage. Reprinted with permission from Ushiyama et al. [28]. Copyright 2024, American Chemical Society

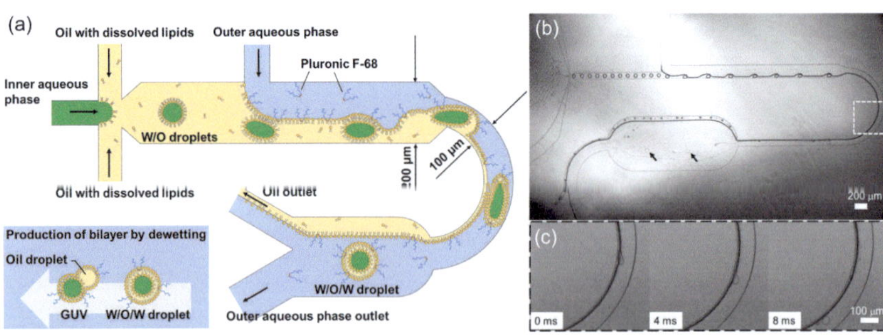

Fig. 14.2 **a** Schematic of the production process of monodisperse W/O/W droplets in the microfluidic channel by the W/O interface transfer and GUV generation, followed by dewetting of excess oil. **b** Microscopic image of the entire channel during successful production of W/O/W droplets. The black arrows indicate the W/O/W droplets. **c** Sequential close-up images of a W/O droplet transferring across the W/O interface to form a W/O/W droplet. Reprinted with permission from Ushiyama et al. [25] Copyright 2022, Elsevier

squalene, the interface of the W/O laminar flow is unstable. Adding 10–30% v/v 1-octanol to squalene significantly improved the stability of the interface.

Furthermore, adding a triblock copolymer surfactant (Pluronic F-68) to the external aqueous phase led to the successful transfer of W/O droplets across the W/O interface. The success rate of droplet transfers (the number of W/O/W droplets transferred successfully divided by the number of W/O droplets generated by flow focusing) was 0% without Pluronic F-68 but reached 94.2% at a concentration of 1% w/v. Pluronic surfactants with high hydrophilic-lipophilic balance (HLB) values, such as Pluronic F-68, have a large hydrophilic portion and impart a positive curvature to the W/O interface, which is thought to facilitate interface transfer [26].

Formation of the lipid bilayer by oil dewetting.

The W/O/W droplets collected from the outlet of the microfluidic channel had a thin oil layer (~1 μm). After a few minutes, the oil spontaneously dewetted and formed oil droplets on top of the vesicles (Fig. 14.3a and b). We confirmed that the degree of oil dewetting can be controlled by increasing the concentration of lipids in the oil phase [27]. As the DOPC and POPC concentrations increased, the contact angle of the dewetted oil increased. Notably, at 6 mg/ml POPC in the oil phase, the oil droplets were detached by gentle pipetting, and GUVs surrounded only by a lipid bilayer were obtained.

GUV Size Control

The size of the W/O droplets could be controlled by adjusting the flow rate of the inner aqueous phase without changing the flow rates of the oil and outer aqueous phases. When a microchannel with a height of 37 μm was used, the diameter of the W/O/W droplets that could be produced was 25–45 μm (Fig. 14.3c). W/O droplets smaller than ~ 25 μm were not sufficiently pressed against the interface of the O-W laminar flow and remained in the oil stream. On the other hand, W/O droplets larger than ~ 45 μm were significantly deformed and split into two droplets when

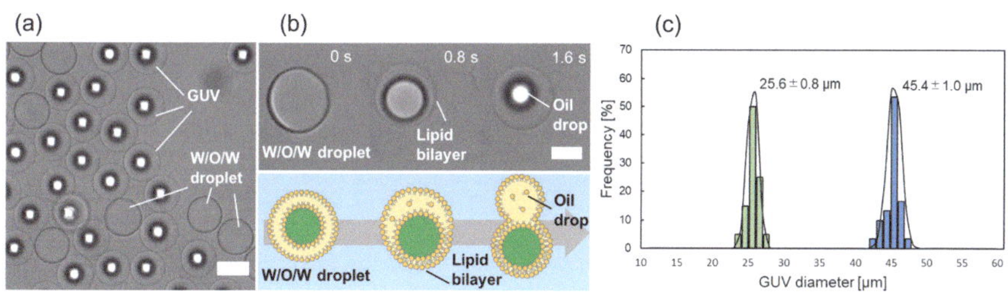

Fig. 14.3 **a** Microscopic image of the monodisperse vesicles during the dewetting process. Scale bar is 40 μm. Sequential close-up images and a schematic of the dewetting process. Scale bar is 40 μm. Histograms of GUV diameters showing the monodispersity of the GUVs generated under two flow conditions. Solid lines are Gaussian fits for each histogram (n = 30, n = 20). Reprinted with permission from Ushiyama et al. [25]. Copyright 2022, Elsevier

pressed against the interface. The GUVs produced were fairly monodisperse, with CVs ranging from 2 to 3%.

14.3 In Vitro Translation/Transcription (IVTT) in GUVs Toward Construction of Artificial Cells and Molecular Robots

To utilize GUVs as artificial cells or molecular robots, the implementation of complex and functional molecular systems is required. To demonstrate one of these possibilities, we investigated protein synthesis, an important biochemical reaction in cells, in the GUVs produced by our microfluidic device [28]. Using a commercially available IVTT system, the PURE system (PUREfrex 1.0, Gene Frontier Corp.) was used to investigate the conditions under which protein synthesis proceeds efficiently, while maintaining vesicle integrity. When all the components included in the IVTT system kit were encapsulated as the internal solution of GUVs, all W/O/W droplets collapsed. We then investigated the components that affect the formation and stability of the vesicles in detail. We found that cations such as potassium glutamate (K^+) and magnesium acetate (Mg^{2+}) and charged substances such as spermidine and phosphocreatine caused membrane rupture. However, this problem was resolved by adding 0.5% polyvinyl alcohol (PVA) to the external solution. Glycerol caused membrane deformation of GUVs made of POPC; however, the vesicles maintained their spherical shape when the lipids were changed to eggPC and cholesterol. Based on these results, we tested the synthesis of green fluorescent protein (GFP) in the GUVs using the IVTT system (Fig. 14.4).

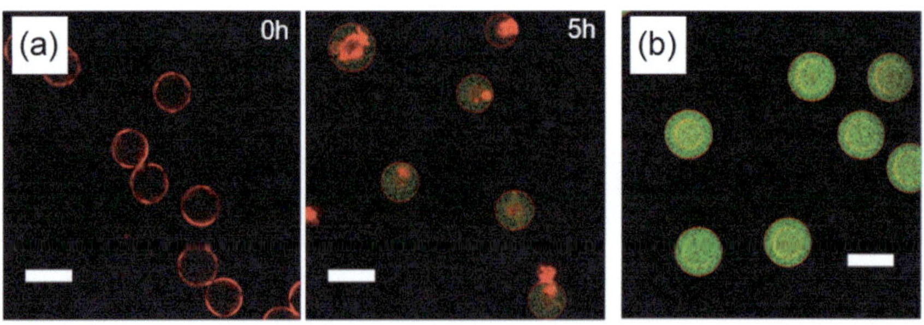

Fig. 14.4 Fluorescent microscope images of vesicles. Green fluorescence from GFP and red fluorescence from the membrane marker. (scale bars 50 μm) **a** GUVs with *in vesicle* synthesis experiment at 0 h and 5 h incubation, **b** GUVs encapsulating the PURE system solution after 5 h GFP synthesis in a test tube. Reprinted with permission from Ushiyama et al. [28]. Copyright 2024, American Chemical Society

Next, to verify if the present system can be extended to the synthesis and reconstitution of membrane proteins, in situ synthesis of α-hemolysin (αHL), a nanopore-forming peptide, was performed. The function of the nanopores was monitored using a fluorescent dye leakage assay. In the GUV containing the IVTT system, the GUV collapsed after the start of αHL synthesis, likely because of the combined destabilizing effect of the PURE system components and αHL. Therefore, the concentrations of all components of the internal solution were halved, and the reaction was allowed to proceed. As a result, it was observed that the fluorescent-tagged protein (Alexa Fluor 647-conjugated transferrin: TA 647), which is larger than the pore size of αHL heptamer, remained inside, while the Alexa Fluor 488 dye, which is smaller in size, gradually decreased (Fig. 14.5a, b and c). In addition, when the IVTT system was added to the external solution of the GUV and αHL was synthesized outside the GUV, the leakage rate was higher than that of internal synthesis (Fig. 14.5d, e and f). Thus, it was confirmed that the synthesized αHL formed a pore. The pore-forming activity of αHL also confirmed the unilamellarity of the vesicle membrane.

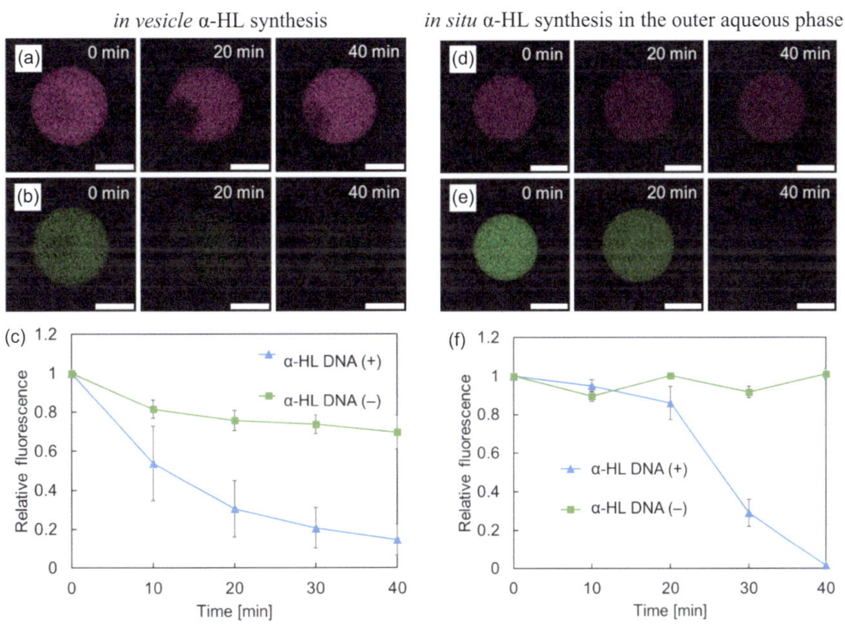

Fig. 14.5 Results of fluorescent dye leakage assay with **a, b, c** *in vesicle* α-HL synthesis and **d, e, f** in situ α-HL synthesis in the outer aqueous phase. **a, b, d, e** Fluorescent confocal images of a vesicle at various time points during incubation. Fluorescence of **a, d** TA647 and **b, e** Alexa Fluor 488, respectively. Scale bars are 20 μm **c, f** Time variation of fluorescence intensity of Alexa Fluor 488 relative to TA 647 (blue triangle, with α-HL DNA; green square, without α-HL DNA; n = 3). Reprinted with permission from Ushiyama et al. [28]. Copyright 2024, American Chemical Society

14.4 Conclusion

We established a simple and highly reproducible method for generating monodisperse GUVs using a microfluidic device. The W/O droplets transfer across the W/O interface, and this approach allows the production of monodisperse GUVs with diameters of 25–45 μm with a yield of ~100% without partial surface treatment in the channel or precise flow control. In addition, the operation can be started immediately after connecting the PDMS device to a vial containing the solution with a tube, and the stable state of W/O/W droplet generation can be achieved in a short time (1–2 min), so that only about 20 μL of the internal solution is required. The prepared GUVs had monolayer and semipermeable membrane properties, and protein synthesis inside and insertion of membrane proteins into the membrane were possible. This technology is expected to contribute to the creation of artificial cells that reproduce the structures and functions of living cells [31–33], and molecular robots using membrane protein-based biosensors.

References

1. Walde, P.: Building artificial cells and protocell models: experimental approaches with lipid vesicles. BioEssays **32**, 296–303 (2010)
2. Dzieciol, A.J., Mann, S.: Designs for life: protocell models in the laboratory. Chem. Soc. Rev. **41**, 79–85 (2012)
3. Miller, D.M., Gulbis, J.M.: Engineering protocells: prospects for self-assembly and nanoscale production-lines. Life (Basel) **5**, 1019–1053 (2015)
4. Peng, Z., et al.: Lipid vesicle-based molecular robots. Lab Chip **24**, 996–1029 (2024)
5. Walde, P., et al.: Giant vesicles: preparations and applications. ChemBioChem **11**, 848–865 (2010)
6. Van de Cauter, L., et al.: Exploring giant unilamellar vesicle production for artificial cells - current challenges and future directions. Small Methods **7**, e2300416 (2023)
7. Reeves, J.P., Dowben, R.M.: Formation and properties of thin-walled phospholipid vesicles. J. Cell. Physiol. **73**, 49–60 (1969)
8. Rideau, E., et al.: Self-assembly of giant unilamellar vesicles by film hydration methodologies. Adv Biosyst **3**, e1800324 (2019)
9. Angelova, M.I., Dimitrov, D.S.: Liposome electroformation. Faraday Discuss. **81**, 303–311 (1986)
10. Boban, Z. et al.: Giant Unilamellar vesicle electroformation: What to Use, What to Avoid, and How to Quantify the Results. Membranes (Basel) **11**, 860 (2021)
11. Pautot, S., et al.: Engineering asymmetric vesicles. Proc. Natl. Acad. Sci. U.S.A. **100**, 10718–10721 (2003)
12. Zhang, Y. et al.: A practical guide to preparation and applications of giant unilamellar vesicles formed via centrifugation of water-in-oil emulsion droplets. Membranes (Basel) **13**, 440 (2023)
13. Van Swaay, D., deMello, A.: Microfluidic methods for forming liposomes. Lab Chip **13**, 752–767 (2013)
14. Martino, C., deMello, A.J.: Droplet-based microfluidics for artificial cell generation: a brief review. Interface Focus **6**, 20160011 (2016)
15. Zhang, G., Sun, J.: Lipid in chips: a brief review of liposomes formation by microfluidics. Int. J. Nanomedicine **16**, 7391–7416 (2021)

16. Funakoshi, K., et al.: Lipid bilayer formation by contacting monolayers in a microfluidic device for membrane protein analysis. Anal. Chem. **78**, 8169–8174 (2006)
17. Shum, H.C., et al.: Double emulsion templated monodisperse phospholipid vesicles. Langmuir **24**, 7651–7653 (2008)
18. Kim, S.H., et al.: Double-emulsion drops with ultra-thin shells for capsule templates. Lab Chip **11**, 3162–3166 (2011)
19. Arriaga, L.R., et al.: Ultrathin shell double emulsion templated giant unilamellar lipid vesicles with controlled microdomain formation. Small **10**, 950–956 (2014)
20. Deshpande, S., et al.: Octanol-assisted liposome assembly on chip. Nat. Commun. **7**, 10447 (2016)
21. Deng, N.N., et al.: Monodisperse uni and multicompartment liposomes. J. Am. Chem. Soc. **138**, 7584–7591 (2016)
22. Hu, P.C., Li, S., Malmstadt, N.: Microfluidic fabrication of asymmetric giant lipid vesicles. ACS Appl. Mater. Interfaces **3**, 1434–1440 (2011)
23. Matosevic, S., Paegel, B.M.: Stepwise synthesis of giant unilamellar vesicles on a microfluidic assembly line. J. Am. Chem. Soc. **133**, 2798–2800 (2011)
24. Deshpande, S., Dekker, C.: On-chip microfluidic production of cell-sized liposomes. Nat. Protoc. **13**, 856–874 (2018)
25. Ushiyama, R., et al.: Plug-and-play microfluidic production of monodisperse giant unilamellar vesicles using droplet transfer across Water-Oil interface. Sensor Actuat B-Chem **355**, 131281 (2022)
26. Komaiko, J.S., McClements, D.J.: Formation of food-grade nanoemulsions using low-energy preparation methods: a review of available methods. Compr Rev Food Sci Food Saf **15**, 331–352 (2016)
27. Bao, P., et al.: Production of giant unilamellar vesicles and encapsulation of lyotropic nematic liquid crystals. Soft Matter **17**, 2234–2241 (2021)
28. Ushiyama, R., et al.: Identifying conditions for protein synthesis inside giant vesicles using microfluidics toward standardized artificial cell production. ACS Synth. Biol. **13**, 68–76 (2024)
29. Tsugane, M., Suzuki, H.: Reverse transcription polymerase chain reaction in giant unilamellar vesicles. Abstr. Sci. Reports **8**(1) (2018).
30. Tsugane, M., Suzuki, H.: Elucidating the membrane dynamics and encapsulation mechanism of large DNA molecules under molecular crowding conditions using giant unilamellar vesicles. ACS Synth. Biol. **9**(10), 2819–2827 (2020).
31. Nanjo, S., Tsugane, M., Matsuura, T., Suzuki, H.: Protein synthesis via in vitro transcription and translation system inside monodisperse vesicles fabricated by microfluidics. Abst. Electr. Eng. Jpn. **218**(2) (2025).
32. Yoneyama, R., Morikawa, N., Ushiyama, R., Maruyama, T., Sato, R., Tsugane, M., Takinoue, M., Suzuki, H.: Controlled formation of DNA condensates as model nuclei in monodisperse giant vesicles. JACS Au. **5**(7), 3533–3544 (2025).
33. Yoneyama, R., Ushiyama, R., Maruyama, T., Sato, R., Tsugane, M., Takinoue, M., Suzuki, H.: Formation dynamics of patchy/Janus DNA condensates in monodisperse giant vesicles generated using microfluidics. RSC Appl. Interf. **2**(6),1599–1606 (2025).
34. Katsuta, S., Okano, T., Koiwai, K., Suzuki, H.: Ejection of large particulate materials from giant unilamellar vesicles induced by electropulsation. Langmuir **35**(40), 13196–13204 (2019).
35. Sakakura, T., Nishimura, K., Suzuki, H., Yomo, T.: Statistical analysis of discrete encapsulation of nanomaterials in colloidal capsules. Anal. Meth. **4**(6), 1648–1655 (2012).

Chapter 15
Evaluating Mechanical Properties of Liposomes

Kaoru Uesugi

15.1 Introduction to Evaluating Liposome Mechanical Properties

Sensors, processors, and actuators are indispensable components of robots. Conventionally, these components have been constructed from materials such as metals, semiconductor materials (e.g., Si and dopant materials), and resins. These materials are typically rigid and chemically stable, and they are often fabricated using top-down methods, resulting in relatively large structures. In contrast, recent developments have introduced "molecular robots" that are constructed from nanoscale components, including biomolecules, using bottom-up approaches. These molecular robots are characterized by their wet and mechanically flexible nature.

A notable example of such molecular robots is the "amoeba-like molecular robot." The concept of this robot is to integrate essential elements, such as sensors, processors, and actuators, within a compartment separated by a membrane like vesicles or liposomes, which are lipid bilayer vesicles. These molecular robots are expected to mimic the behavior of amoebae, being capable, in the future, of receiving external stimuli, processing the received information, transmitting signals, and performing mechanical work on the external environment or enabling movement (via actuation) accordingly [1].

A distinctive feature of the amoeba-like molecular robot is the dynamic deformation of its lipid bilayer compartments. For the amoeba-like molecular robot to achieve dynamic deformation and movement, it is essential to transmit forces to the membrane via the action of molecular motors and the polymerization of molecules, thereby deforming the lipid bilayer vesicle membrane from within. In fact, cells also dynamically deform their membranes through intracellular molecular motors and

K. Uesugi (✉)
Department of Robotics, Ritsumeikan University, 1-1-1 Noji-higashi, Kusatsu, Shiga 525-0058, Japan
e-mail: k-uesugi@fc.ritsumei.ac.jp

S. Murata et al. (eds.), *Molecular Robotics II*,
https://doi.org/10.1007/978-981-95-2394-8_15

cytoskeletons [2, 3], and the driving principles of amoeba-like molecular robots are likely similar.

For instance, Sato et al. [4] successfully attached a microtubule-kinesin system to the inner side of a liposome membrane, enabling the transmission of forces generated by biological molecular motors to the membrane, thus inducing shape changes. Hayashi et al. [5] achieved liposome deformation by polymerizing tubulin encapsulated within the liposome, where the elongated microtubules exerted pressure from the inside. Furthermore, Tanaka et al. [6] demonstrated liposome deformation by polymerizing F-actin encapsulated within the liposome and inducing its nematic liquid crystal formation.

In the context of the amoeba-like molecular robot, liposomes can be considered analogous to the "casing" of conventional robots. It is necessary to consider the question "What mechanical properties are necessary for the development of liposome-based molecular robots?" For artificial cell-type molecular robots to undergo flexible and dynamic deformations and movements, understanding the required forces to achieve sufficient liposome deformation is crucial.

For example, when dynamically moving, reshaping, and relocating like cells, or extending nanotubes for communication between adjacent units, the liposome membrane undergoes bending and stretching. Thus, mechanical properties are critical for achieving arbitrary deformations. Even if the robot does not spontaneously deform, the mechanical properties of the entire liposome become crucial when it invades confined spaces, akin to red blood cells traversing capillaries.

The mechanical properties of liposomes have been characterized by various parameters, including membrane tension, bending rigidity, overall Young's modulus of the liposome, membrane fragility, and the forces required to induce deformation. For instance, it has been reported that the membrane composition of liposomes—such as the type and composition ratio of the lipids [7] or the addition of cholesterol [8, 9]—or the presence of actin cytoskeletal proteins [10] that constitute structures binding to the membrane can alter the membrane tension, bending rigidity, and the forces necessary for deformation, as well as the overall stiffness of the liposomes. Discussions on the mechanical properties of liposomes have also extended beyond the perspective of lipid type: they consider the structural aspects such as the number of hydrocarbons (chain length) in lipids and the number of unsaturated bonds in lipids [11–13]. Additionally, it is known that bending rigidity exhibits temperature dependency [14] and salt concentration dependency [15]. Furthermore, it was reported that in nanovesicles, the degree of lamellarity affects the stiffness of the vesicle [16]. This suggests that the presence of internal structures within liposomes may similarly influence the overall mechanical properties of the liposomes.

However, reports on the mechanical property evaluation of liposomes with the aim of designing molecular robots are still scarce. This chapter introduces the evaluation of the mechanical properties of liposomes or liposome-like samples (single cells) in reference to the development of liposome-based molecular robots. The previously reported typical methods for evaluating the mechanical properties of liposomes are reviewed: atomic force microscopy, laser trapping, image analysis, micropipette aspiration, and microfluidic channels.

15.2 Evaluation of Liposome Mechanical Properties Using Atomic Force Microscopy

Figure 15.1 shows the principle of atomic force microscopy (AFM). The AFM enables simultaneous imaging and force measurement of target objects at the microscale using a cantilever. The cantilever is fabricated by wet anisotropic etching of single-crystal silicon, silicon nitride, or silicon oxide, and a probe with a tip radius of less than 10 nm is formed at its end. A laser beam is directed at the back of the cantilever, and as the probe approaches the sample, the cantilever deflects. The displacement of the laser beam reflected from the back of the cantilever is detected by a photodetector, and displacement of the laser beam is converted to deflection of the cantilever z. The force f_{AFM} is calculated by Eq. (15.1) according to Hooke's law.

$$f_{AFM} = k_c z. \tag{15.1}$$

Here, k_c represents the spring constant of the cantilever, and k_c of the rectangular cantilever is derived from Eq. (15.2) [17].

$$k_c = \frac{E_c w_c t_c^3}{4L_c^3}. \tag{15.2}$$

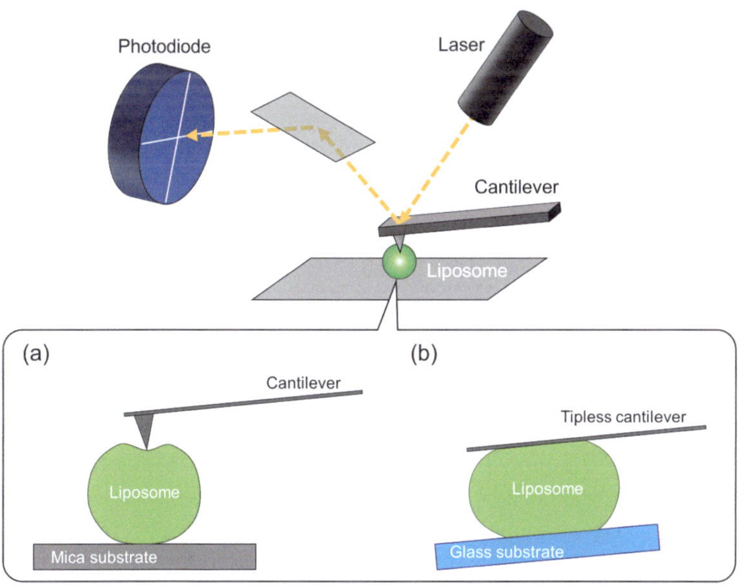

Fig. 15.1 Principle of AFM for evaluation of liposome mechanical properties. (**a**) A cantilever with a sharp tip is pressed against a liposome to derive its mechanical properties. (**b**) The liposome is compressed by sandwiching it between a tipless cantilever and a glass substrate, thereby determining the mechanical properties of the liposome

Here, E_c is Young's modulus of the cantilever material, and w_c, L_c, and t_c represent the width, length, and thickness of the cantilever. This principle of the "optical lever" allows for scanning the sample surface with sub-nanometer precision while measuring forces at the piconewton level. Young's modulus E of the sample can be obtained by fitting the force-indentation curve to the Hertz model (assuming a spherical tip, Eq. (15.3)) or the Sneddon model (assuming a conical tip, (15.4)) [18].

$$F_{sphere} = \frac{4ER^{1/2}\delta^{3/2}}{3(1-\nu^2)} \tag{15.3}$$

$$F_{cone} = \frac{2E\delta^2 \tan\theta}{\pi(1-\nu^2)} \tag{15.4}$$

Here, F_{sphere}, F_{cone} are indentation force, R is the tip sphere tip probe radius, δ is the indentation depth, ν is the Poisson ratio of sample, and θ is the half opening angle of the conical tip probe.

Leveraging these advantages, Et-Thakafy et al. [19] conducted mechanical property evaluations of liposomes absorbed on flat silicon substrates. These researchers compressed the center of a liposome using a sharp-tipped probe and applied the Hertz model and shell theory to the resulting force curves to determine Young's modulus and bending rigidity of the liposomes (Fig. 15.1a). On the other hand, Schäfer et al. [20] sandwiched liposomes between a tipless cantilever and a glass substrate, compressing them to derive the membrane tension and area compressibility modulus from the contour of a liposome (Fig. 15.1b). The advantage of compressing liposomes between two parallel plates, as opposed to conventional AFM evaluation, is the ability to avoid potentially stochastic rupturing of liposomes at the contact point with a sharp probe tip, thereby enabling more accurate measurements.

15.3 Evaluation of Mechanical Properties Using Laser Trapping

The concept of trapping objects using laser beams was proposed by Ashkin [21]. When a Gaussian laser beam enters and refracts within an object, the photon momentum changes (by incidence, reflection, and refraction), resulting in a radiation pressure force corresponding to this momentum change. This radiation pressure force enables non-contact and non-invasive handling of microscale objects, such as cells and liposomes, with piconewton-level forces. There are two primary methods for laser trapping: using counter-propagating lasers and using a single laser [21].

One methodology employing counter-propagating lasers is the dual-beam laser trap (optical stretcher) (Fig. 15.2a), which Delabre et al. [22] used to evaluate the bending rigidity and initial tension values of liposomes. When combined with microfluidic channels, the optical stretcher allows for high-throughput evaluation.

For simplicity, beam axis force is focused on here. When a laser beam is directed at a liposome, it enters the optically denser medium (the interior of the liposome has a higher refractive index than the exterior), generating radiation pressure force (due to the change in the momentum of the photons) that acts in the opposite direction of the beam entry to the liposome surface. Conversely, as the laser beam exits the liposome, radiation pressure force pushes the liposome surface in the direction of the beam exit. With a single laser beam, the liposome is simply transported along the beam axis. However, by introducing a counter-propagating laser beam, the liposome is trapped at the point where forces balance, causing it to stretch along the beam axis by radiation pressure force [23, 24]. The mechanical properties of liposomes can be determined using the stretching force and the degree of liposome deformation. In the optical stretcher, because laser beams are not focused within the samples, radiation damage is minimal [24].

Another method using a single laser beam is the optical tweezers. Koster et al. [25] measured the forces required to extract tubes from the liposome membrane using single-beam laser tweezers. When a Gaussian laser beam is directed at a bead that is not on the center beam axis, the radiation pressure force F_2 caused by Ray 2, which is light with a higher intensity in the central region of the laser, is greater than the radiation pressure force F_1 caused by Ray 1, which is light with a lower intensity in the outer region of the laser (Fig. 15.2b right). As a result, the net radiation pressure F_{net} acts as a force directed toward the center of the laser, causing the bead to be drawn toward the center. Furthermore, by focusing the laser beam with a lens,

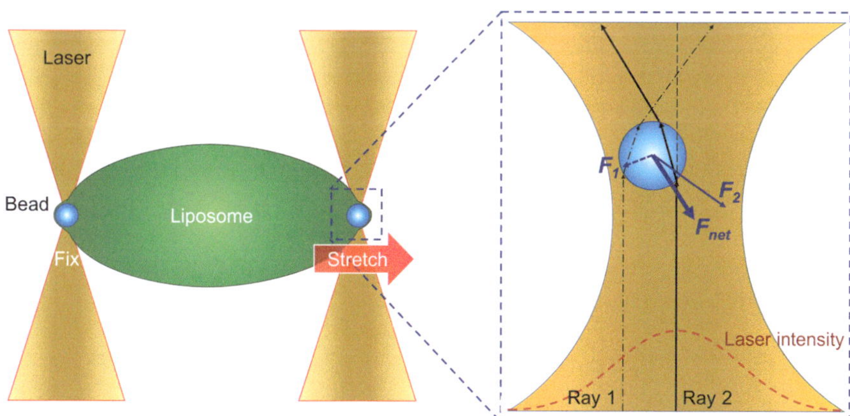

Fig. 15.2 Schematic illustrations explaining the principle of the dual-beam laser trap (optical stretcher) and the optical tweezers. (**a**) The Gaussian laser beam passes through the liposome and the radiation pressure force is loaded to the liposome in the direction away from the light source. By using a counter-propagating laser beam, the liposome is caused to stretch along the beam axis. (**b**) When a Gaussian laser beam is directed at a bead that is not on the center beam axis, the stronger light intensity at the center of the laser beam causes the bead to be drawn toward the center due to the radiation pressure force. A stretching test can be performed by holding one laser-trapped bead stationary and moving the other along the pulling axis of a liposome

the bead is constrained in the axial direction and trapped near the focal point. This means the bead also confined in the vertical direction. Beads attached to the surface of a liposome are trapped and pulled using a single-beam laser tweezer, forming tubes from the liposomes, and forces required to extract tubes are evaluated from the holding force of the laser tweezers. Additionally, Kato et al. [7] and Inaba et al. [26] measured the forces required to deform liposomes into lemon-shaped structures by using two single-beam laser tweezers (Fig. 15.2b left). Two polystyrene beads were encapsulated within a liposome and optically trapped using single-beam lasers. One bead was kept fixed, and the other was displaced away from it to exert force on the inner membrane surface, resulting in membrane stretching. The mechanical properties of liposomes were evaluated based on the extent of its deformation and pulling force, which was estimated from the trapping force of the laser beams.

15.4 Evaluation of Mechanical Properties by Image Analysis

When red blood cells are observed by optical microscopy, red blood cells seem to flicker. This flicker is thought to arise from significant cell shape fluctuations due to thermal fluctuation of the membrane [27] that results in a dynamically changing interference (phase contrast) [28]. Because membrane fluctuations are intimately associated with membrane mechanical properties, it can be considered that the flicker is also related to the cell membrane mechanical properties [28]. In other words, by analyzing these thermal fluctuations, flicker serves as a useful probe for evaluating membrane mechanical properties. Similar membrane fluctuations occur in liposomes and the mechanical properties of liposomes also can be evaluated.

A representative analytical method for this purpose is the flicker noise spectroscopy technique. This method analyzes temporal fluctuations in vesicle shape. Flicker noise spectroscopy is a non-invasive technique, and it has been used for evaluating liposomes. By determining the membrane contour location, the membrane fluctuation spectrum can be extracted, allowing evaluation of the membrane tension and membrane bending modulus [29, 30].

15.5 Evaluation of Mechanical Properties by the Micropipette Aspiration Method

When applying suction pressure to a liposome using a narrow glass pipette with a suction pore less than half the diameter of the liposome, the liposome is aspirated into the pipette. By measuring the aspirated length of liposome or the curvature of the membrane surface, the mechanical properties of the liposome can be evaluated. The micropipette aspiration (MPA) method has long been used to evaluate the mechanical properties of cells due to its relatively simple experimental setup. Because the MPA method enables the evaluation of various physical quantities using a single system, it has become a standard technique for assessing the mechanical properties of cells and vesicles. Therefore, a detailed explanation of this method is provided here. It should be noted in MPA experiments that the hypothesis of a constant vesicle volume during an experimental run (attributable to the low permeability of the lipid bilayer to the solvent over the few minutes of the run) is assumed [31].

In MPA, atmospheric pressure is often utilized to generate the suction pressure. A micropipette is connected to a reservoir, which is open to the air, using a tube filled with liquid. Coating glass pipettes with substances such as bovine serum albumin (BSA) prevents the adsorption and rupture of the liposome membrane on the glass surface. The reservoir is adjustable in height, allowing fine-tuning. The suction pressure (ΔP) is regulated by the height difference (h) between the tip of the micropipette and the liquid surface in the reservoir when the liquid surface of the reservoir is positioned below the capillary tip (Fig. 15.3). The suction pressure ΔP can be derived using a simple equation involving h, and for experiments lasting several tens of seconds or longer, it can be approximated by the relationship $\Delta P = \rho g h$ (where ρ is liquid density and g is gravitational acceleration). However, to obtain more accurate values, a pressure sensor is required. Alternatively, suction pressure can also be generated using dedicated microchannel pump systems. The deformation of the aspirated liposome is observed using a phase-contrast microscope or a fluorescence microscope. To minimize the effects of evaporation and the resulting convection within the liposome suspension droplet, it is preferable to fill the suspension into a chamber, such as one constructed from two coverslips, which allows for microscope observation while reducing evaporation.

The following describes examples of evaluating various mechanical properties of cells and liposomes (vesicles) using MPA method. For example, the relationship between the length L of the aspirated region of a lipid bilayer capsule (e.g., a red blood cell) under a suction pressure ΔP and the membrane shear modulus μ is expressed by Eq. (15.5) [32].

$$\Delta P = \frac{\mu}{R_p}\left[\frac{2L}{R_p} - 1 + \ln\left(\frac{2L}{R_p}\right)\right] \tag{15.5}$$

Here, R_p represents the radius of the micropipette.

From the suction pressure ΔP, the radius of the aspirated area liposome R_p aspirated by the suction pipette, and the radius of the outer spherical segment of the cell or liposome R_c, the membrane tension σ of the cell or liposome can be obtained as Eq. (15.6) [12, 33] (Fig. 15.4).

$$\Delta P = 2\sigma\left(\frac{1}{R_p} - \frac{1}{R_c}\right) \tag{15.6}$$

Furthermore, the apparent change in surface area (specifically, the increase) ΔA can also be obtained from the change in aspirated length ΔL [12, 34] as given by

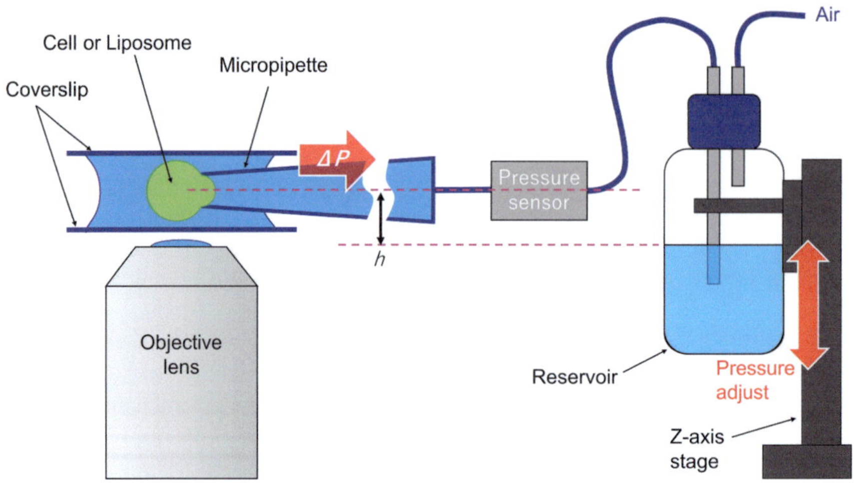

Fig. 15.3 Schematic illustration of the MPA system consisting of a microscope for observing cell or liposome deformation, a chamber composed of two coverslips for holding the cell or liposome suspension, a micropipette for aspirating the cell or liposome, a pressure sensor for measuring the aspiration pressure, and a water-filled reservoir with a height adjustable by a Z-axis stage for generating the aspiration pressure (ΔP)

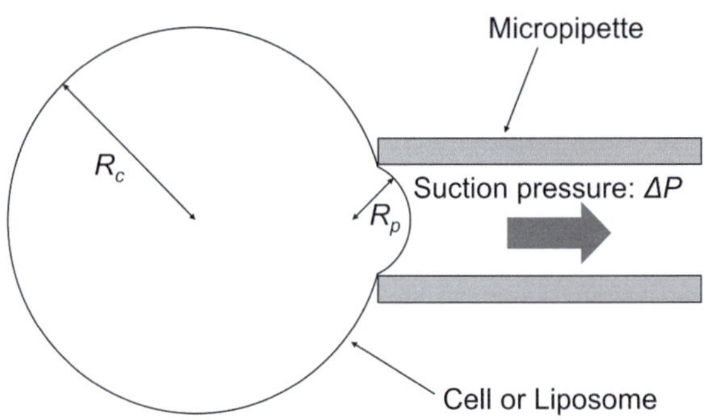

Fig. 15.4 Schematic illustration of a membrane tension model for the MPA of cells or liposomes. The membrane tension is derived from the curvature of the cell or liposome (R_c and R_p) and the suction pressure (ΔP)

Eq. (15.7).

$$\Delta \mathrm{A} = 2\pi R_p \Delta L\left(1 - \frac{R_p}{R_v}\right) \tag{15.7}$$

Here, R_v represents the radius of the vesicle.

The acquired membrane tension allows for the calculation of the forces necessary for the formation of microtubes or nanotubes from liposomes (which are analogous to lamellipodia or nanotubes in cells). Additionally, this tension can also be used to discuss morphology of the shell (the membrane of liposomes and cells).

By using Eq. (15.8) that expresses the membrane surface area change ΔA as a function of membrane tension σ, both the bending rigidity (κ_b) and the area compressibility modulus (K_A) can be determined from a single type of experiment [12, 35]. $k_B T$ is the thermal energy (where k_B is the Boltzmann constant and T is absolute temperature).

$$\frac{\Delta A}{A} = \frac{k_B T}{8\pi \kappa_b} \ln\left(1 + \frac{\sigma A}{\pi^2 \kappa_b}\right) + \frac{\sigma}{K_A} \tag{15.8}$$

Specifically, the bending and stretching moduli (κ_b, K_A) are deduced by fitting the experimental curve relating the apparent area expansion to membrane tension.

According to Theret et al. [36], Young's modulus E of the liposome can be expressed as shown in Eq. (15.9).

$$E = \frac{3a\Delta P}{2\pi L}\Phi \tag{15.9}$$

Here, a represents the radius inside the pipette, L denotes the aspiration length inside the pipette, and Φ is a function related to the external and internal radii of the micropipette tip, which varies depending on whether the "force problem" or the "punch problem" is applied. In the force problem, the applied force is determined by a simple force balance. However, since it assumes uniform stress not only within the pipette but also in the contact region at the pipette tip, the shape of the sample in the contact area varies depending on the applied stress. In contrast, the punch problem is considered a boundary-value problem in which the sample is flat in the contact region and remains in continuous contact with the pipette tip (assuming friction is negligibly small). This model, which assumes that the sample is flat in the contact region, is more realistic and has been widely used in many studies employing MPA.

Liposomes and cells possess not only elastic but also viscous components. Sato et al. [37] considered cells as a three-element model, as shown in Fig. 15.5, and they adapted Theret et al.'s model (Eq. (15.9)) to account for the viscoelastic properties of cells as provided by Eq. (15.10). Sato et al. introduced a value of $\Phi = 2$, and in most cases, Φ values in the range of 2.0–2.1 are used.

$$L(t) = \frac{\Phi a \Delta P}{\pi k_1}\left\{1 - \frac{k_2}{k_1 + k_2} exp\left(-\frac{t}{\tau}\right)\right\} h(t) \tag{15.10}$$

Here, $L(t)$ is the aspiration length inside the pipette at time t, k_1 and k_2 are Young's moduli, τ is the time constant related to viscosity, and $h(t)$ is the unit step function (Heaviside function). Additionally, the initial elastic modulus E_0, steady-state elastic modulus E_∞, and viscosity coefficient η are expressed as follows: (Eqs. (15.11)–(15.13)).

$$E_0 = \frac{3}{2}(k_1 + k_2) \tag{15.11}$$

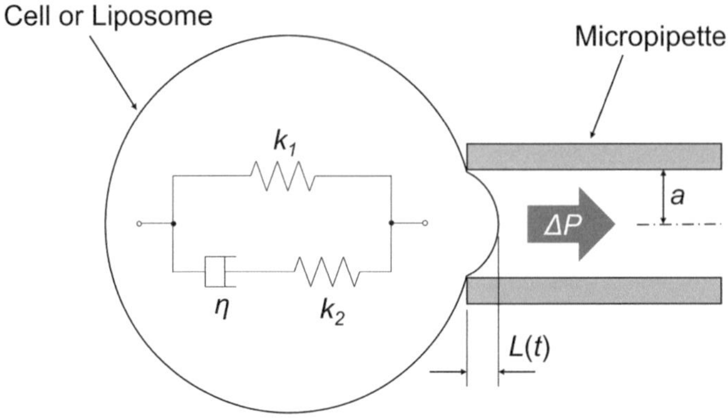

Fig. 15.5 Schematic illustration of a viscoelastic model for the MPA of cells or liposomes. The elastic moduli (k_1, k_2) and viscosity (η) can be evaluated using the MPA method

$$E_\infty = \frac{3}{2}k_1 \tag{15.12}$$

$$\eta = \frac{k_1 k_2}{k_1 + k_2}\tau \tag{15.13}$$

.

The evaluation of mechanical properties of liposomes using the MPA method has been reported, particularly regarding membrane properties [12, 38]. However, reports on the mechanical properties of entire liposomes, including both the membrane and the internal fluid, are rare. Given that amoeba-like molecular robots significantly deform their overall shape, it is crucial to enhance understanding of the mechanical properties of entire liposomes.

The MPA method provides a comprehensive approach for evaluating the mechanical properties of liposomes. While the elastic modulus derivation model is primarily used for cells, its application to liposomes enables the assessment of not only mechanical properties of the membrane but also the overall mechanical characteristics of the system. This is considered to include the influence of internal components (e.g., the mechanical properties of filaments [5], the stiffness associated with the alignment forces related to the nematic liquid crystal behavior of filaments [6], the mechanical properties of the inner solution, and the effects of multilamellar structures [16]), as well as the effect of proteins [10] or other molecular structures (e.g., DNA structures [4] or cholesterol [8, 9]) or functions of these molecular structures (motion or conformational change) adsorbed on the membrane surface [4] (Fig. 15.6).

Additionally, the viscoelastic parameters obtained through the MPA method reflect the viscous friction between the membrane and the surrounding fluid, and the viscoelasticity of the inner and outer solutions. These parameters characterize the liposome as a composite material, representing the properties of the entire structure rather than just the membrane. Such evaluations are particularly valuable

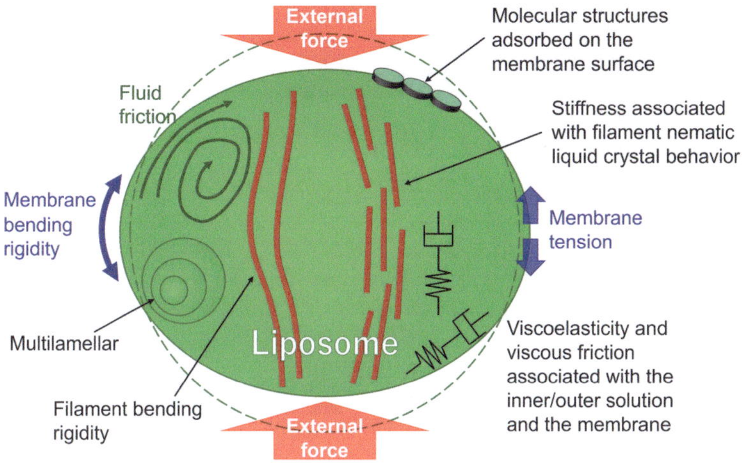

Fig. 15.6 Schematic diagram illustrating the mechanical properties that can be assessed using the MPA method which enables evaluation of the overall mechanical characteristics of the liposome system. These characteristics include the influence of internal components, the membrane structure, as well as the viscoelasticity and viscous friction associated with the membrane and the inner and outer solutions

for analyzing dynamic deformations of the liposome at the whole-structure scale and time-dependent changes in forces. This approach may also prove effective for designing molecular robots that require dynamic control of the entire system.

15.6 Evaluation of Mechanical Properties of Liposomes Using Microfluidic Channels

The MPA method allows for detailed examination of the mechanical properties of individual liposomes. However, it requires complex manipulations (particularly to hold the liposome, necessitating the experimenter be highly skilled and experienced) and a complex experimental setup (e.g., a micromanipulator). Additionally, only one liposome can be evaluated per aspiration, leading to a time-consuming process for increasing the sample size. To address these issues, a structure mimicking the MPA system within microfluidic channels represents a valid and feasible methodology.

In this approach, a liposome suspension is directed through microfluidic channels using constrictions that mimic the suction part of the micropipette. By analyzing the deformation of liposomes trapped at these constrictions, the mechanical properties of the liposomes are evaluated. This method does not require the experimenter to have any advanced techniques, as it simply involves flowing the suspension through the channels. Furthermore, multiple constrictions within the channels allow for simultaneous evaluation of several liposomes, improving time efficiency. These advantages have already led to the microfluidic channel-based MPA system being used to evaluate the mechanical properties of cells [31, 39]. In recent years, microchannels have also been utilized for evaluating the mechanical properties of liposomes [31, 40].

The basic design of a microchannel mimicking a micropipette is shown in Fig. 15.7 [39] (Although the referenced microfluidic device was originally reported for cell trapping, the following explanation assumes the trapping of liposomes). Generally, microchannels are fabricated using photolithography techniques and consist of polydimethylsiloxane (PDMS)-based channels bonded to a glass substrate. In many cases, microchannels are equipped with multiple constricted regions to trap liposomes. Additionally, to eliminate constraints in the height direction, microchannels are sometimes fabricated using a multilayer construction method. In the microchannel sideview of Fig. 15.7, the top layer and bottom layer act as a trapping region and they also provide the main microfluidic channel, and the middle layer mimics the cross section of the micropipette as the constricted region and acts as a trapping structure and also provides the main microfluidic channel. Applying a coating, such as BSA, to the microchannel surface helps prevent the adsorption and rupture of the liposome membrane.

When a liposome suspension is introduced into the microchannel, some liposomes are trapped in the constricted regions. This situation is analogous to the trapping of

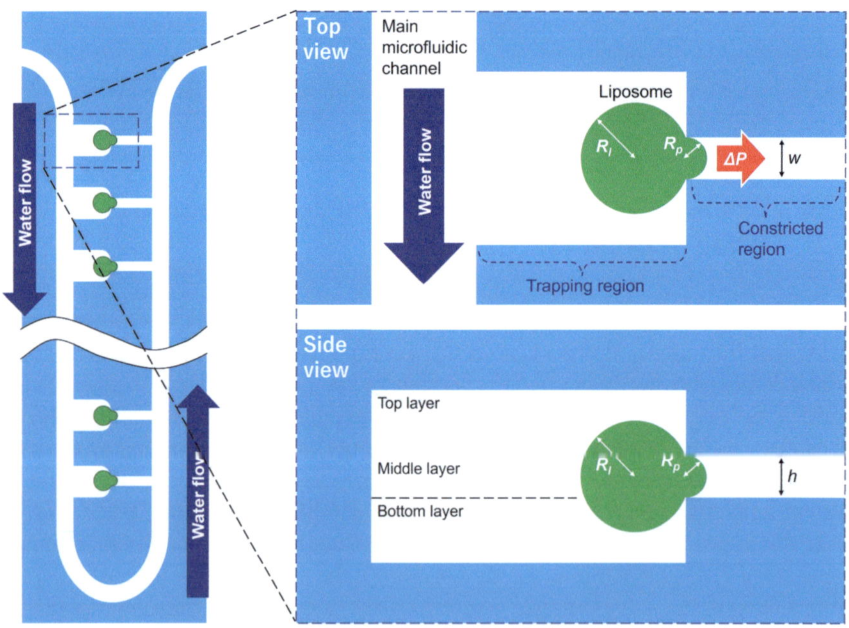

Fig. 15.7 Schematic illustration of a microchannel used for evaluating the membrane tension of liposomes. The microchannel has multiple constricted regions mimicking the tip of the micropipette that are able to trap liposomes. The membrane tension is derived from the curvatures of the liposome (R_l and R_p) and the suction pressure (ΔP)

a liposome with micropipette in the MPA method, as described in Sect. 15.5 and can be analyzed similarly. Therefore, the membrane tension σ can be determined using Eq. (15.14) in the same manner as Eq. (15.6).

$$\Delta P = 2\sigma\left(\frac{1}{R_p} - \frac{1}{R_l}\right) \tag{15.14}$$

R_p represents the radius of aspirated area of liposome aspirated by a micropipette, R_l represents the radius of the outer spherical segment of the liposome, and ΔP represents the pressure difference when the liposome is in pressure equilibrium at the suction pore. In this example, the shape of the pipette structure (the constricted region) is rectangular, which causes the radius of the liposome on the aspirated side (R_p) to differ from its actual value. In such cases, the hydraulic radius of an aspiration pore is calculated as $R_p = wh/(w + h)$, and when the shape of the pipette structure is square ($w = h$), R_p is simply regarded as $w/2$ (or $h/2$) [39]. However, an analytical formula may also be applied [40], or the constricted region may be designed with a circular shape from the outset [31, 41].

For delivering liposome suspensions, microchannel-specific pump systems or syringe pumps are commonly used (Fig. 15.8). Additionally, placing a flow sensor or pressure sensor between the pump system and the microchannel allows for precise monitoring of flow rate and pressure, as well as enabling feedback control of flow and pressure. The pressure difference (ΔP) can be set using various methods, including simply using the pressure value at the inlet [31], deriving partial pressures based on fluid resistance within the microchannel [40], or calculating ΔP from flow velocity using the Darcy–Weisbach equation [39]. The microchannel is mounted on a microscope stage, and the deformation of liposomes is observed using phase-contrast microscopy or fluorescence imaging. (When observing liposomes in a microfluidic channel using phase-contrast microscopy, the liposome shape is often unclear due to multiple reflection interference. Therefore, fluorescence observation is recommended for morphological observation.) In this example, the derivation of membrane tension has been explained; however, similar to MPA, it is also possible to evaluate the elastic modulus and viscosity.

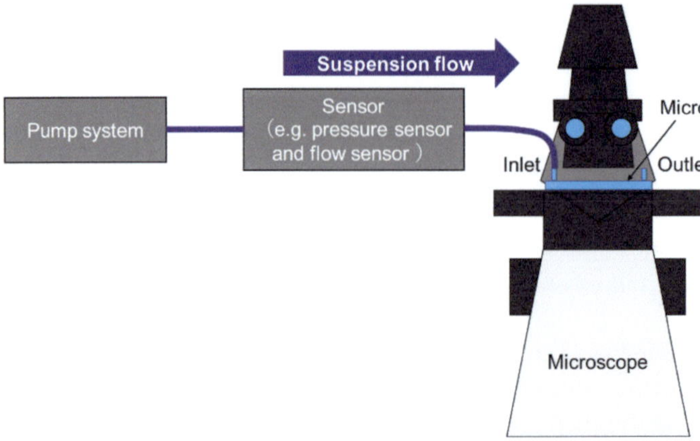

Fig. 15.8 Schematic illustration of the microchannel system for evaluating the mechanical properties of liposomes. The system consists of a microscope for observing liposome deformation, a pump system for introducing the suspension into the microchannel, various sensors for monitoring flow rate and pressure, and the microchannel itself

References

1. Hagiya, M., et al.: Molecular robots with sensors and intelligence. Acc. Chem. Res. **47**, 1681–1690 (2014)
2. Sens, P., Plastino, J.: Membrane tension and cytoskeleton organization in cell motility. J. Phys. Condens. Matter **27**, 273103 (2015)
3. Clark, A.G., et al.: Stresses at the cell surface during animal cell morphogenesis. Curr. Biol. **24**, R484–R494 (2014)
4. Sato, Y. et al.: Micrometer-sized molecular robot changes its shape in response to signal molecules. Sci. Robot. **2**, eaal3735 (2017)
5. Hayashi, M., et al.: Reversible morphological control of tubulin-encapsulating giant liposomes by hydrostatic pressure. Langmuir **32**, 3794–3802 (2016)
6. Tanaka, S. et al.: Repetitive stretching of giant liposomes utilizing the nematic alignment of confined actin. Commun. Phys. **1** (2018)
7. Kato, N., et al.: Effects of lipid composition and solution conditions on the mechanical properties of membrane vesicles. Membranes **5**, 22–47 (2015)
8. Henriksen, J., et al.: Vesicle fluctuation analysis of the effects of sterols on membrane bending rigidity. Eur. Biophys. J. **33**, 732–741 (2004)
9. Arriaga, L.R., et al.: Stiffening effect of cholesterol on disordered lipid phases. A combined neutron spin echo + dynamic light scattering analysis of the bending elasticity of large unilamellar vesicles. Biophys. J. **96**, 3629–3637 (2009)
10. Luo, T., et al.: Mimicking the mechanical properties of the cell cortex by the self-assembly of an actin cortex in vesicles. Appl. Phys. Lett. **104**, 153701 (2014)
11. Olbrich, K., et al.: Water permeability and mechanical strength of polyunsaturated lipid bilayers. Biophys. J. **79**, 321–327 (2000)
12. Rawicz, W., et al.: Effect of chain length and unsaturation on elasticity of lipid bilayers. Biophys. J. **79**, 328–339 (2000)
13. Hossain, M.I. et al.: Effects of hydrocarbon chain on the vesicle size distribution, kinetics of average size, bending modulus, and elastic modulus of lipid membranes. Eur. Phys. J. E. **45** (2022)

14. Lee, C.H., et al.: All-optical measurements of the bending rigidity of lipid-vesicle membranes across structural phase transitions. Phys. Rev. E **64**, 020901 (2001)
15. Claessens, M.M.A.E., et al.: Charged lipid vesicles: effects of salts on bending rigidity, stability, and size. Biophys. J. **87**, 3882–3893 (2004)
16. Vorselen, D., et al.: Multilamellar nanovesicles show distinct mechanical properties depending on their degree of lamellarity. Nanoscale **10**, 5318–5324 (2018)
17. Albrecht, T.R., et al.: Microfabrication of cantilever styli for the atomic force microscope. J. Vac. Sci. Technol. A **8**, 3386–3396 (1990)
18. Liang, W., et al.: Recent advances in AFM-based biological characterization and applications at multiple levels. Soft Matter **16**, 8962–8984 (2020)
19. Et-Thakafy, O., et al.: Mechanical properties of membranes composed of gel-phase or fluid-phase phospholipids probed on liposomes by atomic force spectroscopy. Langmuir **33**, 5117–5126 (2017)
20. Schäfer, E., et al.: Mechanical properties of giant liposomes compressed between two parallel plates: impact of artificial actin shells. Langmuir **29**, 10463–10474 (2013)
21. Ashkin, A.: Acceleration and trapping of particles by radiation pressure. Phys. Rev. Lett. **24**, 156–159 (1970)
22. Delabre, U., et al.: Deformation of phospholipid vesicles in an optical stretcher. Soft Matter **11**, 6075–6088 (2015)
23. Guck, J., et al.: Optical deformability of soft biological dielectrics. Phys. Rev. Lett. **84**, 5451–5454 (2000)
24. Guck, J., et al.: The Optical Stretcher : a novel laser tool to micromanipulate cells. Biophys. J. **81**, 767–784 (2001)
25. Koster, G., et al.: Membrane tube formation from giant vesicles by dynamic association of motor proteins. Proc. Natl. Acad. Sci. U. S. A. **100**, 15583–15588 (2003)
26. Inaba, T., et al.: Formation and maintenance of tubular membrane projections require mechanical force, but their elongation and shortening do not require additional force. J. Mol. Biol. **348**, 325–333 (2005)
27. Brochard, F., Frequency, J.F.L.: Frequency spectrum of the flicker phenomenon in erythrocytes. J. Phys. **36**, 1035–1047 (1975)
28. Yoon, Y.Z., et al.: Flickering analysis of erythrocyte mechanical properties: Dependence on oxygenation level, cell shape, and hydration level. Biophys. J. **97**, 1606–1615 (2009)
29. Pécréaux, J., et al.: Refined contour analysis of giant unilamellar vesicles. Eur. Phys. J. E **13**, 277–290 (2004)
30. Doskocz, J., et al.: Statistical analysis of bending rigidity coefficient determined using fluorescence-based flicker-noise spectroscopy. J. Membr. Biol. **251**, 601–608 (2018)
31. Landiech, S., et al.: Parallel on-chip micropipettes enabling quantitative multiplexed characterization of vesicle mechanics and cell aggregates rheology. APL Bioeng. **8**, 026122 (2024)
32. Waugh, R., Evans, E.: Thermoelasticity of red blood cell membrane. Biophys. J. **26**, 115–131 (1979)
33. Evans, E., Yeung, A.: Apparent viscosity and cortical tension of blood granulocytes determined by micropipet aspiration. Biophys. J. **56**, 151–160 (1989)
34. Piontek, M.C., et al.: Active probing of the mechanical properties of biological and synthetic vesicles. Biochim. Biophys. Acta - Gen. Subj. **1865**, 129486 (2021)
35. Evans, E., Rawicz, W.: Entropy-driven tension and bending elasticity in condensed-fluid membranes. Phys. Rev. Lett. **64**, 2094–2097 (1990)
36. Theret, D.P., et al.: The application of a homogeneous half-space model in the analysis of endothelial cell micropipette measurements. J. Biomech. Eng. **110**, 190–199 (1988)
37. Sato, M., et al.: Application of the micropipette technique to the measurement of cultured porcine aortic endothelial cell viscoelastic properties. J. Biomech. Eng. **112**, 263–268 (1990)
38. Needham, D., Nunn, R.S.: Elastic deformation and failure of lipid bilayer membranes containing cholesterol. Biophys. J. **58**, 997–1009 (1990)

39. Lee, L.M., Liu, A.P.: A microfluidic pipette array for mechanophenotyping of cancer cells and mechanical gating of mechanosensitive channels. Lab Chip **15**, 264–273 (2015)
40. Elias, M., et al.: Microfluidic characterization of biomimetic membrane mechanics with an on-chip micropipette. Micro Nano Eng. **8**, 100064 (2020)
41. Abdelgawad, M., et al.: A fast and simple method to fabricate circular microchannels in polydimethylsiloxane (PDMS). Lab Chip **11**, 545–551 (2011)

Chapter 16
Microhand and Technology for Manipulating and Measuring Micro-objects

Masaru Kojima

To realize a methodology for constructing large-scale molecular systems, technology for the complex arrangement of different functions, such as sensors, processors, and actuators, is implemented in artificial cells. Furthermore, the technology for evaluating these functions is essential. Therefore, a microhand system capable of the advanced manipulation and measurement of a local mechanical stimulus response is introduced.

16.1 Micromanipulation

Micromanipulation is a technique used to handle objects on the order of micrometers. In recent years, micromanipulation has been widely used in the fields of medicine and biology for gene transfers into cells, the measurement of cellular and tissue characteristics, and other microscopic tasks. First, representative methods developed for manipulating micro-objects are introduced.

Various methodologies have been proposed for micromanipulation. Manipulation using the force of a fluid flowing through a microfluidic channel has been reported [1]. In this method, micromanipulation is completed in the channel. Therefore, it has attracted attention, as it can solve the problem of contamination and requires technical expertise. For example, a fluidic chip that automatically performs embryo manipulation during cloning has been reported. In this report, embryo manipulation involved the following processes: (a) oocyte cutting, (b) the sorting of anucleated oocytes, and (c) the coupling and fusion of anucleated oocytes and donor cells. The authors developed a fluidic chip for each of these processes and integrated them to enable embryo manipulation within a single chip [2]. Another example

M. Kojima (✉)
The University of Osaka, Osaka, Japan
e-mail: kojima.masaru.es@osaka-u.ac.jp

S. Murata et al. (eds.), *Molecular Robotics II*,
https://doi.org/10.1007/978-981-95-2394-8_16

is a report of high-speed and high-precision cell manipulation by constraining the movement of cells in a fluidic channel in one dimension [3]. Microfluidic chips are often fabricated for specific purposes and lack versatility; however, they are capable of precise environmental manipulation.

Many manipulators have been developed to handle objects. For example, a magnetically driven on-chip robot with a grasping mechanism has been developed. This robot is intended for use in microfluidic chips, and it is capable of grasping and releasing movements. In this study, the on-chip enucleation of oocytes was achieved using a microknife [4]. There has also been a report of the development of microfingers, in which two glass rods with tips fabricated to a diameter of approximately 100 μm are used to grasp, move, and rotate micro-objects [5]. In these methods, the manipulator is in direct contact with the object.

On the microscale, various surface forces, such as electrostatic and intermolecular forces (called "scale effects") are dominant, making it difficult to release objects. In addition, because of the direct contact with the object, there is a risk that the physical and chemical properties of the object's surface may be changed. This risk presents a major challenge in contact manipulation.

However, noncontact methods use various forces to manipulate an object without physical contact. Although noncontact manipulation does not require the influence of intersurface forces, it is difficult to achieve highly accurate positioning. Depending on the method, the working environment and objects that can be manipulated are limited. Noncontact methods include laser manipulation, which uses the radiation pressure of laser light [6] and dielectrophoresis [7]. In addition, a manipulation method using a thermosensitive gel has been proposed [8]. In this method, when a probe-shaped heater is heated in a solution containing a thermoresponsive gel, a hardened gel mass is generated around the heater. When the heater was turned off, the gel returned to a liquid state, and the object could be grasped and released with the gel.

16.2 Two-Fingered Microhand

Two-fingered microhand systems were also developed for contact manipulation [9]. The two-fingered microhand is a manipulator that mimics the human manipulation of chopsticks and has achieved multiscale manipulation at submicro- to several hundred micro-orders of magnitude.

An example of a microhand using the 3PRS mechanism for a parallel mechanism is shown in (Fig. 16.1). PRS corresponds to a prismatic linear joint, revolute rotational joint, and spherical ball joint, thus indicating the type of joint used. The 3PRS mechanism consists of a base plate to which three sets of linkage mechanisms are fixed and end plates to which end effectors are mounted. Rotary joints use an arc-hinge structure, and ball joints are flexible joints that act as passive joints. A piezoelectric element, which is a linear actuator, is used to operate the end effector. The 3PRS mechanism has three degrees of freedom (DOFs): one DOF of translation

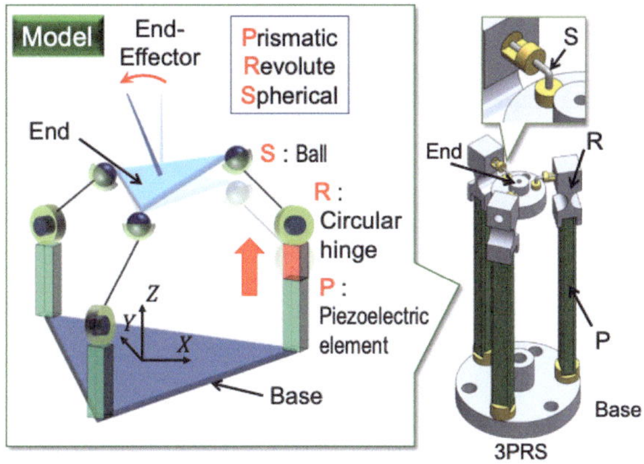

Fig. 16.1 Two-fingered microhand model (3PRS mechanism)

and two DOFs of rotation in the end effector. The use of piezoelectric elements as actuators enables high-speed and precise operations [10].

An example of using the parallel mechanism as upper and lower modules with a glass needle attached to the tip of each module as an end effector has been reported [9] (Fig. 16.2a). The upper and lower modules corresponded to chopstick on the index finger and thumb sides, respectively. During the object-grasping operation, the lower module was fixed, and only the upper module moved. During the object transfer operation, only the lower module moved, and the relative positions of the lower and upper modules were fixed such that the object could be maintained in the grasping state. By combining the three-dimensional motion of the two end effectors, various operations can be expected, such as cutting, stretching, pressing, drilling, stirring, and splashing, as well as grasping and transferring objects. Recently, we developed a compact microhand, as shown in Fig. 16.2b [11]. The lower module of this microhand consists of a 3-axis stage, and only the upper module employs a parallel mechanism. Compere with previously developed model, its smaller size makes it easier to install in an environment under a microscope, and the improved rigidity reduces vibrations during high-speed transportation.

16.3 Configuration of the Microhand System

The system consisted of a two-fingered microhand, an automated stage, a microscope, and an automated focusing device with an objective piezoelectric actuator. The main body of the micromanipulator is fixed to the microscope and has a limited range of motion (Fig. 16.3a) [12, 13]. Therefore, when manipulating an object outside the movable range of the micromanipulator, the object to be grasped is moved to the field of view of the microscope using a 3-DOF automated stage. The microscopic image of the object is acquired using a high-speed camera, and the microhand and automated stages are controlled based on this image. Hence, the high-speed automatic

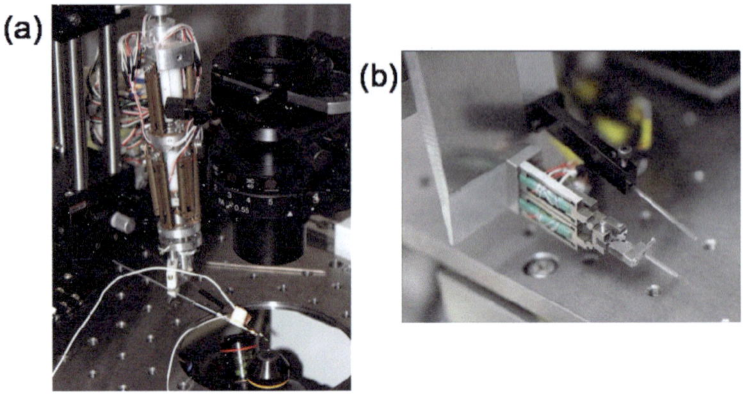

Fig. 16.2 Example of a two-fingered microhand

manipulations of grasping and placement are enabled (Fig. 16.3b). The high-speed automated placement of microbeads has been successfully performed using this system. Furthermore, automatic stiffness measurements of cells (mouse fibroblasts) were performed using a system with a microforce sensor mounted on the tip of the end effector.

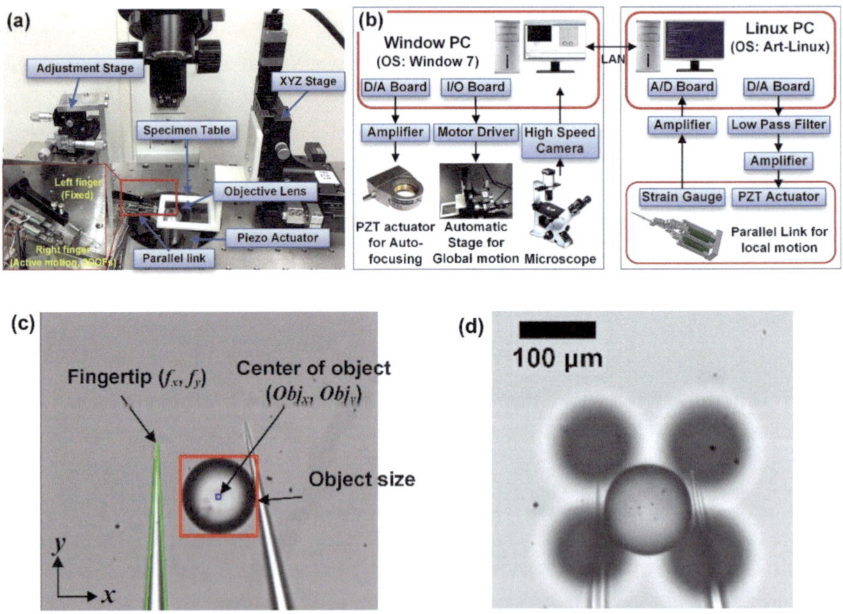

Fig. 16.3 Configuration of the microhand system and an example of high-speed manipulation [13]

16.4 High-Speed Manipulation Using a Microhand System

Fully automated pick-and-place manipulation using a microhand system has been reported for the rapid and accurate assembly of micro-objects, such as cells. Micromanipulation in liquids presents challenges, such as strong adhesion. Reliable manipulation systems and special release techniques are essential for solving these problems. A report on high-speed automated manipulation addressed these issues [12]. In this study, a dexterous microhand was used to hold the object steady, and an automated stage was used to quickly transport the object. If an object adheres to one of the end effectors, two release methods are used to remove it: (1) release using a local stream of flow that is locally caused by the high-speed motion of the end effector or (2) dynamic release, in which the end effector is directly vibrated to pull off the adhered object. The system recognizes objects adhering to the two fingertips and uses vision-based technology to distinguish between these release methods and achieve automated release (Fig. 16.3c). This technique dramatically increases the manipulation speed and has been successfully used to assemble microbeads and spheroids (Fig. 16.3d).

This high-speed placement method is applicable to various micro-objects and is thought to be useful for placing molecular robot systems that use artificial cells as units. For example, using BAM-modified substrates (BAM: biocompatible anchor for cell membranes) [14] that are inserted into lipid membranes and used to fix cells to substrates, it is possible to place and fix liposomes in arbitrary positions, thus making it possible to construct and verify a wide variety of molecular systems.

16.5 Microhand System for Stiffness Measurement

In the field of life sciences, a wide variety of analytical experiments have been conducted on cellular tissues and single cells for development in the medical field and technological innovation in the pharmaceutical field. As an example of cellular property analysis, experiments analyzing stiffness based on the reaction forces from cells have attracted particular attention. For example, the stiffness of a cancer cell is 70% lower than that of a normal cell [15]. However, it is difficult to conduct experiments to measure cell stiffness. Experiments that require a wide variety of repetitive manipulations and long durations require the experimenter to perform delicate and highly precise manipulations. Because the experimenter is forced to learn the required skills empirically from manipulations over a long period, robotics technology offers sufficient potential to support delicate and high-precision operations to supplement these experiential skills. Hence, micro/nanorobot systems have been constructed to support microscopic operations. In this context, a method for measuring the stiffness of micro-objects using the proposed microhand system was developed.

16.6 Cell Stiffness Measurement Using a Microhand System

As an example of stiffness measurement using a microhand, cell stiffness was measured by mounting a microforce sensor on an end effector with a needle-shaped tip. A cell was grasped using a two-fingered microhand, the reaction force from the cell was recorded when the end effector was pushed into the cell, and the stiffness of the cell was evaluated based on the obtained force curve. Then, the nanoindentation method was used to measure and calculate the stiffness. This system was also used to measure the change in stiffness during the process of viral infection, and the change in stiffness of the cells due to viral infection was confirmed [16].

As another example of the application in basic research, the stiffness measurements of ameloblast-like cell spheroids have been reported [17]. Ameloblasts are cells that form enamel on tooth surfaces. When these cells are cultured in a calcification medium, accelerated hardening is expected to accompany calcification. Therefore, hardening associated with the culture was evaluated using stiffness measurements collected via a microhand system. The results showed that stiffening was more stable when the cells were cultured in a calcified medium than when they were cultured in a non-calcified medium. In other words, the promotion of hardening was revealed via direct cell measurement by microhand system.

Various studies have indicated that the stiffness of a cell changes, depending on its state, which has attracted attention as a new method by which to evaluate cells. A two-fingered microhand system can directly measure such parameters. Hence, further research and development are expected.

16.7 Attempt to Measure the Stiffness of Micro-objects with Complex Structures Using a Microhand

The aforementioned methods have not been able to cause large deformations in the structures within cells. The method of placing cells between two microplates and compressing them is suitable for large cell deformations and has been used to measure cell nuclei [18]. Furthermore, elongation can be achieved by modifying the probe surface with a cell adhesive. Another feature of this method is that it is highly scalable, owing to its simplified design. Therefore, we incorporated the method of using a plate-shaped end effector into the microhand. To apply a large strain to the cell and measure its stiffness, reaction force measurements were performed using a plate-shaped end effector [19, 20] rather than a needle-shaped end effector [16]. Consequently, it is easier to measure reaction forces, including that of the intracellular structure, and a more stable measurement is possible because uniform loading can be applied to a spherical object. Stiffness measurements of cells and spheroids have been successfully performed using this plate-shaped end effector [21, 22].

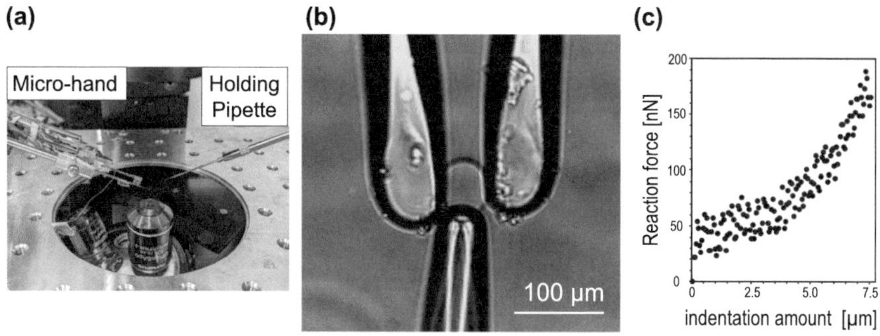

Fig. 16.4 Liposome stiffness measurements taken using a microhand system

There have also been reports measuring the force spontaneously generated by cells. Hydras are well-known regenerative organisms. When a hydra is cut into small pieces, it forms a spherical cell mass during regeneration, which then changes its shape and regenerates. This cell mass oscillates periodically, and quantitative measurements of the force generated by this periodic movement have been achieved [23].

Attempts have been made to develop a system for measuring liposomes and liquid crystal capsules (Kawakami and Kojima, unpublished). When taking these measurements, the objects slipped and moved when a load was applied, making the measurements impossible. To solve this problem, a fixation technique using aspiration was applied. A thermally processed pipette used to hold egg cells was used for aspiration, and a small amount of negative pressure was applied to the pipette to fix the object by aspirating it (Fig. 16.4 a, b).

This experimental system successfully measured the stiffness of the cell-sized liposomes and cells. Although experimental data are still preliminary, this method successfully measured the stiffnesses of liposomes during deformation (Fig. 16.4c). The collected measurements were comparable to the stiffness values measured using atomic force microscopy (AFM) and other techniques [24]. By using this method, the differences in the stiffnesses of liposomes with various membrane compositions can be clarified. In addition, it is possible to evaluate the stiffnesses of liposomes encapsulating the cytoskeleton and measure the forces generated by the deformation of liposomes encapsulating molecular motor proteins. Therefore, this method is also expected to enable the in situ quantitative evaluation of the output of molecular robotic (cybernetic) systems.

16.8 Summary

This section introduces a microrobot system that can freely manipulate microparticles. A specialized system for evaluating the mechanical properties of micro-objects and their applications is also presented. To realize the construction of large-scale molecular robot (cybernetic) systems, it is necessary to have technology for the complex arrangement of different functions, such as sensors, processors, and actuators, implemented in artificial cells as well as technology with which to evaluate these functions. Although the methods introduced in this chapter are difficult to parallelize because they involve a single object, they may be useful for prototyping molecular robotic (cybernetic) systems, owing to their high flexibility. In addition, there is a possibility of developing a system that integrates microfluidic systems, and this technology is sure to contribute to the realization of molecular robotics and cybernetics.

References

1. Murphy, T.W., et al.: Recent advances in the use of microfluidic technologies for single cell analysis. Analyst **143**, 60–80 (2018)
2. Yamanishi, Y., et al.: Design and fabrication of all-in-one unified microfluidic chip for automation of embryonic cell manipulation. J. Robot. Mech. **22**, 371–379 (2010)
3. Hirose, Y. et al.: Online red blood cell stiffness evaluation system by utilizing a high speed vision and micro channel. Trans. Society of Instrument. Control. Eng. **47**, 221–229 (2011)
4. Ichikawa, A., et al.: On-chip enucleation of an oocyte by untethered microrobots. J. Micromech. Microeng. **24**, 095004 (2014)
5. Inoue, K., et al.: Micromanipulation using micro hand with two rotational fingers. J. Micro-Nano Mech. **7**, 33–44 (2012)
6. Ashkin, A., et al.: Observation of a single-beam gradient force optical trap for dielectric particles. Opt. Lett. **11**, 288–290 (1986)
7. Rosenthal, A., et al.: Dielectrophoretic traps for single-particle patterning. Biophys. J. **88**, 2193–2205 (2005)
8. Takeuchi, M., et al.: Handling of micro objects using phase transition of thermoresponsive polymer. J. Micro-Nano Mech. **8**, 53–64 (2013)
9. Tanikawa, T., et al.: Developmant of a micro-manipulation system having a two-fingered micro-hand. IEEE Trans. Robotics and Automation **15**, 152–162 (1999)
10. Ramadan, A., et al.: Developmental process of a chopstick-like hybrid-structure two- fingered micromanipulator hand for 3-D manipulation of microscopic objects. IEEE Trans. Ind. Electro. **56**, 1121–1135 (2009)
11. Avci, E., et al.: Piezo-actuated parallel mechanism for biological cell release at high speed. Biomed. Microdevice **17**, 98 (2015)
12. Avci, E., et al.: High-speed automated manipulation of microobjects using a two-fingered microhand. IEEE Trans. Ind. Electron. **62**, 1070–1079 (2015)
13. Kim, E., et al.: High-speed manipulation of microobjects using an automated two-fingered microhand for 3D microassembly. Micromachines **11**, 534 (2020)
14. Kato, K., et al.: Rapid protein anchoring into the membranes of mammalian cells, using oleyl chain and poly(ethylene glycol) Derivatives. Biotechnol. Prog. **20**, 897–904 (2004)
15. Cros, S., et al.: Nanomechanical analysis of cells from cancer patients. Nature Nanotech. **2**, 780–783 (2007)

16. Ohara, K., et al.: Dextrous cell diagnosis using two-fingered microhand with micro force sensor. J. Micro-Nano Mech. **7**, 13–20 (2012)
17. Tadaki, M., et al.: A 3D culture model study monitoring differentiation of dental epithelial cells into ameloblast-like cells". RSC Adv. **6**, 62109–62118 (2016)
18. Goswami, R., et al.: Mechanical shielding in plant nuclei. Curr. Biol. **30**, 2013–2025 (2020)
19. Khangai, N. et al.: Cell stiffness measurement using two-fingered Micro- hand with micro force sensor equipped with plate-shaped end effector. In: Proceeding 11th International Conference Ubiquitous Robots Ambient Intelligence, pp. 517–521 (2014)
20. Kawakami, M., et al.: Automated microhand system for measuring cell stiffness by using two plate end-effectors. IEEE Robot. Autom. Lett. **7**, 2385–2390 (2022)
21. Yabugaki, H. et al.: Automated stable grasping with two-fingered microhand using micro force sensor. In: Proceeding IEEE International Conference Robotics and Automation, pp. 2771–2776 (2013)
22. Kojima, M. et al.: Cell stiffness measurement by two-fingered micro-hand system with plate shaped end effector. In: Proceeding International Symposium Micro-Nano Mechatronics Human Science, pp. 1–2 (2018)
23. Veschgini, M., et al.: Tracking mechanical and morphological dynamics of regenerating hydra tissue fragments using a two fingered micro-robotic hand. Appl. Phys. Lett. **108**, 103702 (2016)
24. Takechi-Haraya, Y., et al.: Improved atomic force microscopy stiffness measurements of nanoscale liposomes by cantilever tip shape evaluation. Anal. Chem. **91**, 10432–10440 (2019)

Chapter 17
Synthetic Polymers for the Modulation of Membrane Structures and Functions

Kazuma Yasuhara

The essential function of biomembranes is to serve as barriers that compartmentalize cells and intracellular organelles from the external environment. They also provide functional interfaces responsible for intercellular communication, substance transport, and energy production. These complex functions are achieved through the dynamic formation of supramolecular assemblies of lipid bilayers and membrane proteins working in concert. Establishing methods to control the physical properties of lipid membranes and modulate their functions is indispensable for realizing molecular robotics using artificial cell membranes. In this chapter, we present examples of methods to control the properties and related functions of lipid membranes using engineered synthetic membrane-active polymers. Specifically, we designed and synthesized a series of amphiphilic polymers that mimic the membrane-binding motifs of natural proteins or peptides to induce pore formation in lipid membranes, lipid domain formation, and lipid nanodisc formation.

17.1 Molecular Design of Membrane-Active Synthetic Polymers Inspired by Natural Peptides

In nature, there are numerous proteins and peptides that modify the structure and function of lipid membranes to induce various cellular functions such as cell membrane deformation, cell division, membrane fusion, and membrane permeation. A common structural feature among these membrane-interacting proteins and peptides is the amphiphilic motif, which is formed by the localization of hydrophobic

K. Yasuhara (✉)
Division of Materials Science, Graduate School of Science and Technology, Center for Digital Green-Innovation, Data Science Center, Medilux Research Center, Nara Institute of Science and Technology, Ikoma, Nara, Japan
e-mail: yasuhara@ms.naist.jp

S. Murata et al. (eds.), *Molecular Robotics II*,
https://doi.org/10.1007/978-981-95-2394-8_17

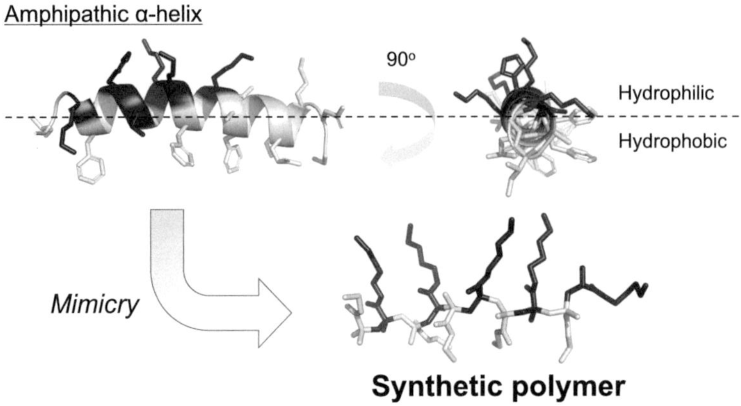

Fig. 17.1 Design of membrane-active polymers inspired by the natural membrane-binding domain of proteins and peptides

and hydrophilic amino acid residues on the same molecular backbone. Here, we introduce Magainin-2, an antimicrobial peptide isolated from the skin of the African clawed frog *Xenopus laevis* known to form pores in lipid membranes, as an example of a membrane-active peptide. Magainin-2 peptide, consisting of 23 amino acid residues, forms an α-helix upon interacting with lipid membranes, whereas it adopts a random coil structure in water [1]. By forming an α-helix, hydrophilic (cationic) amino acid residues such as lysine and histidine are localized on one side of the helix, while hydrophobic amino acid residues are localized on the opposite side, resulting in the formation of an amphiphilic structure (Fig. 17.1).

Inspired by the amphiphilic structures of membrane-binding proteins and peptides, several synthetic polymers have been designed to exhibit functions through the interaction with lipid membranes [2]. Synthetic polymers have several advantages, such as large-scale production at low cost and ease of modification by synthetic approaches. However, there are considerable differences in molecular structure between natural peptides/proteins and synthetic polymers. While proteins and peptides have specific amino acid sequences and exist as single molecular species, synthetic polymers are always obtained as mixtures of multiple molecular species due to the difficulty in the synthesis with fully controlled monomer sequences and molecular weight. Furthermore, synthetic polymers hardly form higher-order structures like α-helices and β-sheets, which are commonly found in proteins and peptides. Therefore, it is challenging to completely replicate the structures of proteins and peptides with synthetic polymers. However, by focusing on the physicochemical characteristics of peptides, several functions related to biomembranes have been successfully induced using synthetic polymers.

17.2 Formation of Nanopores in Membranes by a Synthetic Polymer

In nature, many amphiphilic peptides spontaneously form pore structures and increase membrane permeability through interactions with lipid membranes. These peptides are utilized by living organisms as toxins or antimicrobial agents [3]. We previously developed artificial molecules to mimic these pore-forming peptides, using a synthetic polymethacrylate molecular framework [4]. The amphiphilic polymethacrylate derivatives that form membrane pores were synthesized by free radical polymerization of methacrylate monomer units with cationic aminoethyl groups and hydrophobic n-butyl groups as side chains (Fig. 17.2a). The molecular weight (MW) of the polymer was controlled by the addition of methyl 3-mercaptopropionate (MMP) as a chain transfer agent. The number-averaged molecular weight and the average fraction of the hydrophobic unit were found to be 2,100 g·mol^{-1} and 0.36, respectively.

Membrane permeation induced by the pore-forming polymer was directly visualized using cell-sized giant unilamellar vesicles (GUVs). Specifically, GUVs were prepared in an aqueous solution of sucrose, and the outer aqueous phase was replaced with a glucose solution of the same osmotic pressure. Due to the difference in refractive index between the sugar solutions, the contrast or darkness inside and outside the vesicles differs, allowing molecular leakage from the vesicles to be monitored by

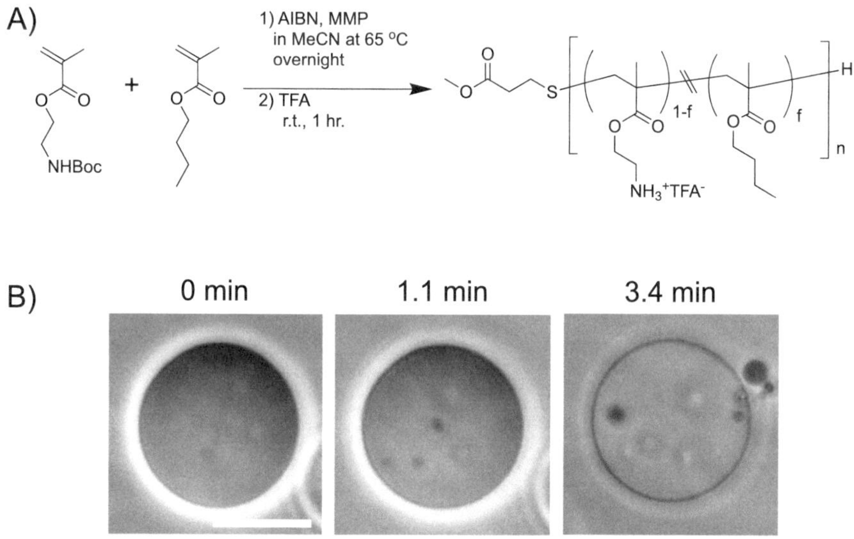

Fig. 17.2 **a** Synthesis of a pore-forming amphiphilic polymethacrylate derivative **b** The microscopic visualization of the membrane permeation on a POPE / POPG GUV induced by the polymer at a concentration of 6.67 μg·mL^{-1}. Bar = 10 μm. The images were reproduced from the reference [4] with permission

phase-contrast microscopic observation [5]. Addition of the polymer at a concentration of 6.67 μg·mL^{-1} induced a decrease in the darkness of the inner aqueous phase of a 1-palmitoyl-2-oleoyl-*sn*-glycero-3-phosphoethanolamine (POPE) / 1-palmitoyl-2-oleoyl-*sn*-glycero-3-phospho-(1'-*rac*-glycerol) (sodium salt) (POPG) (8:2) GUV without an obvious morphological change, indicating the leakage of entrapped sucrose to the outer aqueous phase (Fig. 17.2b). The membrane permeation induced by the polymer was found to be highly dependent on the size of the marker molecule. In particular, the permeation of sucrose (MW = 342 g·mol^{-1}, hydrodynamic diameter (D_{hy}) of 0.98 nm) was observed, whereas Rhodamine B isothiocyanate–Dextran (MW = 70 kDa, D_{hy} of 12 nm) did not leak, suggesting that the polymer forms size-limited pores in the membrane. The pore-forming polymer displayed concentration-dependent interaction with the lipid membrane. At low polymer concentrations, the polymer adsorbed to the membrane surface without inducing permeation, whereas high polymer concentrations induced solubilization of the membrane. Additionally, pore formation behavior was not observed when a methyl group was introduced as a hydrophobic side chain, which instead induced rupture of the giant vesicles. These results suggest that the polymer needs to be inserted into the hydrophobic core of the lipid membrane by introducing sufficiently long hydrophobic chains to form a pore structure.

17.3 Lipid Domain Formation Induced by a Synthetic Polymer

As suggested by the lipid raft hypothesis, the clustering of specific lipid molecules and membrane proteins within membranes is thought to play important roles in various cellular functions such as signal transduction [6]. Several proteins and peptides have been found to promote the formation of domain structures in lipid membranes. Here, we introduce a synthetic polymer that induces domain formation in lipid molecules [7]. The domain-forming polymer is a polymethacrylate derivative with cationic aminobutyl groups and hydrophobic ethyl groups as side chains, synthesized by atom transfer radical polymerization (Fig. 17.3a). The number-averaged molecular weight and the average fraction of the hydrophobic group were found to be 5,600 g·mol^{-1} and 0.44, respectively.

To visualize the domain structure in the membrane, POPE/POPG GUVs labeled with Texas Red® 1,2-dihexadecanoyl-*sn*-glycero-3-phosphoethanolamine, triethylammonium salt (TR-DHPE) were prepared by a gentle hydration method. Before the addition of the polymer, a uniform fluorescence signal was observed throughout the GUV surface, indicating that the two lipid components were well miscible. Upon addition of the polymer, both dark and bright regions appeared on the vesicular membrane, suggesting changes in the localization of TR-DHPE associated with domain formation (Fig. 17.3b). It is likely that the polymer, having multiple cationic

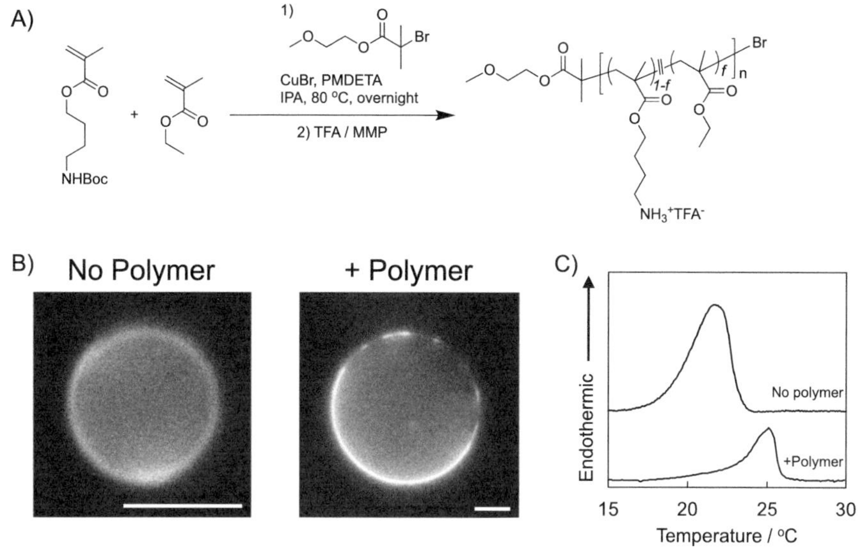

Fig. 17.3 **a** Synthesis of a lipid domain-forming amphiphilic polymethacrylate derivative **b** The microscopic visualization of the domain formation on a POPE/POPG GUV induced by the polymer at a concentration of 0.067 μg·mL^{-}. The GUV was stained with TR-DHPE. Bar = 10 μm **c** DSC thermograms of POPE / POPG (8:2) vesicles in the absence and presence of the polymer. [Lipid] = 1 mM, [Polymer] = 0.1 mM in 10 mM HEPES buffer (pH = 7.4, [NaCl] = 100 mM). The images were reproduced from the reference [7] with permission

side chains, induced the clustering of the negatively charged lipid, POPG. Differential scanning calorimetry (DSC) measurements also confirmed the formation of POPE-rich domains due to the exclusion of POPG molecules upon the addition of the polymer (Fig. 17.3c). Such domain formation through multi-point interactions between charged groups of macromolecules and oppositely charged lipid molecules has also been reported for antimicrobial peptides [8]. Furthermore, in a binding study using a polymer with a fluorescent dye introduced at the terminal, a sigmoidal time dependence was observed, indicating cooperative binding. Taken together, the domain formation promotes further adsorption of the polymers to the membrane by creating new binding sites.

17.4 Lipid Nanodisc-Forming Synthetic Polymer

Molecular assemblies formed by lipids are not limited to hollow vesicles but can take various forms depending on their molecular structure and composition. Here, we introduce a synthetic polymer that spontaneously forms lipid nanodiscs through the fragmentation of membranes. High-density lipoprotein (HDL), which is responsible for cholesterol transport in the body, forms nanometer-scale disc-shaped molecular assemblies in its early stages [9]. HDL consists of apolipoprotein A-I (Apo A-I) and lipid molecules, where the amphiphilic α-helices of Apo A-I cover the hydrophobic edges of the lipid bilayer, thereby stabilizing the nanodisc structure in water. We developed amphiphilic polymers that spontaneously form nanodiscs by mimicking the amphiphilic structure of Apo A-I [10]. The nanodisc-forming polymer was obtained by random copolymerization of methacryloyl choline chloride as a hydrophilic unit and butyl acrylate as a hydrophobic unit through free radical polymerization (Fig. 17.4a). The degree of polymerization (n) was controlled using methyl 3-mercaptopropionate (MMP) as a chain transfer agent, and the fraction of the hydrophobic monomer unit (f) was adjusted by varying the monomer feed ratio. Negative-stain TEM observation confirmed that the addition of the polymer with 60% of the hydrophobic monomer unit and a number-averaged molecular weight of 4.7 $kg \cdot mol^{-1}$ to 1, 2-dipalmitoyl-*sn*-glycero-3-phosphocholine (DMPC) liposomes resulted in the fragmentation of the lipid membrane and the subsequent formation of disc-shaped molecular assemblies with diameters of several tens of nanometers (Fig. 17.4b). A systematic screening of the polymer library with varying molecular weights and hydrophobic monomer contents revealed that the structural requirements for polymers forming lipid nanodiscs include (i) an appropriate amphiphilic balance (hydrophobic monomer content, f of about 0.4 to 0.6) and (ii) an appropriate molecular weight range (3,000 to 9,000 $g \cdot mol^{-1}$) (Fig. 17.4C).

The lipid nanodiscs formed by this polymer do not show strong absorption in the UV–visible region and have uniform small sizes, providing a powerful platform for analyzing the interactions of membrane-active molecules with lipid membranes using various spectroscopic methods. We have evaluated the interactions between several amyloidogenic proteins and lipid membranes using lipid nanodiscs. Thioflavin-T fluorescence and circular dichroism measurements revealed that nanodiscs inhibit the amyloid fibril formation of human islet amyloid polypeptide (hIAPP), which is known to be involved in type 2 diabetes [10]. Similarly, for amyloid-β, which causes Alzheimer's disease, nanodiscs have been shown to inhibit amyloid fibril formation and reduce neurotoxicity [11]. In these amyloidogenic protein systems, nanodiscs with limited surface areas can restrict the number of proteins that can bind and accumulate. As a result, unlike in aqueous solutions, it is likely that protein aggregation can be trapped in an oligomeric state.

Furthermore, nanodiscs can also be used as nanocarriers for molecular delivery in living systems [12]. We evaluated the uptake behavior of nanodiscs formed with 1,2-dipalmitoyl-*sn*-glycero-3-phosphocholine by HeLa cells by labeling the nanodiscs with Lissamine rhodamine B 1,2-dihexadecanoyl-*sn*-glycero-3-phosphoethanolamine, a fluorescent lipid. Quantitative analysis revealed that nanodiscs are taken up with significantly higher efficiency compared to liposomes. It was also found that nanodiscs formed with higher polymer content, which form smaller aggregates, display improved cellular uptake efficiency. The intracellular destination of the nanodiscs depends on their size, with larger nanodiscs remaining

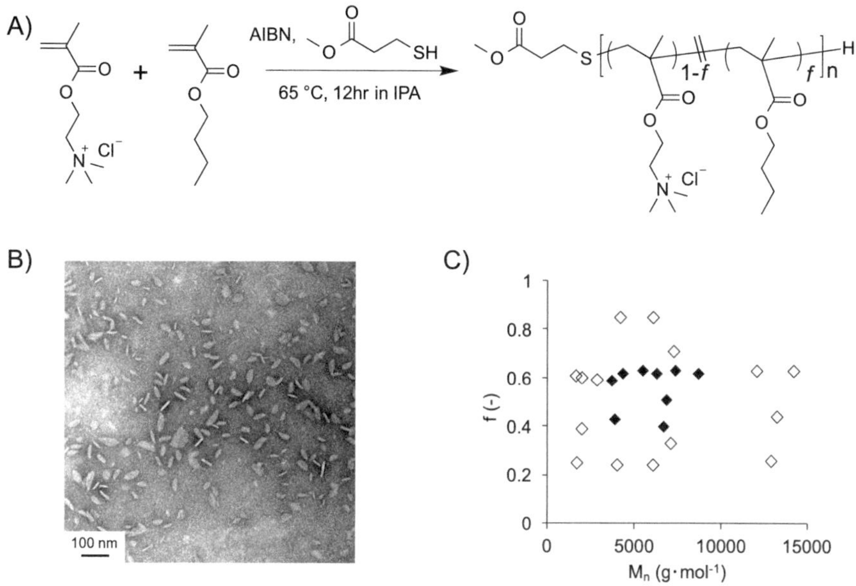

Fig. 17.4 **a** Synthesis of lipid nanodisc-forming amphiphilic polymethacrylate derivatives **b** Negative-stain TEM image of the DMPC nanodiscs formed with the polymer **c** The effect of number-averaged molecular weight (M_n) and the fraction of hydrophobic side chains (f) on the formation of nanodiscs. Filled diamonds correspond to the polymers that can spontaneously form nanodiscs. The images were reproduced from the reference [10] with permission.

in the cytoplasm, while smaller nanodiscs are transported near the nucleus. Additionally, the pharmacological activity of the anticancer agent paclitaxel was significantly enhanced when encapsulated in nanodiscs compared to conventional liposomes.

17.5 Conclusion

In this chapter, we have introduced several synthetic membrane-active polymers inspired by the amphiphilic structure of membrane-binding domains of proteins and peptides. A common aspect of these examples is the approach of capturing the physicochemical characteristics of the molecule, rather than strictly replicating the molecular structure (amino acid sequence) of proteins and peptides. This strategy is expected not only to provide molecular tools for artificial cell membrane-based molecular robotics but also to contribute to the reconstructive understanding of biological systems composed of complex intermolecular interactions.

Acknowledgements The author deeply acknowledges Prof. Kenichi Kuroda and Prof. Ayyalusamy Ramamoorthy (University of Michigan) for their collaboration on a part of this work.

References

1. Zasloff, M., et al.: Antimicrobial peptides of multicellular organisms. Nature **415**, 389–395 (2002)
2. Ergene, C., et al.: Biomimetic antimicrobial polymers: recent advances in molecular design. Polym. Chem. **9**, 2407–2427 (2018)
3. Shai, Y., et al.: Mode of action of membrane active antimicrobial peptides. Biopolymers **66**, 236–248 (2002)
4. Tsukamoto, M., et al.: Mechanistic study of membrane disruption by antimicrobial methacrylate random copolymers by the single giant vesicle method. Langmuir **37**, 9982–9995 (2021)
5. Tamba, Y., et al.: Single giant unilamellar vesicle method reveals effect of antimicrobial peptide magainin 2 on membrane permeability. Biochemistry **44**, 15823–15833 (2005)
6. Lingwood, D., et al.: Lipid rafts as a membrane-organizing principle. Science **327**, 46–50 (2010)
7. Yasuhara, K., et al.: An antimicrobial peptide-mimetic methacrylate random copolymer induces domain formation in a model bacterial membrane. J. Membr. Biol. **255**, 513–521 (2022)
8. Epand, R., et al.: Lipid domains in bacterial membranes and the action of antimicrobial agents. Biochim. Biophys. Acta Biomembr. **1788**, 289–294 (2009)
9. Cavigiolio, G., et al.: The Interplay between size, morphology, stability, and functionality of high-density lipoprotein subclasses. Biochemistry **47**, 4770–4779 (2008)
10. Yasuhara, K., et al.: Spontaneous lipid nanodisc fomation by amphiphilic polymethacrylate copolymers. J. Am. Chem. Soc. **139**, 18657–18663 (2017)
11. Sahoo, B.R., et al.: Alzheimer's amyloid-beta intermediates generated using polymer-nanodiscs. Chem. Commun. **54**, 12883–12886 (2018)
12. Hao, J., et al.: Lipid nanodiscs spontaneously formed by an amphiphilic polymethacrylate derivative as an efficient nanocarrier for molecular delivery to intact cells. RSC Adv. **14**, 6127–6134 (2024)

Chapter 18
Membrane Deformation and Transport by Self-Assembling Peptides

Kazunori Matsuura

Nano-architectures self-assembled from rationally designed peptides and proteins are essential components of molecular robots [1]. By engineering self-assembly based on peptide secondary structures, protein–ligand interactions, and protein–protein interactions, various nanostructures can be constructed, including fibers, tubes, rings, and capsules [2–5]. For these nano-architectures to function as elements of molecular robots, they should interact with liposomes—the compartments of molecular robots—and perturb them in response to stimuli [6]. Studies have investigated morphological changes in liposomes accompanying structural changes in peptide assemblies in response to stimuli, as well as substance transfer and signal transduction across the liposome membrane. This chapter outlines recent studies on liposome deformation and substance transport across membranes based on interactions between peptide assemblies and lipid membranes.

18.1 Dramatic Deformation of Giant Vesicles Induced by Photoresponsive Peptide Nanofiber

Cytoskeletons, such as microtubules and actin filaments, are natural protein assemblies that dynamically polymerize and depolymerize to control cell morphology and intracellular/extracellular movements. The polymerization and depolymerization of microtubules and actin filaments are controlled by the hydrolysis of guanosine triphosphate (GTP) and adenosine triphosphate (ATP), respectively. Replicating these dynamics using artificial materials has proven challenging. Recently, the polymerization and depolymerization of various peptide and protein fibers have been controlled by external stimuli such as light [7, 8]. Additionally, it has been reported

K. Matsuura (✉)
Tottori University, Tottori, Japan
e-mail: ma2ra-k@tottori-u.ac.jp

S. Murata et al. (eds.), *Molecular Robotics II*,
https://doi.org/10.1007/978-981-95-2394-8_18

that the polymerization and depolymerization of natural cytoskeletons and physical perturbations to giant unilamellar vesicles (GUVs) induce deformations similar to those observed in natural cells [9, 10].

We developed the first artificial cytoskeleton-like system, where the reversible polymerization and depolymerization of spiropyran (SP)/merocyanine (MC)-modified peptide nanofibers in GUVs dramatically altered their morphology [11]. We designed SP-modified β-sheet-forming peptides FKFEC^{SP}KFE, which can be reversibly converted to the MC form by visible/UV light irradiation. Confocal laser scanning fluorescence microscopy (CLSM) with thioflavin T staining and transmission electron microscopy (TEM) showed that the SP-peptide formed β-sheet nanofibers, whereas photoisomerization to the MC-peptide almost completely dissociated the nanofibers (Fig. 18.1). Interestingly, the morphology of GUVs encapsulating the MC-peptide dramatically changed into worm-like vesicles upon photoisomerization to the SP-peptide and then reversibly changed back to spherical GUVs upon photoisomerization to the MC-peptide (Fig. 18.2). This photo-controlled system, mimicking the polymerization and depolymerization of the natural cytoskeleton, can be applied as a component of spatiotemporal-controlled molecular robots and artificial cell systems.

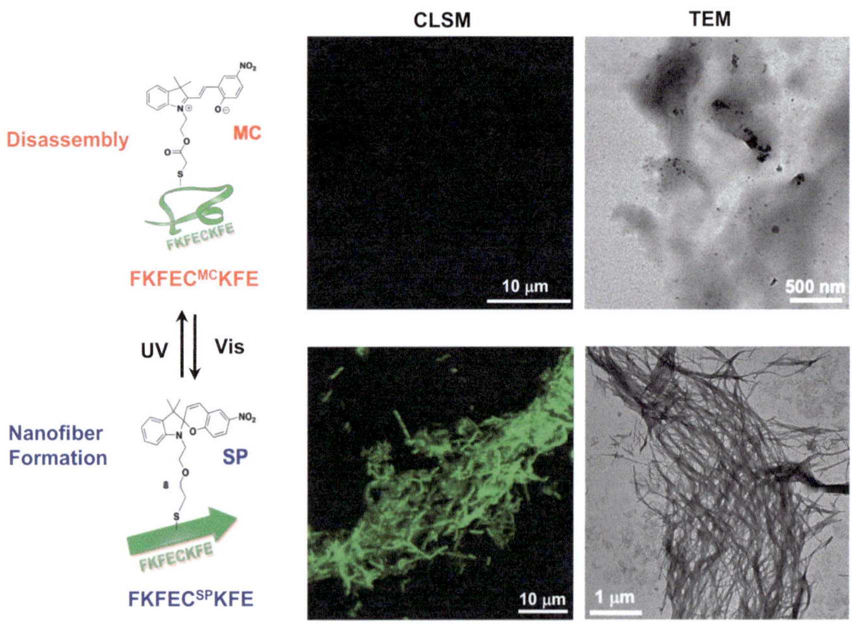

Fig. 18.1 Reversible peptide nanofiber formation and dissociation by photoisomerization of SP/MC. Confocal laser scanning fluorescence microscopy (CLSM) and transmission electron microscopy (TEM) images of 100 μM FKFEC$^{SP/MC}$KFE peptide and 100 μM Thioflavin T during photoisomerization to form and dissociate peptide nanofibers in a 10 mM phosphate buffer (pH 7.4) at 25 °C [11]

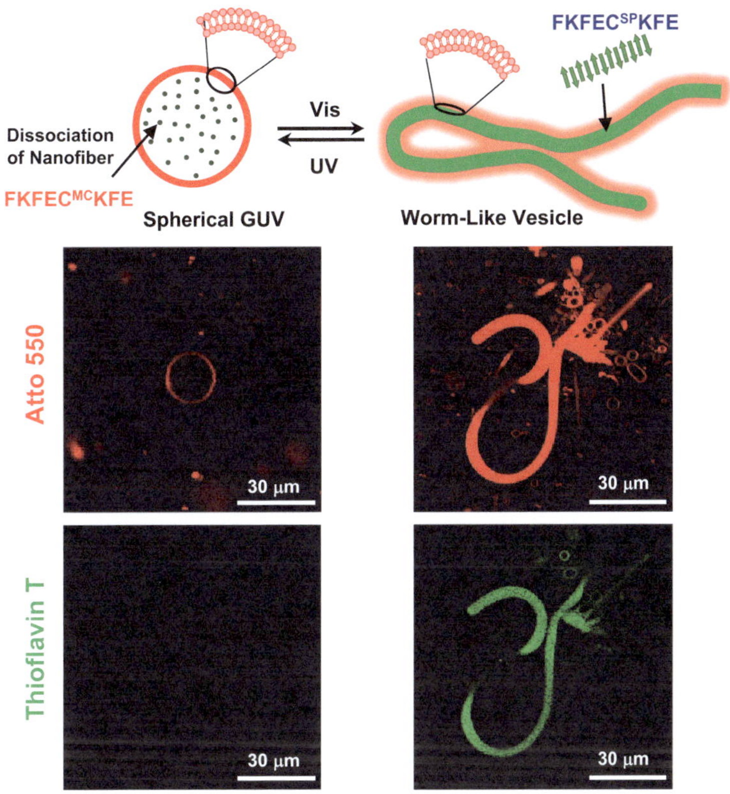

Fig. 18.2 Dramatic deformation of GUV induced by photoisomerization of peptides. CLSM images of GUV (5 µM Atto550-labeled 1,2-dioleoyl-*sn*-glycero-3-phosphoethanolamine (Atto550-DOPE), 1 mM 1-palmitoyl-2-oleoyl phosphatidylcholine (POPC)) encapsulating 10 µM FKFEC^{MC}KFE and 100 µM Thioflavin T during photoisomerization in a 10 mM phosphate buffer (pH 7.4) at 25 °C [11]

18.2 Budding of Giant Vesicles by Anchoring Artificial Viral Capsid Self-Assembled from β-Annulus Peptide

A spherical virus is a supramolecular complex, several tens of nanometers in size, consisting of genome nucleic acids encapsulated in self-assembled coat proteins called capsids. Spherical viral capsids with a discrete size and internal space are valuable natural supramolecular materials that can be used as nanoreactors, drug delivery carriers, and base materials for vaccines [12]. Recently, it has become possible to construct artificial virus-like nanocapsules by designing the self-assembly of proteins and peptides [13]. We developed an artificial viral capsid by self-assembling a 24-residue β-annulus peptide (INHVGGTGGAIMAPVAVTRQLVGS), which is involved in the formation of the dodecahedral internal skeleton of the tomato bushy stunt virus [14]. This artificial viral capsid can encapsulate various guest macromolecules through N-terminal modification of the β-annulus peptide, while C-terminal modification enables the display of various functional molecules, such as proteins and DNAs [13, 15–17].

Natural viruses can self-replicate by hijacking biosynthetic systems within an infected host cell, followed by budding out of the cell membrane to infect the next host cell. Viral capsid proteins self-assemble around the cell membrane by anchoring various signaling proteins to the membrane, and the virus is enveloped by the cell membrane and released outside the cell. For example, the budding of the human immunodeficiency virus (HIV) from an infected host cell is induced by the modification of structural proteins bearing long-chain fatty acids, followed by their anchoring to the cell membrane [18]. By mimicking the budding mechanism of HIV, we designed an alkyl anchor-modified artificial viral capsid to enable the transport of capsids to the lipid membrane and subsequent budding out [19]. The C-terminus of the β-annulus peptide was modified using an octyl chain as an alkyl anchor via a disulfide bond and co-assembled with tetramethylrhodamine (TMR)-labeled β-annulus peptide to form alkyl anchor-modified capsids. Adding the β-annulus peptide bearing an octyl chain to the outer aqueous phase of GUVs induced the budding of daughter vesicles encapsulating capsids from outside-to-inside the mother GUV (Fig. 18.3a). Conversely, encapsulating the β-annulus peptide bearing an octyl chain in the inner aqueous phase of GUVs induced the budding of daughter vesicles encapsulating capsids from inside-to-outside the mother GUV. Time-series CLSM images exhibited the convex deformation of GUVs followed by the detachment of daughter vesicles from the parent vesicles (Fig. 18.3b). The budding of daughter vesicles from inside-to-outside GUVs might be induced by the anchoring of octyl chains to the lipid bilayer inside GUVs, followed by self-assembly to form artificial viral capsids on the membrane. Such an artificial virus budding system can be applied to intercellular molecular transport and communication systems, drug delivery, in-cell component recovery, and the control of cell function.

18.3 Invasion of Enveloped Viral Replica into Giant Vesicles and Cells Through Membrane Fusion

The nucleocapsids of natural envelope-type viruses, such as the influenza virus, HIV, and coronavirus, are covered with a lipid bilayer membrane equipped with membrane proteins. During the infection of host cells by these viruses, membrane fusion is an essential step often triggered by the binding of viral membrane proteins to the host cell membrane. For instance, glycoprotein gp41 on the envelope of HIV promotes membrane fusion, facilitating entry into the host cell [20].

To develop a supramolecular model system that mimics viral entry into a GUV or cell through membrane fusion, we created an enveloped viral replica by complexing an anionic artificial viral capsid with a cationic lipid bilayer. This

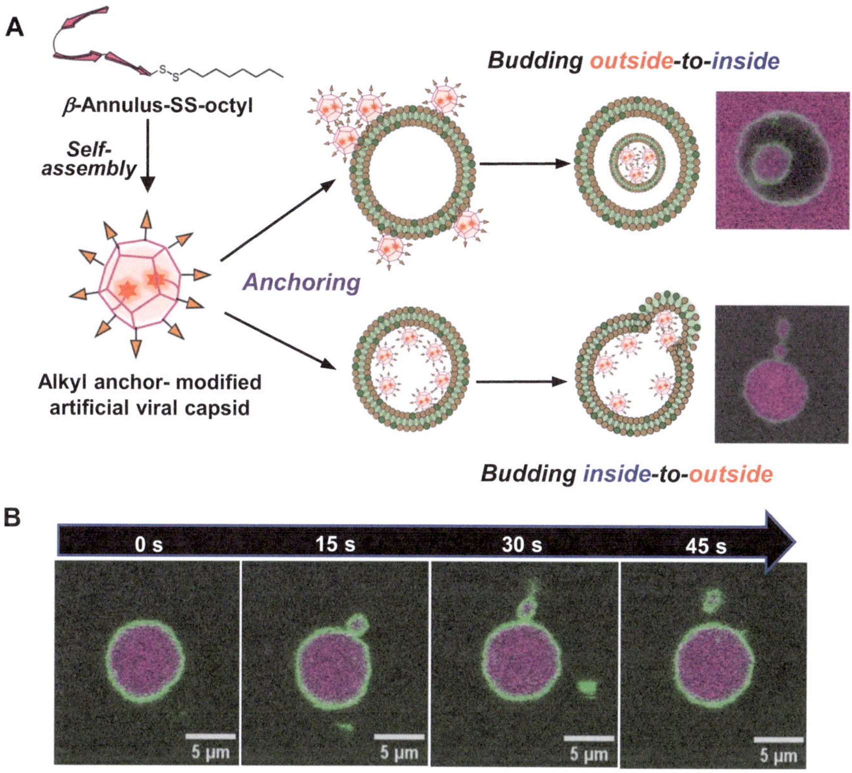

Fig. 18.3 **a** Alkyl anchor-modified artificial viral capsid budding outside-to-inside and inside-to-outside of GUV **b** Time-series CLSM images of nitrobenzoxadiazole (NBD)-labeled GUV (1 mM POPC, 100 μM cholesterol, 10 μM 1,2-dioleoyl-*sn*-glycero-3-phosphoethanolamine-*N*-(7-nitro-2–1,3-benzoxadiazol-4-yl (NBD-PE)) budding inside-to-outside after the addition of 2 μM TMR-β-annulus-SS-octyl inside GUV [19]

enveloped viral replica was constructed using a cationic lipid bilayer (1,2-dioleoyl-3-trimethylammonium-propane (DOTAP)/ 1,2-dioleoyl-sn-glycero-3-phosphocholine (DOPC)/NBD-PE = 1/9.95/0.05) complexed on an anionic artificial viral capsid co-assembled from β-annulus-EE (INHVGGTGGAIMAPVAVTRQLVGSEE) and TMR-labeled β-annulus-EE peptides [21]. The cationic enveloped viral replica electrostatically interacted with the anionic GUV comprising 40% DOPG/DOPC, allowing the capsid to enter the GUV through membrane fusion. In contrast, no entry of the capsid was observed inside the zwitterionic GUV composed of DOPC [22] (Fig. 18.4a). Furthermore, HepG2 cells internalized the cationic enveloped viral replicas through membrane fusion and micropinocytosis (Fig. 18.4b). After 1 h of adding the viral replicas to the cells, yellow fluorescent dots colocalizing TMR and NBD were observed on the cell surface. After 3 h, strong TMR fluorescence was seen throughout the cell. By contrast, when anionic capsids or enveloped capsids coated with DOPC (without DOTAP) were added to HepG2 cells, only weak TMR fluorescence was detected inside the cell. These results suggest that membrane fusion could be induced by interactions between the enveloped viral replica and the cell membrane without external stimuli, providing insights into the infection mechanism of natural enveloped viruses.

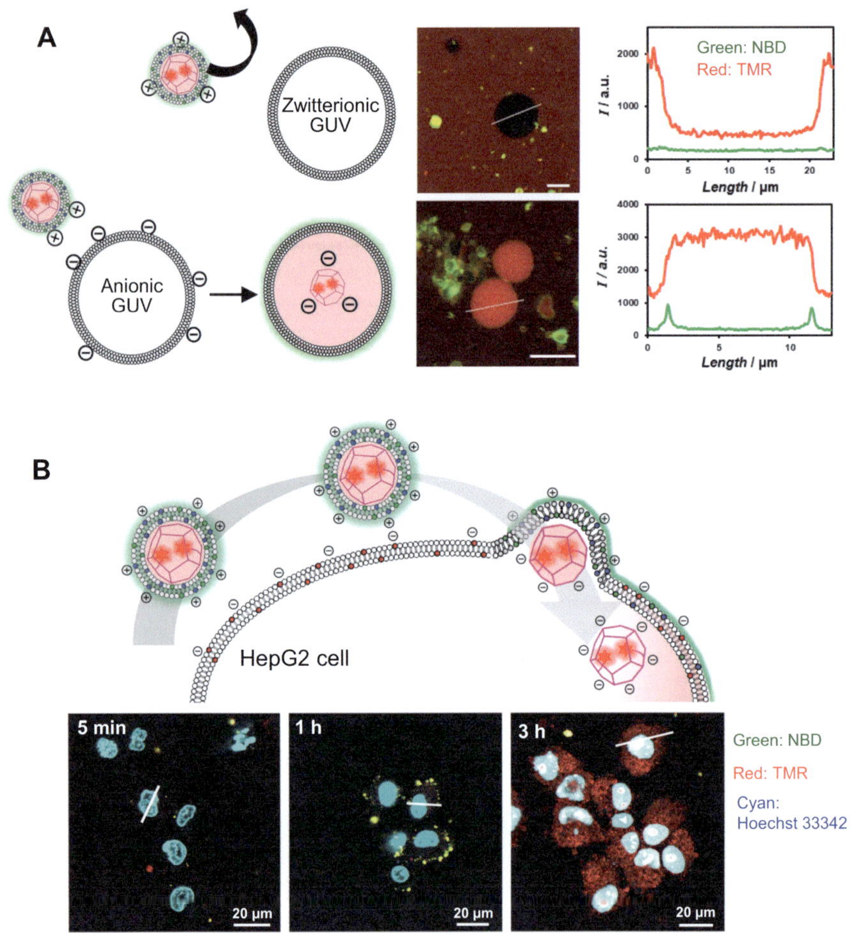

Fig. 18.4 **a** CLSM images of DOPC GUV and 40% DOPG/DOPC interacting with TMR/NBD-labeled enveloped viral replica (45 μM β-annulus-EE peptide, 5 μM TMR-β-annulus-EE peptide, 150 μM DOTAP, 1492.5 μM DOPC and 7.5 μM NBD-PE) **b** CLSM images of HepG2 cells incubated with the TMR/NBD-labeled enveloped viral replica [22]

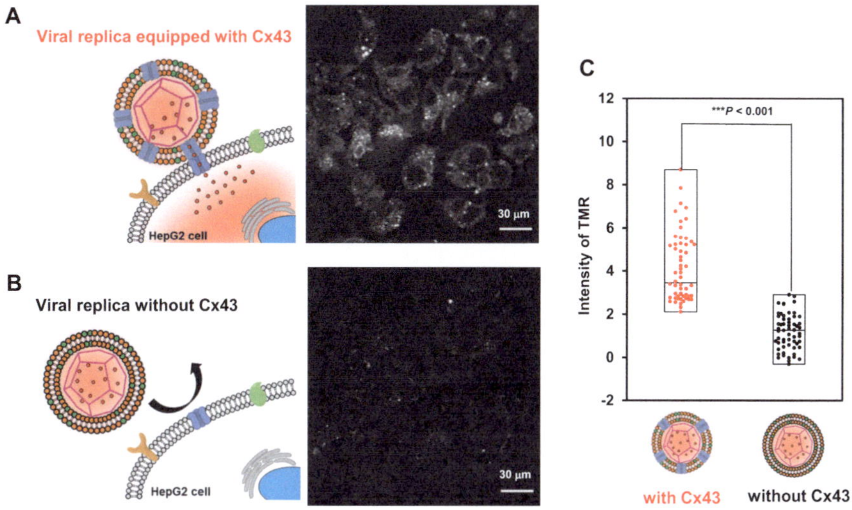

Fig. 18.5 Transport of TMR from Cx43-equipped enveloped viral replicas into HepG2 cells. CLSM images of HepG2 cells incubated with TMR-encapsulated enveloped viral replicas (10 μM β-annulus-EE, 30 μM DOTAP, 300 μM DOPC, 1.7 μM TMR) equipped with Cx43 **a** and without Cx43 **b** Cx43 was expressed from 17.3 nM pURE-Cx43 using the PURE system **c** Box plot of the fluorescence intensity distribution of TMR in HepG2 cells normalized outside of the cells (N = 60) [23]

18.4 Transport of Fluorescent Dye from Connexin-Equipped Viral Replica into Cells Through Gap Junction

The membrane protein connexin-43 (Cx43) forms tubular gap junctions on the cell membrane, with pores approximately 1.6 nm in diameter, facilitating cell-to-cell communication. We successfully equipped Cx43 onto the enveloped viral replicas using a cell-free protein expression system (PURE system) [23]. TEM images of the viral replicas with anti-Cx43 primary antibody and gold nanoparticle-labeled secondary antibody confirmed the expression of Cx43 on the envelope. The transport of the fluorescent dye TMR from the Cx43-equipped enveloped viral replicas into Cx43-expressing HepG2 cells through gap junction formation was also demonstrated (Fig. 18.5). CLSM images showed that fluorescence of TMR was observed inside cells after the addition of TMR-encapsulated Cx43-equipped viral replicas (Fig. 18.5a). In contrast, transport of TMR from viral replicas without Cx43 to the cells was minimal (Fig. 18.5b). Quantitative analyses revealed that the fluorescence intensity in cells receiving TMR from Cx43-embedded viral replicas was significantly higher than that from enveloped capsids without Cx43 (Fig. 18.5c). This technology may be applicable for transferring drugs or signaling molecules using Cx43-equipped viral replicas into Cx43-expressing cells and liposomes.

References

1. Matsuura, K.: 4.9 Trends in the peptide/protein design technology. In Murata, S. (ed.) Molecular Robotics. Springer Nature Singapore, Singapore, pp. 167–173 (2022)
2. Matsuura, K.: Rational design of self-assembled proteins and peptides for nano- and micro-sized architectures. RSC Adv. **4**, 2942–2953 (2014)
3. Luo, Q.J., et al.: Protein assembly: Versatile approaches to construct highly ordered nanostructures. Chem. Rev. **116**, 13571–13632 (2016)
4. Kuan, S.L., et al.: Functional protein nanostructures: A chemical toolbox. Chem. Soc. Rev. **47**, 9069–9105 (2018)
5. Ramakers, B.E., et al.: Molecular tools for the construction of peptide-based materials. Chem. Soc. Rev. **43**, 2743–2756 (2014)
6. Murata, S., et al.: Molecular cybernetics: Challenges toward cellular chemical artificial intelligence. Adv. Funct. Mater. **32**, 2201866 (2022)
7. Sendai, T., et al.: Photoreconfigurable supramolecular nanotube. J. Am. Chem. Soc. **135**, 11509–11512 (2013)
8. Liu, M., et al.: Light-driven dissipative self-assembly of a peptide hydrogel. Chem. Commun. **57**, 13776–13779 (2021)
9. Li, C., et al.: Reversible deformation of artificial cell colonies triggered by actin polymerization for muscle behavior mimicry. Adv. Mater. **34**, e2204039 (2022)
10. Tanaka, S., et al.: Repetitive stretching of giant liposomes utilizing the nematic alignment of confined actin. Commun. Phys. **1**, 18 (2018)
11. Liang, Y., et al.: Dramatic morphological changes in liposomes induced by peptide nanofibers reversibly polymerized and depolymerized by the photoisomerization of spiropyran. Front. Mol. Biosci. **10**, 1137885 (2023)
12. Khudyakov, Y., Pumpens, P. (eds.): Viral Nanotechnology. CRC Press, Boca Raton (2016)
13. Matsuura, K.: Synthetic approaches to construct viral capsid-like spherical nanomaterials. Chem. Commun. **54**, 8944–8959 (2018)
14. Matsuura, K., et al.: Self-assembled synthetic viral capsids from a 24-mer viral peptide fragment. Angew. Chem. Int. Ed. **49**, 9662–9665 (2010)
15. Matsuura, K.: Dressing up artificial viral capsids self-assembled from C-terminal-modified β-annulus peptides. Polym. J. **52**, 1035–1041 (2020)
16. Liang, Y., et al.: Anticancer activity of reconstituted ribonuclease S-decorated artificial viral capsid. ChemBioChem **23**, e202200220 (2022)
17. Sakamoto, K., et al.: An artificial viral capsid decorated with a DNA aptamer internalizing into lymphoma cells. J. Mater. Chem. B **11**, 6053–6059 (2023)
18. Sundquist, W.I., et al.: HIV-1 assembly, budding, and maturation. Cold Spring Harb. Perspect. Med. **2**, a006924 (2012)
19. Matsuura, K., et al.: Alkyl anchor–modified artificial viral capsid budding outside-to-inside and inside-to-outside giant vesicles. Sci. Tech. Adv. Mater. **25**, 2347191 (2024)
20. Doms, R.W., et al.: HIV-1 membrane fusion: Targets of opportunity. J. Cell Biol. **151**, F9–F14 (2000)
21. Furukawa, H., et al.: Enveloped artificial viral capsids self-assembled from anionic β-annulus peptide and cationic lipid bilayer. Chem. Commun. **56**, 7092–7095 (2020)
22. Furukawa, H., et al.: A supramolecular system mimicking the infection process of an enveloped virus through membrane fusion. Sci. Rep. **13**, 19934 (2023)
23. Furukawa, H., et al.: Embedding a membrane protein into an enveloped artificial viral replica. RSC Chem. Biol. **3**, 231–241 (2022)

Chapter 19
Transforming Molecular Robots Learned from Living Cells

Masahito Hayashi and Kingo Takiguchi

Molecular robotics, and chemical AI or molecular cybernetics, have tried to construct by diversifying and organizing artificial cell models, especially those that cause deformation or movement [1, 2 3]. Living cells are "natural molecular robots" that have developed on earth in a long evolutionary process. Learning deformation and movement mechanisms from them and using materials involved in their motility are a reasonable and powerful approach. Here, we introduce the research that uses the liposome that incorporates cytoskeletons and molecular motors that are responsible for morphogenesis and movement of cells as representative demonstrations of constructing the motile artificial cell models.

19.1 Transforming Mechanisms of Living Cells

The cell membrane is the container that separates cell contents from the external environment. The cell membrane is composed of amphiphilic lipids (phospholipids, glycolipids, and sterols) and membrane proteins (ion channels, pumps, receptors, signaling proteins, and proteins that form beneath the membrane as described below) [4, 5]. Amphiphilic lipids spontaneously assemble to form bilayers in aqueous solution. The bilayer closes spherically to form vesicles that separate the inside and

M. Hayashi
Department of Biotechnology and Life Science, Tokyo University of Agriculture and Technology, Koganei-shi, Tokyo, Japan
e-mail: hayashi.bmi@gmail.com

K. Takiguchi (✉)
Graduate School of Science, Nagoya University, Nagoya, Japan
e-mail: j46037a@nucc.cc.nagoya-u.ac.jp

S. Murata et al. (eds.), *Molecular Robotics II*,
https://doi.org/10.1007/978-981-95-2394-8_19

outside of the cell. The bilayer is flexible and easily deforms since the lipids flow two-dimensionally across the membrane. Its flexibility enables the dynamic deformation of the living cell.

Rigid structures, such as the cytoskeleton, which enables the cell to maintain its shape, support the cell membrane from the inside. Cytoskeletal network reconstruction alters the cell shape dynamically. Cytoskeletal elongation and shortening, their sliding movement motorized by molecular motors, and cytoskeletal network reorganization change the morphology of living cells.

Cytoskeletal fibers consist of actin filaments, microtubules, intermediate filaments, and septin [6–9]. Among these, actin filaments and microtubules are responsible for the rapid cellular deformation that occurs within minutes to hours. They can be isolated and purified in large quantities from living tissues (muscle and brain tissue, respectively); thus, they have been widely utilized to construct artificial cell models.

Actin filaments are double helical nanofibers with 6–7 nm thick polymerized from actin monomers. Microtubules are 25 nm thick nanotubes polymerized from heterodimers of α-tubulin and β-tubulin. Actin filament and microtubule polymerization require ATP and GTP, respectively. Both fibers demonstrate polarity, which identifies the direction of the polymerization and molecular motor movement. Actin filaments are flexible fibers with a persistence length of approximately 20 μm. Conversely, microtubules are rigid nanotubes with a persistence length of several hundred micrometers or more.

Molecular motors move in one direction on cytoskeletal fibers with the hydrolytic energy of ATP [6, 10, 11]. Myosin motors move on actin filaments; whereas, kinesin and dynein motors move on microtubules. Myosin causes muscle contraction, cell division, and cell locomotion. Kinesin and dynein cause mitosis, transport cellular cargo, and beat eukaryotic cilia and flagella. The multimer of the molecular motors spans two fibers, causing the fibers to slide and move between each other.

In the case of cell-sized liposomes, the actin filaments can be distributed beneath the membrane, even if there is no specific binding factor, due to the relation between their persistence length and the membrane curvature. Such actin filaments alone are involved in both reinforcement and deformation of the shape of liposomes [12–14]. However, various regulating factors and joints, including binding factors between the lipid membrane and the cytoskeleton, are required to control the degree of reinforcement or to induce the desired deformations. Living cells achieve these regulations or bindings by using a large number of proteins [6, 7].

These proteins are orchestrated by being further controlled as each of them works in a desired part of multiple cascades composed of several hierarchies. The most upstream of these cascades is the acceptance of extracellular stimuli by receptor proteins as input, and information processing is conducted via such as protein phosphorylation on the way, and cytoskeleton assembly/disassembly or movement occurs in the downstream, causing the cellular deformation as the final output. The molecular mechanism itself could be referred to or each protein extracted from cells may be used, in order to induce the deformation or movement of the artificial cell models described in the next section. The binding and regulatory proteins of the cytoskeleton, especially of actin, have already been revealed in hundreds of types, and there is plenty of knowledge about them; thus, they are very promising as a target to use [7].

Two more types of cytoskeletal protein families, named intermediate filament and septin, exist in addition to the actin filament and microtubule. These are durable protein filaments, involved in stabilizing cell morphology, reinforcing plasma membranes, or probably changing membrane shape, which is only once required. Notably, the intermediate filaments and septin filaments do not have polarity, i.e., the properties of the two ends of a filament are the same, and their filaments have no direction; thus, the molecular motors (i.e., myosins, kinesins, or dynein) that work together with the intermediate filaments and septin filaments are absent. At present, intermediate filaments and septin are not expected to be utilized in molecular robotics or molecular cybernetics because they, especially intermediate filaments, are not suitable for causing dynamic changes [8]. Even such intermediate filaments may be useful to inhibit the shape changes via mechanical reinforcement of the liposome membrane or to stabilize the morphology of the area where the shape change has already occurred. Along similar ideas, cell walls, those in plants that are mainly composed of cellulose, or those in bacteria that are made up of peptidoglycan, may be used as containers or parts of artificial cell models, although they are not the cytoskeleton.

19.2 Morphogenesis of Artificial Cell Models Using Natural Materials Such as the Cytoskeleton and Molecular Motor

The construction of artificial cell models is indispensable for molecular robotics, which is the technology underlying the SPA system that is the core of molecular cybernetics (Fig. 19.1) [1–3, 15]. In particular, an artificial cell model capable of deformation or movement is required to prepare the A-unit, which is intended to cause large deformations.

In the early stages of the construction of artificial cell models, gentle hydration was used, which had low liposome efficiency; the solution conditions were frequently restricted, and the cytoskeletal protein encapsulation depended on chance. Hence, the main results obtained were liposome deformations caused by actin filament or microtubule polymerization inside, and only a qualitative analysis enabled the deformation mechanism [16, 17]. The cross-linkers between actin filaments or actin filament and lipid membrane, or the stabilizing factors of microtubules, such as microtubule-associated proteins, were mainly utilized when examining the effects of the coencapsulation of their binding/regulating factors, since these could cause a large effect with a small amount [12, 18].

Such early efforts indicated that liposomes containing cytoskeletal filaments take on many distinctive morphologies. Microtubules, rigid nanofilaments, push the

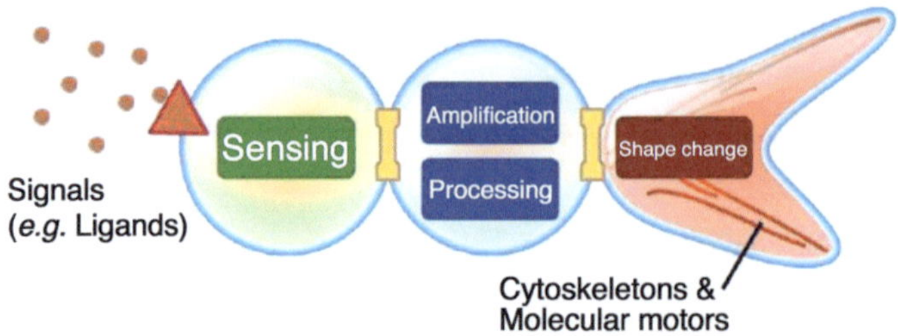

Fig. 19.1 Role of the actuator unit (A-unit) in the SPA system. The SPA system consists of three components, including the S-, P-, and A-units, which play the roles of sensor, processor (amplification and processing), and actuator, respectively. The A-unit changes shape under the P-unit control. Small DNA strands will be utilized for communication between the S-, P-, and A-units. This is a modification of Fig. 19.2 previously published in IEEE Transactions on Molecular, Biological and Multi-Scale Communications 9, 354–363 (2023) that is licensed under a CC BY 4.0 [3]

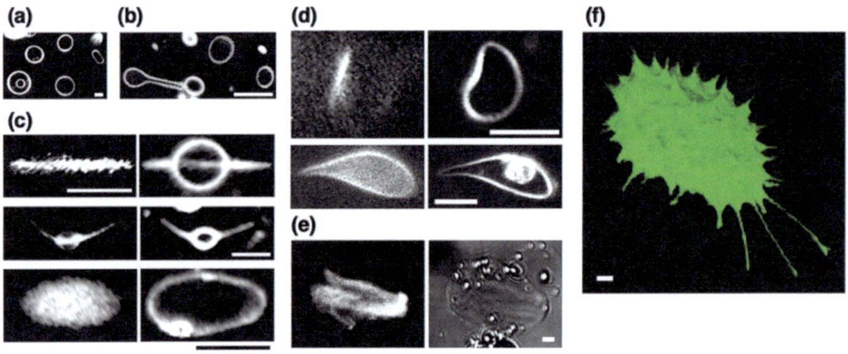

Fig. 19.2 Large-scale shape changes of the molecular robots with biomolecular nanofibers. Liposomes without the cytoskeleton (**a**) and with actin filaments (**b**)–(**e**) or microtubules (**f**). Actin filament encapsulation alone changes the shape of the liposome to a flat disk or racket-like shape (**b**), whereas that with cross-linking proteins (in order from the top, fascin, α-actinin, and filamin) forms bundles of actin filaments and makes membrane protrusions or causes a distorted shape (**c**). Additionally, shape changes occur when a molecular motor, such as BBMI, a myosin that connects the actin filament and the membrane (**d**), or HMM, a myosin fragment that has multiple motor domains (**e**), is coencapsulated. The left column of (**c**)–(**e**) are fluorescence images demonstrating the actin filament distribution. Many protrusions are simultaneously elongated from a liposome when microtubules are encapsulated at a high concentration (**f**). Scale bars indicate 5 μm. Panels (**a**)–(**e**) are modifications of Figures previously published in Methods in Enzymology 464, 31–53 (2009) with permission (panel (**f**) is unpublished data) [13]

membrane from inside with both ends to protrude long membrane tubes on the liposome, causing the deformation into a Φ shape. Conversely, actin filaments, which are soft nanofilaments, cannot push the membrane at both ends but bend along the inner surface of the liposome to make thick circular bundles spontaneously. The circular bundles push the membrane with their side and deform the liposome to a disk-like shape.

The cross-linker between actin filaments induces more dynamic morphological alterations. Rigid actin bundles with fascin or α-actinin make membrane protrusions on the liposomes, and durable actin gel with filamin makes the liposomes a certain shape other than a sphere (Fig. 19.2a, b, and c) [12]. Myosin I isolated from the brush border (Brash Border Myosin I, BBMI) plays the role of the cross-linker between the actin filament and the lipid membrane in intestinal epithelial cells.[1] Actin filament became able to interact with the membrane due to BBMI coencapsulation; thus, actin filament, even in a low concentration, affected membrane curvature, causing a shape change of the liposome (Fig. 19.2d) [13].

Based on the preparation method and the conditions, rather than liposomes desired, multilamellar membranes (also called onions) that do not have an inside space that can be freely utilized, or lipid aggregates (also called lipid droplets) that are not made of lipid bilayer, are sometimes obtained.[2] These are distinguishable by observation with phase-contrast or fluorescent microscopes, by determining whether the lipids are distributed only in the outermost region and/or the internal space consists of the aqueous solution. Currently, methods to form lipid bilayers by passing water-in-oil droplets created by emulsifying an aqueous solution into a lipid-dissolved oil through an oil–water interface that forms a lipid monolayer are predominantly utilized. These "interface passing methods" have dramatically improved the quality and yield of the resulting liposomes, making the encapsulation of the molecules of interest easier [19, 20]. These methods have enabled the encapsulation of high concentrations of target substances and of the complexes that were previously hard to encapsulate because of their large size or a large number of constituents, as well as the quantitative analysis of the deformation.

Myosin, including its fragment that possesses multiple motor domains, is a molecular motor that slides the actin filament and assembles a large-scale meshwork of actin filaments. Inducing deformations easily to even large-sized liposomes became possible by encapsulating such assemblies (Fig. 19.2e). Furthermore, their characteristic is that the entire liposome is deformed and significantly varies from a spherical shape.

A large number of protrusions could be extended from a liposome when a large number of microtubules were polymerized inside by encapsulation at a high concentration (Fig. 19.2f). The number of protrusions is two on each pole (or only one on one pole) in the case of the abovementioned Φ-shaped deformation. Thus, this increase in the number of protrusions is drastic.

---

[1] Among the proteins named as molecular motors by estimation from their amino acid sequence or three-dimensional structure, whether it works as a motor remains unknown. The BBMI is one of them. It connects the actin filament and lipid membrane, but it remains unclear whether it works as a motor. Additionally, some kind of kinesin works to shorten a microtubule by causing depolymerization from its one end rather than as a motor sliding along the microtubule.

[2] Papers that contain incorrect information and interpretation have frequently been reported in the field of biology by confusing lipid droplets with liposomes and using them for experiments. Noteworthily, scientific journals, which are widely read, are no exception.

However, the deformations of liposomes mentioned so far were difficult to reverse and/or repeat or transform into a different shape. Thus, further ingenuity was required to achieve such.

Inside the liposomes that encapsulate high concentrations of actin filaments, dense actin filaments spontaneously align in one direction to form a nematic liquid crystal. The dense actin filaments deform the liposomes into a spindle shape that was elongated along the alignment direction since the nematic liquid crystal has bending rigidity on the axis of alignment (Fig. 19.3a and b) [21]. Importantly, this liposome morphology was successfully manipulated using the dependency of nematic liquid crystal formation on filament concentration and length. Not only did the liposome itself tend to deform, but also the concentration of the internal actin filament could be decreased or increased by increasing or decreasing the water in the liposome through the change in the osmolarity of the outer solution, respectively. The aspect ratio of the spindle-shaped liposomes was changed reversibly by this simple method, as expected (Fig. 19.3a). Strong excitation light on fluorescently labeled actin filaments severs the filaments by light-induced free radicals to shorten the actin filaments and increase their number. Conversely, the spontaneous annealing of actin filaments increases their length and decreases their number after the light is turned off. Therefore, a giant liposome encapsulating fluorescent-labeled actin filaments changes its shape from spindle-like to spherical when irradiated with strong excitation light, whereas it changes its shape from spherical to spindle-like when the light is turned off. The process is reversible and can be repeated several times (Fig. 19.3b). Notably, a similar repetitive deformation was realized by the encapsulation of microtubules, not only actin filaments (Fig. 19.3c) [22]. Microtubules are highly sensitive to changes in temperature and static water pressure. Thus, by decreasing the temperature or increasing the static water pressure, the microtubules inside depolymerize, shortening the Φ-shaped liposome protrusions. Conversely, by increasing the temperature or decreasing the static water pressure, the microtubules polymerized, thereby elongating these protrusions. Additionally, the process is reversible and can be repeated until all the coencapsulated GTP, which is essential for microtubule activity, is hydrolyzed.

The elongation and shortening of the nanofilaments generate dynamic motions as described above. Microtubules and actin filaments are polymerized or depolymerized and change the shape of living cells in response to the surrounding conditions, including temperature, static and osmotic pressure, and light stimulation, through a large number of regulatory factors. Alternatively, aligned microtubules with multimers of kinesin, as well as aligned actin filaments with double-headed myosins or myosin filaments, generate dynamic motions by active sliding between the microtubules or actin filaments. Focusing on this point, these naturally derived factors involved in cell motility have been attempted to cause a continuous or oscillating flow of cytoskeletons in liposomes, by developing an artificial molecular clutch or crosslinker created from DNA fragments for the force transmission between microtubule and membrane, or for aligning microtubules, or progressing the simultaneous encapsulation of multiple factors. However, some are "hybrids" with artificial materials

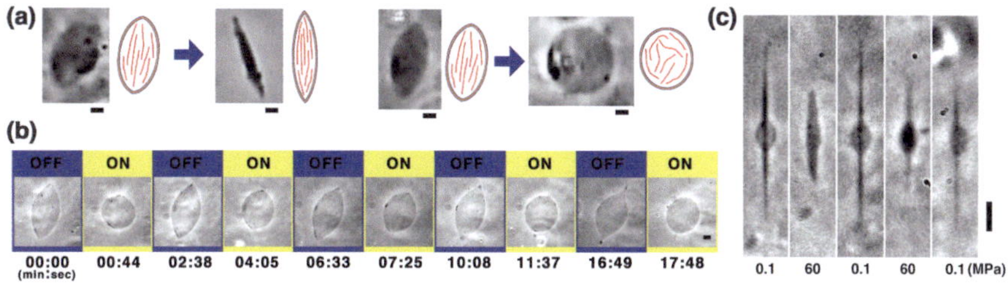

Fig. 19.3 Manipulation of the liposome shape change. Dense actin filaments take the nematic liquid crystal phase and change the shape of the encapsulating liposome to a spindle-like shape. This shape change is repeatedly manipulated by increasing or decreasing the actin filament concentration by changing the liposomal volume via altering the osmolarity (**a**), or by severing the actin filaments caused by irradiation of excitation light against fluorescently labeled actin and their spontaneous annealing (**b**). Microtubule-encapsulating liposomes that change into a bipolar shape can be repeatedly manipulated by inducing the polymerization/depolymerization of microtubules by changing the hydrostatic pressure (**c**). Scale bars indicate 5 μm. Panels (**a**), (**b**), and (**c**) are modifications of Figures previously published in Communications Physics 1, 18 (2018), and Langmuir 32, 3794–3802 (2016), respectively [21, 22]

described in the next section [23–25]. These efforts caused the liposome to successfully reproduce the amoeba-like movement, and the construction of a microtubule assembly that leads to in vitro mitotic spindle formation was achieved.

As useful natural materials, this section has focused on the cytoskeleton and molecular motor, which are involved in the motility of eukaryotes. Of course, proteins that form the motility system of prokaryotes, for example, flagellin that polymerizes to the bacterial flagellar, the protein complex for the sliding of *Mycoplasma*, or the proteins that polymerize to the spiral filaments required for the swimming of *Spiroplasma*, are also promising candidates [26–28]. These may be used not only by encapsulating inside the liposomes but also by attaching to the outside.

19.3 Morphogenesis of Artificial Cell Models Using Artificial Materials Such as Synthesized DNA, Peptide, or Polymer

We have described methods that directly use materials isolated and purified from living organisms. Additionally, artificially designed and synthesized nucleic acids (mainly for DNA origami) [29], peptides, and other polymers may be used to mimic the roles of naturally derived materials, or even to provide much better functions, involved in inducing movements, including polymerization/depolymerization, elongation/shortening, or bending/straightening [10, 30–32]. In the future, the spatiotemporal regulation of morphological changes and movement generation needs to be enabled for the artificial cell models made of liposomes that will play the role of the A-unit, when considering assembling multiple regulatory systems and simultaneously making their operation easier. In particular, when, where, and how to induce what kind of membrane shape change needs to be considered. The entire system which is composed of artificial cell models and consists of naturally derived materials, poses a risk that the difficulty derived from the complex system of life will be brought with the merits of the utilization. Hence, the advantage of using artificially synthesized materials is that the entire mechanism of action of the molecule is within the scope of understanding, or that its action is designed to be simple and clear, including ON/OFF switching.

Of course, making everything from artificially synthesized materials is not necessary. The "hybrid" made from both synthetic and natural materials will also suffice, as mentioned in the previous section. Native systems, which are life, should provide a good reference even if everything were to be made from artificial materials.

19.4 Perspectives

The system has already succeeded in causing large changes in the overall shape of liposomes, which play a role as the container of artificial cell models. Furthermore, some of them have been repeatable several times, not once.

The first indispensable issue is the molecular system for regulating the artificial cell models. In the future, assuming that the regulatory systems are assembled to create artificial cell models, it will be possible to perform more complex operations, the spatiotemporal regulation of the liposomal morphology and movement, that is, when, where, and how to induce or put back what kind of change into the membrane of the liposomes, need to be considered [21, 22].

Another concern is how to integrate various artificial cell models into one multi-cell system. In molecular robotics, developing an artificial cell model in which shape changes or movements are individually inducible is sufficient. However, the task of "constructing an array" of multiple artificial cell models is essential to respond to the demands of molecular cybernetics, which require molecular systems for sensing, processing, and transmission of information [3, 33]. Specifically, artificial cell models that play different roles, i.e., sensor, processor, and actuator, should be arranged in the correct order and direction following each role. There are characteristics that actin filaments and microtubules have a polarity, and molecular motors move in only a certain direction along them [11]. Thus, the process of overcoming the first issue mentioned in the previous paragraph, i.e., realizing "the spatiotemporal regulation," has progressed, and the task of "constructing an array" may be solved if those characteristics can be used. Rather, the regulation of the connection between the artificial cell models is more difficult. It is preferable to be tighter and looser to improve the exchange of signals and molecules and to make deformation and movement freer, respectively, of which adjustment will be the key.

In any case, the challenge to realize the higher-order regulations has just begun, and development is possible through various measures. Life acquired the cytoskeleton during evolution. It is considered the result of the convergence of many primitive dynamic protein systems that are presumed to be their "ancestors," and of the development and refinement of their functions, as well as their regulation systems, over a long period.[3] They would be certainly realized in artificial molecular robotics and molecular cybernetics using the achievements of life evolution.

References

1. Hagiya, M., et al.: Molecular robots with sensors and intelligence. Acc. Chem. Res. **47**, 1681–1690 (2014)
2. Murata, S. (ed.): Molecular Robotics. Springer Nature, London (2022)
3. Kuzuya, A., et al.: From molecular robotics to molecular cybernetics: the first step toward chemical artificial intelligence. IEEE Trans. Mol. Biol. Multi-Scale Commun. **9**, 354–363 (2023)
4. Singer, S.J., Nicolson, G.I.: The fluid mosaic model of the structure of cell membranes: cell membranes are viewed as two-dimensional solutions of oriented globular proteins and lipids. Science **175**, 720–731 (1972)
5. Bruce, A., Alexander, J., Julian, L. (eds.).: Molecular Biology of the Cell (6 th ed.). Garland Publishing, New York (2014)
6. Hohmann, T., Dehghani, F.: The Cytoskeleton-a complex interacting meshwork. Cells **8**, 362–416 (2019)
7. Pollard, T.D., Goldman, R.D.: Overview of the cytoskeleton from an evolutionary perspective. Cold Spring Harb. Perspect. Biol. **10**, a030288 (2018)
8. Fuchs, E., Cleveland, D.W.: A structural scaffolding of intermediate filaments in health and disease. Science **279**, 514–519 (1998)
9. Tanaka-Takiguchi, Y., Kinoshita, M., Takiguchi, K.: Septin-mediated uniform bracing of phospholipid membranes. Curr. Biol. **19**, 140–145 (2009)
10. Inoue, D., et al.: In Vitro Synthesis and Design of Kinesin Biomolecular Motors by Cell-Free Protein Synthesis. ACS Synth. Biol. **12**, 1624–1631 (2023)
11. Tanaka-Takiguchi, Y., et al.: The elongation and contraction of actin bundles are induced by double-headed myosins in a motor concentration-dependent manner. J. Mol. Biol. **341**, 467–476 (2004)
12. Honda, M., et al.: Morphogenesis of liposomes encapsulating actin depends on the type of actin-crosslinking. J. Mol. Biol. **287**, 293–300 (1999)
13. Takiguchi, K., et al.: Construction of cell-sized liposomes encapsulating actin and actin-cross-linking proteins. Methods Enzymol. **464**, 31–53 (2009)
14. Miyazaki, M., et al.: Cell-sized spherical confinement induces the spontaneous formation of contractile actomyosin rings in vitro. Nat. Cell Biol. **17**, 480–489 (2015)
15. Lasic, D.D.: Applications of Liposomes. In Handbook of Biological Physics: Structure and Dynamics of Membranes (Lipowsky, R., Sackmann, E., eds.). Elsevier Science B. V., Amsterdam 1A, 491–519 (1995)

[3] Cytoskeletal proteins are greatly conserved among eukaryotes. In particular, the amino acid sequence of actin is 100% homologous within mammals and aves and >90% homologous even in fish, yeast, and slime molds. This may be because they have acquired extreme multifunctionality and versatility, and have continued to maintain them. Conversely, many different proteins, which have similarities to cytoskeletal ones and so are thought to be the "ancestors," are observed in prokaryotes such as bacteria. Each plays its specific role in several tasks, such as separating DNA or distributing enzymes.

16. Miyata, H., Hotani, H.: Morphological changes in liposomes caused by polymerization of encapsulated actin and spontaneous formation of actin bundles. Proc. Natl. Acad. Sci. U.S.A. **89**, 11547–11551 (1992)
17. Hotani, H., Miyamoto, H.: Dynamic features of microtubules as visualized by dark-field microscopy. Advan. Biophys **26**, 135–156 (1990)
18. Kaneko, T., Itoh, T.J., Hotani, H.: Morphological transformation of liposomes caused by assembly of encapsulated tubulin and determination of shape by microtubule-associated proteins (MAPs). J. Mol. Biol. **284**, 1671–1681 (1998)
19. Natsume, Y., Toyota, T.: Giant vesicles containing microspheres with high volume fraction prepared by water-in-oil emulsion centrifugation. Chem. Lett. **42**, 295–297 (2013)
20. Archer, R.J., et al.: Scalable synthesis of planar macroscopic lipid-based multi-compartment structures. Langmuir **39**, 4863–4871 (2023)
21. Tanaka, S., Takiguchi, K., Hayashi, M.: Repetitive Stretching of Giant Liposomes Utilizing the Nematic Alignment of Confined Actin. Commun. Phys. **1**, 18 (2018)
22. Hayashi, M., et al.: Reversible morphological control of tubulin-encapsulating giant liposomes by hydrostatic pressure. Langmuir **32**, 3794–3802 (2016)
23. Nakajima, D., et al.: A unified purification method for actin-binding proteins using a TEV-cleavable his-strep-tag. MethodsX **13**, 102884 (2024)
24. Sato, Y. et al.: Micrometer-sized molecular robot changes its shape in response to signal molecules. Sci. Robotics **2**, aal3735 (2017)
25. Kawamata, I. et al.: Autonomous assembly and disassembly of gliding molecular robots regulated by a DNA-based molecular controller. Sci. Adv. **10**, eadn4490 (2024)
26. Kinosita, Y., Sowa, Y.: Flagellar polymorphism-dependent bacterial swimming motility in a structured environment. Biophys Physicobiol. **20**, e200024 (2023)
27. Miyata, M., Hamaguchi, T.: Prospects for the gliding mechanism of mycoplasma mobile. Curr. Opin. Microbiol. **29**, 15–21 (2016)
28. Kiyama, H. et al.: Reconstitution of a minimal motility system based on spiroplasma swimming by two bacterial actins in a synthetic minimal bacterium. Sci. Adv. **8**, eabo7490 (2022)
29. Endo, M. (ed.): DNA Origami for Molecular Robotics. Wiley, New York (2022)
30. Kurokawa, C., et al.: DNA Cytoskeleton for stabilizing artificial cells. Proc. Nat. Acad. Sci. U.S.A. **114**, 7228–7233 (2017)
31. Yingbing, L., et al.: Dramatic morphological changes in liposomes induced by peptide nanofibers reversibly polymerized and depolymerized by the photoisomerization of spiropyran. Front. Mole. Biosci. **10**, 1137885–1137885 (2023)
32. Liu, Q., Kuzuya, A., Wang, Z.-G.: supramolecular enzyme-mimicking catalysts self-assembled from peptides. iScience **26**, 10583 (2023)
33. Murata, S., et al.: Molecular cybernetics: challenges toward cellular chemical artificial intelligence. Adv. Funct. Mater. **32**, 2201866 (2022)

Chapter 20
Control of Topological Defect Generation for Deformation of Molecular Robots

Hiroki Miyazako

Abstract In the field of molecular robotics, many studies have applied cytoskeletal proteins and motor proteins for the actuation and deformation of molecular robots by confining compounds of the proteins in giant vesicles or microdroplets. Such confined cytoskeletal proteins exhibit orientational order of nematic liquid crystals, and the deformation of the molecular robots is controlled by singular points of the orientational angle, called topological defects. The aim of this chapter is to give theoretical tools for analyzing the behavior of topological defects to predict where the deformation of molecular robots starts. To this end, we introduce explicit formulae of nematic liquid crystals in arbitrary-shaped simply- and doubly-connected domains and then explain the method for determining defect positions by numerical minimization of the elastic energy of liquid crystals. Finally, we show some numerical examples of defect localization by asymmetric geometries and discuss the design principle of topographical guides for efficient control of topological defects.

20.1 Introduction

In the field of molecular robotics, many studies have applied cytoskeletal proteins and motor proteins for the actuation and deformation of molecular robots by confining compounds of the proteins in giant vesicles or microdroplets. The movement of kinesin and myosin can generate directional force onto microtubules and actin filaments, respectively, and then the generated force can be converted into the deformation of giant vesicles via mechanical interaction between the cytoskeletal proteins and boundaries. This strategy enabled the dancing of giant vesicles [1] and the self-locomotion of droplets [2], which implies the use of motor and cytoskeletal proteins is one of the fundamental mechanisms for the actuation of molecular robots. However, physical and mathematical theories for the actuation have not been well established.

H. Miyazako (✉)
Department of Information Physics and Computing, Graduate School of Information Science and Technology, The University of Tokyo, Tokyo, Japan
e-mail: hiroki_miyazako@ipc.i.u-tokyo.ac.jp

S. Murata et al. (eds.), *Molecular Robotics II*,
https://doi.org/10.1007/978-981-95-2394-8_20

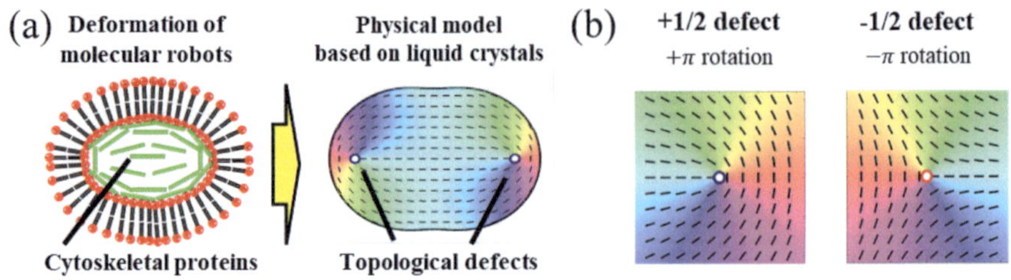

Fig. 20.1 Mathematical model for molecular robots deformed by cytoskeletal proteins. **a** Physical model of molecular robots as confined nematic liquid crystals. **b** Topological defects in nematic liquid crystals

From the physical point of view, cytoskeletal proteins have the characteristics of nematic liquid crystals, which are rod-shaped molecules and have orientational order for some directions (Fig. 20.1a). Since cytoskeletal proteins can be moved by the motor proteins, compounds of motor and cytoskeletal proteins can behave like self-driven complex fluid. Such fluidic behavior is known as "active nematics" in the fields of nonequilibrium physics and biophysics and has been widely studied in the last decade.

When considering the behaviors of active nematics as well as conventional nematic liquid crystals, topological defects have a vital role in the prediction of the alignment of nematic liquid crystals. Topological defects are singular points of nematic alignment around which direction of nematic liquid crystals cannot be defined and can be characterized by topological charges, which are obtained by dividing the total rotation angle around a defect by 2π (Fig. 20.1b). Since nematic liquid crystals are non-polar molecules, the orientational angle of nematic liquid crystals is defined between $-\pi/2$ and $\pi/2$. Thus, the minimum topological charges are $\pm 1/2$, as shown in Fig. 20.1b. In general, topological charges take a half- or full-integer value except for zero.

The topological defects with topological charges of $\pm 1/2$, which are called $\pm 1/2$ defects in later, are widely observed in many biological liquid crystals. However, the symmetries of the alignment around $\pm 1/2$ defects are quite different from each other: the alignment around a $-1/2$ defect has $2\pi/3$ rotational symmetry, whereas a $+1/2$ defect does not have such rotational symmetry. Due to the mechanical asymmetry of $+1/2$ defects, $+1/2$ defects can actively move in nematic liquid crystals, which can be viewed as driving forces of active nematics. Furthermore, topological defects can be starting points of deformation of molecular robots. When cytoskeletal proteins are polymerized in closed containers such as giant vesicles or microdroplets, the polymerized cytoskeletal proteins show nematic alignment order, and defects are generated around surfaces. Then, the shapes of the containers change from sphere- to spindle-shaped structures and the defects on the surfaces become apices of the spindles [3, 4] (Fig. 20.2a). Such spindle-shaped structures are referred to as tactoids in the field of liquid crystals. Thus, it is important to control spatial positions of topological defects in molecular robots for controlling the shapes of the robots.

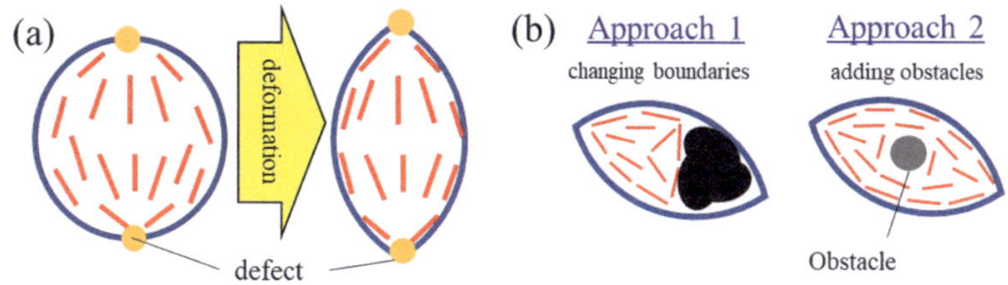

Fig. 20.2 Control of deformation of molecular robots by topological defects. **a** Deformation of molecular robots by topological defects near the surface. **b** Control strategies for deformation of molecular robots

Though there are a lot of studies for controlling topological defects using light [5], this approach would not be appropriate for controlling molecular robots at a large scale because external equipment is necessary for complex light patterns.

Topographical guides inside or outside molecular robots will be one of the promising methods for spatial control of topological defects without adding external physical fields. Especially, there are two strategies for enabling such topography-based control: (i) changing the external boundaries of molecular robots and (ii) adding obstacles like microbeads in molecular robots (Fig. 20.2b). Both approaches can induce asymmetry of nematic alignment order, which can enable spatial localization of topological defects in molecular robots. In addition, the two approaches can be modeled as the alignment of nematic liquid crystals in simply- or doubly-connected domains, which can be mathematically analyzed based on complex function theories according to our recent studies [6, 7].

The aim of this chapter is to give theoretical tools for analyzing the behavior of topological defects to predict where the deformation of molecular robots starts. To this end, we introduce explicit formulae of nematic liquid crystals in arbitrary-shaped simply- and doubly-connected domains and then explain the method for determining defect positions by numerical minimization of the elastic energy of liquid crystals [6, 7]. Then, we show some numerical examples of defect localization by asymmetric geometries and discuss the design principle of topographical guides for efficient control of topological defects.

20.2 Physical Model and Topological Constraints of Confined Nematic Liquid Crystals

In this section, we consider that cytoskeletal proteins exist in a confined two-dimensional domain Ω for the simplicity of discussion. In addition, the cytoskeletal proteins are modeled as nematic liquid crystals, which are aligned along a specific direction denoted by a unit vector $\boldsymbol{n} = \left[\cos\phi(x, y)\ \sin\phi(x, y)\right]^T$, where $\phi(x, y)$ represents the alignment angle of the nematic liquid crystals. Since $\phi(x, y)$ is defined between $-\pi/2$ and $\pi/2$, we identify $\phi(x, y) + n\pi$ with $\phi(x, y)$, where n is an integer.

When liquid crystals are confined, the alignment of the liquid crystals is distorted due to the forced alignment along the boundary of the domain $\partial\Omega$. Thus, the elastic energy of the distortion, called the Frank elastic energy, should be considered for calculating the alignment of the liquid crystals. According to the theory of liquid crystals [8], the Frank elastic energy with one-constant approximation is calculated as follows:

$$F = \frac{K}{2} \iint_{\tilde{\Omega}} |\nabla\phi(x, y)|^2 dxdy,$$

where K is the Frank elastic constant and $\tilde{\Omega}$ is a subdomain of Ω which does not have any singularities of the alignment. It follows from the variation of the Frank elastic energy F that the alignment angle $\phi(x, y)$ should satisfy the Laplace equation $\nabla^2\phi(x, y) = 0$.

Another important point of the confinement is a topological constraint related to topological charges of topological defects. Let us consider the domain Ω have M holes ($M \geq 0$) and there are N topological defects in Ω. For the k-th defect located at (x_k, y_k) ($k = 1,2, \cdots, N$), the total rotation angle around the defect is given by $2\pi q_k$, where q_k is the topological charge of the k-th defect. Since the total rotation angle along each boundary of the domain Ω is 2π, we can show that the summation of the topological charges should satisfy the following constraint according to Green's theorem:

$$\sum_{k=1}^{N} q_k = 1 - M.$$

The above constraint is critical for considering the number and the topological charges of defects in the domain. If there is no hole ($M = 0$), the summation of the topological charges is equal to 1, which means there always exist some topological defects with positive topological charges in the domain. In contrast, defects do not always exist when there is one hole ($M = 1$) because the summation of the charges should be zero. In this case, pairs of a $+1/2$ and $-1/2$ defects tend to be generated. When there are multiple holes ($M \geq 2$), there always exist some topological defects with negative topological charges in the domain. Since the holes in the domain can be considered as obstacles in molecular robots, the topographical control approach with obstacles is reasonable for the point that the number and sign of the topological charges can be controlled by adding holes.

20.3 Explicit Formula Approach for Predicting Alignment of Confined Nematic Liquid Crystals

This section devotes to explain how the alignment of confined nematic molecules can be calculated with an example for simply-connected domains ($M = 0$) based on our previous study [6, 7]. According to Sect. 20.2, the summation of the topological charges is equal to 1 in this case and the alignment angle $\phi(x, y)$ of the confined nematic liquid crystals should satisfy the following conditions:

(I) $\phi(x, y)$ is harmonic on $\tilde{\Omega}$.
(II) $\phi(x, y)$ changes by $2\pi q_k$ around the k-th defect located at $z_k = x_k + iy_k$.
(III) On the boundaries of the domain Ω, $\phi(x, y)$ is equal to the tangential angle of the boundaries.

To calculate the alignment angle $\phi(x, y)$ which satisfies the above three conditions (I)–(III), it is natural to consider the use of complex potentials because it is guaranteed that real and imaginary parts of complex potentials are always harmonic. Let us consider a conformal mapping $z = g(\zeta)$ from a unit disk $\mathbb{D} := \{ \zeta \in \mathbb{C} \mid |\zeta| = 1\}$ in ζ plane to the domain Ω in z plane, where $\mathbb{C}$ is a set of complex numbers and the coordinate of Ω is denoted by $z = x + iy$. The existence of the conformal mapping $g(\zeta)$ is guaranteed by Riemann's mapping theorem. Then, the real part of the following potential $f_\Omega(z)$ is equal to the alignment angle $\phi(x, y)$ which satisfies the conditions (I)–(III) [6]:

$$f_\Omega(z) = f_{\mathbb{D}}\left(g^{-1}(z)\right) - i\log g\prime\left(g^{-1}(z)\right),$$

where

$$f_{\mathbb{D}}(\zeta) = -\sum_{k=1}^{N} iq_k\left(\log(\zeta - \zeta_k) + \log\left(1 - \overline{\zeta_k}\zeta\right)\right) - \frac{\pi}{2},$$

and $\zeta_k = g^{-1}(z_k)(k = 1,2, \cdots, N)$. The first and second terms in $f_\Omega(z)$ represent the alignment of the liquid crystals in the unit disk $\mathbb{D}$ and the complementary term due to the conformal mapping, respectively. It is noted this explicit formula has been extended for doubly-connected domains ($M = 1$) [7].

The real part of the complex potential $f_\Omega(z)$ satisfies the conditions (I)–(III) for any defect position and arbitrary-shaped simply-connected domain. Thus, it is necessary to detect optimal defect configurations in some way. One approach is to minimize the Frank elastic energy with respect to the defect position since the value of the Frank elastic energy depends on the defect positions. By using Green's theorem, the Frank elastic energy can be explicitly calculated by the following contour integral [6]:

$$F = \frac{K}{2i}\oint_{g^{-1}(\partial\tilde{\Omega})} \frac{\partial f_\Omega(g(\zeta))}{\partial\zeta}\overline{f_\Omega(g(\zeta))}d\zeta.$$

Since the Frank elastic energy F is explicitly expressed with respect to defect positions ζ_k $(k = 1,2, \cdots, N)$, its gradient $-\partial F/\partial \overline{\zeta_k}$ is also explicitly calculated. Thus, the energy F can be numerically minimized with respect to ζ_k by the steepest descent method. After the minimization, the optimal defect configuration is obtained by calculating $z_k = g(\zeta_k)$.

20.4 Numerical Investigation of Optimal Defect Configurations and Geometries

In this section, we show numerical examples of the calculation of the optimal defect configurations explained in Sect. 20.3 and discuss design principles of geometries of boundaries and obstacles for efficient control of defect positions in molecular robots.

First, we consider how the geometry of boundaries affects the optimal defect positions for simply-connected domains [6]. Figure 20.3 shows the optimal defect positions for various boundaries, which are parametrically represented by

$$\partial\Omega_p = \left\{(1 + 0.3\cos p\theta)e^{i\theta} | \theta \in [0,2\pi)\right\}(p = 2,3).$$

When $p = 2$, the two $+1/2$ defects are located on the real axis at the final state, which means that the position of the $+ 1/2$ defects can be controlled by making the boundary asymmetric. On the contrary, two optimal defect configurations can

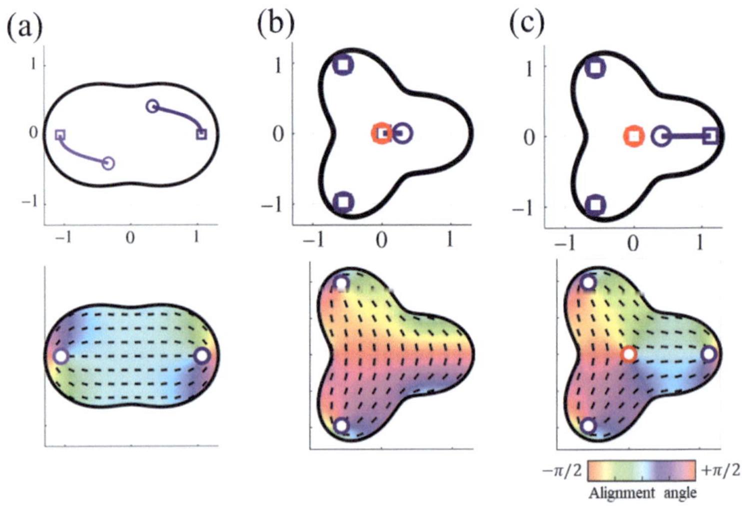

Fig. 20.3 Optimal defect configurations for Ω_p. **a** $p = 2$, **b**, **c** $p = 3$. (Top) Defect trajectories during the minimization of the Frank elastic energy. (Bottom) Final states after the minimization. Adapted with permission from [6] under CC-BY license

be realized when $p = 3$: (i) three $+1/2$ defects (Fig. 20.3b) and one $-1/2$ defects, (ii) two $+1/2$ defects (Fig. 20.3c). This result means that rotational symmetry of the boundary can realize multiple defect configurations. Thus, the rotational symmetry should be broken if the experimenters want to achieve a unique defect configuration for molecular robots.

Next, we discuss the effect of obstacles for defect configurations by considering doubly-connected domains [7]. Contrary to simply-connected domains, the summation of the topological charges should be zero in this case. Thus, we first consider whether pairs of $+1/2$ and $-1/2$ defects can stably exist or not in a concentric annulus, whose outer and inner radii are 1 and $\rho(< 1)$, respectively (Fig. 20.4). According to the numerical optimization in [7], a pair of $\pm 1/2$ defects can exist when $\rho \leq 0.15$ and the two defects are located along the same radial direction. On the other hand, the two defects are annihilated during the minimization process when $\rho \geq 0.16$. Furthermore, the defect pair can be localized by making rounded corners on inner or outer boundaries (see [7] for details). These results indicate that the size of the obstacles should be small and corners should be made on boundaries for controlling a defect pair in molecular robots.

As discussed above, numerical investigations based on explicit alignment formulas can provide design principles for geometries of molecular robots for controlling topological defects, which means that predictive design of geometries or obstacles can be achieved in advance of real experiments. Thus, the explicit formula approach can save the design and experiment costs and accelerate studies of molecular robots.

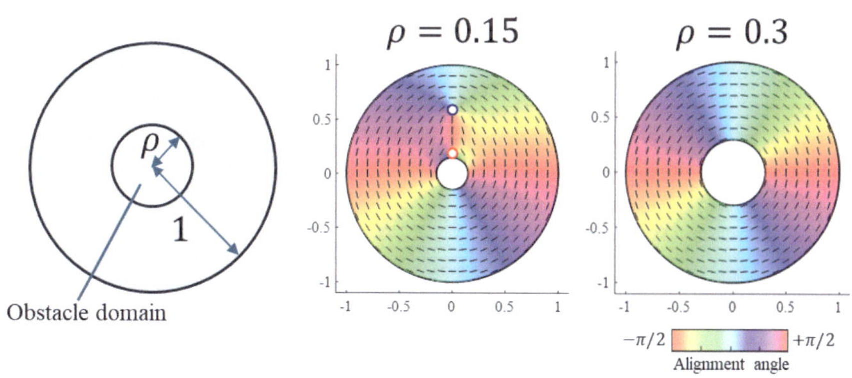

Fig. 20.4 Optimal defect configurations for concentric annulus

20.5 Conclusion

This section discussed the role and control strategies of topological defects in molecular robots. By combining nematic liquid crystals theories with complex function theories, the alignment and topological defects in molecular robots can be mathematically calculated and optimal defect configurations can be predicted by numerical minimization of the Frank elastic energy. According to the numerical examples, the defect positions can be controlled by designing various geometries of molecular robots: for example, the symmetry of the boundaries, size and shape of obstacles in molecular robots.

References

1. Sato, Y., Hiratsuka, Y., Kawamata, I., Murata, S., Nomura, M.S.: Micrometer-sized molecular robot changes its shape in response to signal molecules. Sci. Robot. **2**, eaal3735 (2017)
2. Sanchez, T., Chen, D., DeCamp, S., Heymann, M., Dogic, Z.: Spontaneous motion in hierarchically assembled active matter. Nature **491**, 431–434 (2012)
3. Tanaka, S., Takiguchi, K., Hayashi, M.: Repetitive stretching of giant liposomes utilizing the nematic alignment of confined actin. Commun. Phys. **1**, 18 (2018)
4. Edozie, B., Sahu, S., Pitta, M., Englert, A., Fermino do Rosário, C., Ross, J.L.: Self-organization of spindle-like microtubule structures. Soft matter **15**, 4797–4807 (2019)
5. Zhang, R., Redford, S.A., Ruijgrok, P.V., et al.: Spatiotemporal control of liquid crystal structure and dynamics through activity patterning. Nat. Mater. **20**, 875–882 (2021)
6. Miyazako, H., Nara, T.: Explicit calculation method for cell alignment in non-circular geometries. R. Soc. Open Sci. **92**, 11663 (2022)
7. Miyazako, H., Sakajo, T.: Defect pairs in nematic cell alignment on doubly connected domains. Proc. R. Soc. A. **480**, 20230879 (2024)
8. de Gennes, P.G., Prost, J.: The Physics of Liquid Crystals, 2nd edition. Clarendon Press (1995)

Chapter 21
Emergent Digital Technologies for Evaluating Giant Vesicles

Taro Toyota, Soichiro Hiroi, Shogo Hamada, Keita Abe, and Akihiko Konagaya

Giant vesicles (GVs) which are closed lipid bilayer membranes with diameters greater than 1 μm have garnered attention for the construction of cell-mimetic molecular robots. When evaluating the operation of functionalized GVs that act as sensors, processors, or actuators, or these complexes, the number of GVs that should be tested increases with the number of components and parts involved. This is due to the variation in the yields of their molecules and modules within the GVs and GV complexes. Consequently, functionalized GVs must be evaluated with high precision within a short timeframe. This necessitates continuous improvements in measurement methods, supported by digital technology and high-performance computing. In this section, we review and discuss emerging digital technologies for evaluating GVs.

T. Toyota (✉) · S. Hiroi
The University of Tokyo, Tokyo, Japan
e-mail: cttoyota@mail.ecc.u-tokyo.ac.jp

S. Hiroi
e-mail: hiroi-soichiro1729@g.ecc.u-tokyo.ac.jp

S. Hamada
Institute of Science Tokyo, Yokohama, Japan
e-mail: hamada@comp.isct.ac.jp

K. Abe
Tohoku University, Sendai, Japan
e-mail: keita.abe.e1@tohoku.ac.jp

A. Konagaya
Molecular Robotics Research Institute, Co., Ltd., Tokyo, Japan
e-mail: konagaya@molecular-robot.com

S. Murata et al. (eds.), *Molecular Robotics II*,
https://doi.org/10.1007/978-981-95-2394-8_21

21.1 High-Throughput Image Analysis for GVs

Recent advances in cell measurement experiments in the life sciences can be applied to analyze the properties of GVs. In microscopy, data derived from measuring the properties of individual cells within a specimen often suffer from data errors. Furthermore, recent studies in life science have investigated the properties and functions of cells that are very few in number but have high-level functions (such as stem cells and immune cells) in the body. Thus, the development of pretreatment methods to selectively isolate and enrich these cells is highly desirable. Flow cytometers and cell sorters have been invented for this purpose [1]. A flow cytometer channels a cell dispersion through a capillary, equipped with a laser and multiple photomultipliers to measure scattered light and fluorescence from each cell, and statistically analyzes the characteristics of a large number of cells based on the flow and photonic characteristics of each cell. Based on the analysis data, a cell sorter uses a gradient electric field at the tip of the capillary to select droplets containing cells with specific characteristics. Recently, imaging flow cytometry was developed to statistically analyze a large number of cells passing through a capillary using microscopy images [2].

The shape and structure of GVs have also been analyzed using flow cytometry and cell sorting [3, 4]. However, because GVs, particularly giant unilamellar vesicles (GUVs), are fragile and can deform under shear stress when flowing through capillaries, microscopic observation of GV dispersion in a specimen is desirable to maintain their integrity. Yomo's group realized a combination of fluorescence microscopy image acquisition for thousands of GUVs and statistical analysis of the population of GUVs using the R programming language [5]. In particular, they focused on numerical vectors containing the pixel values of GUV images for statistical analysis rather than individual parameters of GUV properties, such as size, circularity, and lamellarity. Their methodology produced population analysis results consistent with those obtained through imaging flow cytometry. Using phase contrast microscopy images, Zupanc et al. proposed a GV segmentation algorithm based on the Markov random field model [6, 7].

21.2 Evaluation of GV Through Image Processing and Deep Learning

Light scattering and dielectric constants measurements are often used to evaluate the size and structure of vesicles in vesicular dispersions [8]. While these measurements can provide average information about the vesicles in the entire dispersion, they are limited when dealing with multimodal data distributions. Consequently, the size, shape, and inner structure of the GVs are typically identified by incorporating fluorescent molecules into the vesicular membrane and observing the GVs with a fluorescence microscope. In fluorescence microscopy images, GVs appear as closed bright circles in cross-sections, which allows for the application of algorithms using circular Hough transformation [9, 10]. However, this technique is only effective for closed circles. If the GUV membrane undergoes pore formation, phase separation, or localized absorption of fluorescent additives, the fluorescence cross-sectional images show an arc shape of the membrane, making circular Hough transformation unsuitable (in the case of phase separation in the GUV membrane, each domain is stained with a specific fluorescent lipid dye). Riske's group developed a Python-based software (called PoET) for automatic pore edge identification and measurement of pore diameter in GUVs [11]. After producing a binarized image from fluorescence microscopy images of a GUV, PoET utilizes a custom labeling method that searches for white pixels around one spot on the membrane element in the binarized image. If all the detected elements are identical to the spot having white pixels, the pore is closed; if the elements have different labels with black pixels, an open pore is indicated. For morphological analysis of GUVs deformed from spheres, Göpfrich's group provided a Python software package to analyze whole 3D z-stack images of GUVs obtained by confocal laser scanning fluorescence microscopy [12]. This approach uses 3D meshes for quantitative analysis of physical properties, such as curvature and bending energy of GUVs.

Recently, a convolutional neural network (CNN)-assisted software that analyzes entire 3D z-stack fluorescence images of GUVs was provided from Lee's group [13]. This MATLAB-based software is composed of three programs. The input data for the program are fluorescence image z-stacks and section images captured at different GUV heights. The first program performs initial masking, followed by segmentation, to identify individual GUVs. The selection strategy is then applied to isolate GUVs of interest with desired morphologies. The selected valid z-stacks of the GUV undergo further analysis using CNN-assisted programs. The second and third programs used an experimentally trained CNN for GUV selection. The CNN is mainly structured with six different layers: the image input, convolutional, batch normalization, rectified linear unit, pooling, and a fully connected output layer. The research group successfully employed this software to autonomously evaluate lipid-protein interaction and phase separation on the vesicular membrane of the selected GUVs.

The Molecular Robot Research Institute is currently constructing 3-dimensional virtual reality (VR) images for GUVs. Confocal laser scanning fluorescence microscopy images of a GUV are introduced into the VR algorithm developed by the VR co-creation environment project of the Molecular Robot Research Institute [14]. VR images of GUVs allow for both external and internal observation of GUVs (Fig. 21.1).

21.3 Robotization of Experiment Process on GVs

In addition to automated and high-throughput evaluation programs for GVs, digital technologies are facilitating construction of GV-based cell-mimetic molecular robots through the robotization of experimental process. When digitally controlling the

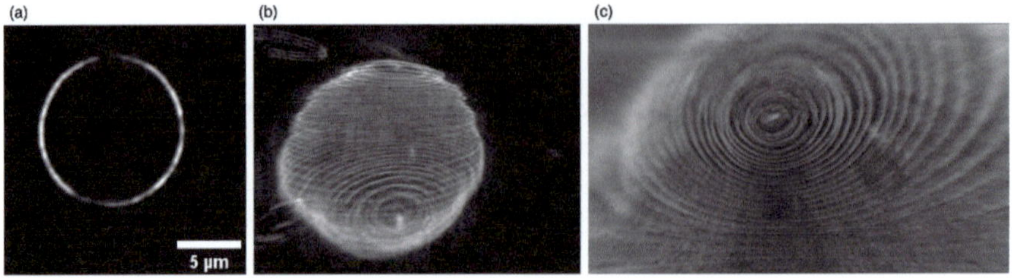

Fig. 21.1 Example of VR image of a GUV. **a** Confocal laser scanning fluorescence microscopy image of a cross-section of a GUV for construction of 3D VR image. **b** 3D VR image of the GUV from the outside. **c** Another 3D VR image of the same GUV from the inside

experimental apparatus, such as microscope, camera acquisition, pump, and valve, is linked to the analysis results of the microscopic observation data, the entire experimental process on GVs is regulated in a closed feedback loop (Fig. 21.2).

Robotization of the experimental process is advantageous for precisely elucidating GV dynamics under various stimuli. Although recent developments in microfluidic devices have expanded the scope of investigations using GVs and enabled the accurate control of experimental conditions [15], exhaustive trials to determine optimal conditions for elucidating functionalized GVs require tremendous effort. There is also a risk of increased systematic errors due to variations in the skill of the experimenters, fatigue, and other factors. Moreover, if the GV processor is designed as a memory device, the hysteresis response to multiple stimuli must be recorded under highly reproducible experimental conditions. Therefore, robotization of the experimental process on GVs is inevitable not only for evaluation but also for functionalization of GVs.

Sugiyama et al. constructed a machine-assisted experimental platform based on a microfluidic device that automatically proceeds with a sequence of protocols: (i)

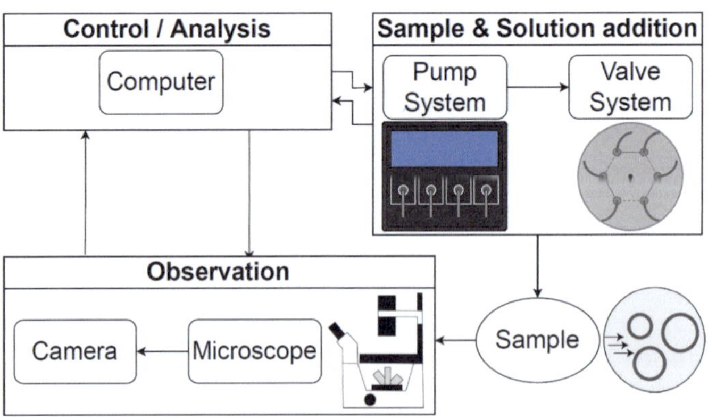

Fig. 21.2 Schematic illustration of a closed feedback loop for the entire experimental apparatus for GV evaluation using microscopy

sorting and trapping GVs, (ii) interactive control of the additive solution linked to image analysis, and (iii) refreshment of the device. This was termed Machine-Assisted, Numerous, Simultaneous, and Interactive Observation of Non-equilibrium self-assembly (MANSION). MANSIONs enable statistical analysis of GV size change under osmotic pressure to estimate membrane permeability to water [16] and water-soluble molecule accumulation in GVs against a concentration gradient across the membrane [17]. This unique accumulation phenomenon in GVs was first found by the research group and has attracted much attention as an intake function for model protocell closely related to origin of life. Various types of GVs, including those containing electrically charged phospholipids, were precisely examined. As a result, the group hypothesized that external flow induces compositional asymmetry of anionic phospholipids between the inner and outer leaflets, facilitating rapid accumulation of water-soluble molecules against the concentration gradient [18].

21.4 Internet of Things (IoT) Technologies for GV Experiments

IoT technology has been used primarily for remote control devices in homes and facilities from the outside, or to automatic device management through inter-device communication. At synchrotron radiation facilities, legal zoning requirements necessitate separating areas where radiation is permissible from those where it is not. Remote operation of experimental apparatus in radiation-permissible areas has expanded research possibilities. Furthermore, the COVID-19 pandemic of 2020–2022 significantly increased interest in digital technologies for remote communication and machine manipulation via the Internet, leading to their rapid adoption globally. Nowadays, advancement of Internet speed and data compression algorithm accelerates the development of telemedicine technologies [19] and the Internet-based co-creation space [20].

Building on this trend and motivated by the desire to enhance collaboration among researchers, we are developing an experimental environment that allows remote control of platforms for observing and manipulating functionalized GVs in microfluidic devices. This environment also automates observation protocols based on user-provided instruction scripts. For inter-device communication over the Internet and real-time handling of GUVs, a server connects computers equipped with control software for each device to the user's computer via Message Queuing Telemetry Transport (MQTT) (Fig. 21.3). The user can send instructions to each device using an easily operable Graphical User Interface created with MATLAB and access experimental data, such as microscopy images and pump pressure data, through the server.

In addition to improving the quality of the research as mentioned above, the remote operation of this experimental platform can engage non-expert researchers, companies, and citizens interested in GV research. They can use protocols developed and shared by GV researchers to obtain reproducible results without specialized skills. As

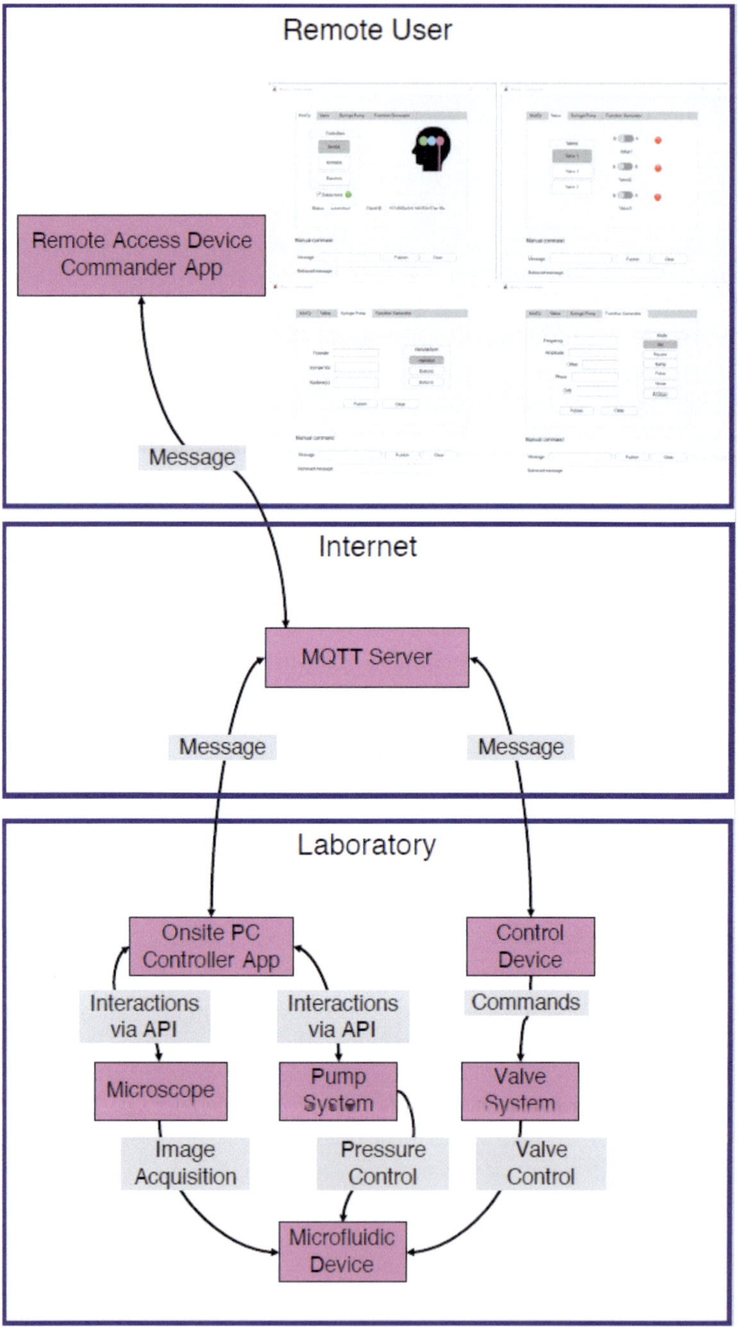

Fig. 21.3 Schematic illustration of the experimental environment for remote control for observing and manipulating GVs in microfluidic devices

an interdisciplinary field, molecular robot research benefits from the customizable nature of GVs for various applications, although the current social implementation of GV technologies remains limited. Reducing barriers to collaboration among researchers and stakeholders through sharing experiences could foster innovation. In outreach activities, operating the experimental platform from venues outside the laboratory could offer participants a unique experience and inspire new ideas for GV research in developing cell-mimetic molecular robots.

21.5 Summary

In this chapter, we have reviewed and discussed recent advances in GV experiments facilitated by digital technologies, from the analysis of microscopy images of GVs to the robotization and remote control of experimental apparatuses. These technological developments will enhance the efficiency of research on GV functionalization and serve as powerful tools for exploring new physicochemical theories of cell-mimetic molecular robots in the future when combined with molecular dynamics simulations.

References

1. Adan, A., et al.: Flow cytometry: basic principles and applications. Crit. Rev. Biotechnol. **37**, 163–176 (2017)
2. Rees, P., et al.: Imaging flow cytometry. Nat. Rev. Methods Primers **2**, 86 (2022)
3. Sato, K., et al.: Quantification of structural properties of cell-sized individual liposomes by flow cytometry. J. Biosci. Bioeng. **102**, 171–178 (2006)
4. Nishimura, K., et al.: Population analysis of structural properties of giant liposomes by flow cytometry. Langmuir **25**, 10439–10443 (2009)
5. Caliari, A., et al.: Quantification of giant unilamellar vesicle fusion products by high-throughput image analysis. Int. J. Mol. Sci. **24**, 8241 (2023)
6. Zupanc, J., Drobne, D., Ster, B.: Markov random field model for segmenting large populations of lipid vesicles from micrographs. J. Liposome Res. **21**, 315–323 (2011)
7. Zupanc, J., et al.: Lipid vesicle shape analysis from populations using light video microscopy and computer vision. PLoS ONE **9**, e113405 (2014)
8. Barenholz, Y., Lasic, D.D.: Handbook of nonmedical applications of liposomes, volume I: theory and basic sciences. CRC Press, Florida (1996)
9. Hermann, E., et al.: Automated analysis of giant unilamellar vesicles using circular Hough transformation. Bioinform. **30**, 1747–1754 (2014)
10. Van Buren, L., et al.: DisGUVery: a versatile open-source software for high-throughput image analysis of giant unilamellar vesicles. ACS Synth. Biol. **12**, 120–135 (2022)
11. Leomil, F.S., et al.: PoET: automated approach for measuring pore edge tension in giant unilamellar vesicles. Bioinform. Adv. 1, vbab037 (2021)
12. Dreher, Y., et al.: GeoV: an open-source software package for quantitative image analysis of 3D vesicle morphologies. Adv. Intell. Sys. **5**, 2300170 (2023)
13. Lee, I.H., et al.: Intelligent fluorescence image analysis of giant unilamellar vesicles using convolutional neural network. BMC Bioinform. **23**, 48 (2022)
14. Konagaya, A., et al.: Co-creation environment with cloud virtual reality and real-time artificial intelligence toward the design of molecular robots. J. Integr. Bioinform. **20**, 20220017 (2022)

15. Göpfrich, K., Platzman, I., Spatz, J.P.: Mastering complexity: towards bottom-up construction of multifunctional eukaryotic synthetic cells. Trends Biotechnol. **36**, 938–951 (2018)
16. Sugiyama, H., et al.: Perfusion chamber for observing a liposome-based cell model prepared by a water-in-oil emulsion transfer method. ACS Omega **5**, 19429–19436 (2020)
17. Sugiyama, H., et al.: Hydrodynamic accumulation of small molecules and ions into cell-sized liposomes against a concentration gradient. Commun. Chem. **3**, 32 (2020)
18. Sugiyama, H., et al.: Role of negatively charged lipids achieving rapid accumulation of water-soluble molecules and macromolecules into cell-sized liposomes against a concentration gradient. Langmuir **38**, 112–121 (2022)
19. Alenoghena, C.O., et al.: Telemedicine: a survey of telecommunication technologies, developments, and challenges. J. Sens. Actuator Netw. **12**, 20 (2023)
20. Gutmann, G., Konagaya, A.: Real-time inferencing and training of artificial neural network for adaptive latency negation in distributed virtual environments. 2020 International Congress on Human-Computer Interaction, Optimization and Robotic Applications (HORA), Ankara (2020)

Part IV
Intelligent Molecular Computation Systems

Chapter 22
DNA Circuits with Memory and Learning Capabilities

Takashi Nakakuki

Abstract We consider DNA circuits in which the circuit function alters according to conditions, unlike typical DNA circuits, which have fixed functions embedded. In a typical design, DNA circuits cannot respond normally in a time-responsive manner to repeatedly applied inputs. However, it is required that DNA circuits have a renewable mechanism to "memorize" input history and to alter circuit functions plastically and adaptively by "learning." Here, we provide some typical design strategies for a renewable mechanism intended for DNA circuits, and a classical conditioning circuit is introduced as an example of DNA circuit design with memory and learning capabilities.

22.1 Introduction

DNA computing provides the theory and technology to rationally design and implement artificial molecular reaction systems that perform the desired information processing [1]. Starting with the realization of DNA reaction systems that mimic simple logical operations, such as tweezer [2], diverse DNA circuits equipping more complex and sophisticated computational capabilities have been studied thoroughly over the years [3, 4]. Notably, DNA circuits have two main model types. The first type is a model built on the design principle of rationally combining basic modules like semiconductor integrated circuits. Typical examples include the DNA logical circuit, in which basic logical gates, such as AND, OR, and NOT, are integrated to design a combinational circuit with complicated Boolean functions [5–7]. The second type is a learning molecular model with memories such as "Neurons" in our brain of which dynamics can be adaptively and plastically altered depending on various conditions in an autonomous manner.

For plastic DNA circuits to have memory and autonomous learning capabilities, they require both a mechanism that stores repeatedly applied input stimuli as inter-

T. Nakakuki (✉)
Kyushu Institute of Technology, Fukuoka, Japan
e-mail: nakakuki@ics.kyutech.ac.jp

S. Murata et al. (eds.), *Molecular Robotics II*,
https://doi.org/10.1007/978-981-95-2394-8_22

nal states and a learning law to update the internal states toward the acquisition of the function to be learned. Regarding the learning law, the error back-propagation method, which is mainly appropriate for a Neumann-type computer, is widely used in the field of machine learning in computer science. However, the learning "algorithms" of neurons and neural networks in our brain are not still fully understood. Although many studies have been addressing to design a variety of learning circuits [8], how to implement memory and autonomous learning capabilities in DNA circuits is an unexplored research area. In this section, we consider the requirements for a DNA circuit with memory and autonomous learning capabilities while introducing the "classical conditioning circuit."

22.2 Renewability: Basic Requirements for DNA Circuits with Memory and Learning Capabilities

Definitions of Memory and Learning

As the terminologies "memory" and "learning" may have different meanings in terms of functionality depending on the problem statement considered, we first define DNA circuits with memory and learning capabilities in this section. Consider a concentration-based DNA circuit designed to have function $\mathscr{A}$ under the initial condition, where the input strand applied to the DNA circuit induces a series of chain reactions based on chemical kinetics, while inducing the output DNA strand. In this context, "information" or "signal" refers to the concentration of a DNA strand. For example, $\mathscr{A}$ might be a Boolean function, such as two-input and one-output AND logic, including cases where no input-output relationship is defined. Let us consider a trial in which the input strands are applied to the circuit and the steady-state concentration of the output strand is observed. When the circuit function autonomously alters to $\mathscr{B}$ by iterating the trials that repetitive inputs are applied to the circuit with the specific input condition and history, we say that the circuit has acquired $\mathscr{B}$ by "learning." In particular, we focus on the learning tasks in which the acquisition of the function $\mathscr{B}$ can be attained autonomously without any help of additional external/auxiliary strands applied to the circuit during the learning process, such as the teaching signal employed in the error back-propagation method in the framework of neural networks in computer science.

Functions of the molecular circuit are determined by the topology, reaction rate constants, and initial concentrations of the reaction system. Because the topology and reaction rate constants are fixed once the base sequences of all DNA strands have been designed, to change the dynamics of the reaction system by learning, the special DNA strand whose concentration increases or decreases depending on the input history is required as a "memory." It should be noted that the above definitions of memory and learning are widely applicable to the designs of both enzyme-free and enzyme-driven DNA circuits.

Renewability as a Fundamental Requirement

Having given definitions of memory and learning, now consider the essential requirements of DNA circuit design. In general, the DNA reaction system proceeds in such a way that the free energy is minimized, and therefore, all DNA strands converge to the steady state in concentration. If the DNA circuit is AND logic, then when the concentrations of the two-input strands are simultaneously applied above a threshold, the concentration of the output strand is also generated above the threshold (hereafter referred to as logically low or high for the concentration below or above the threshold level, respectively). Meanwhile, after the first-time response reaches the steady state, what if the second-time input is applied to the circuit? In a typical design of an AND circuit [5], a logically high output as the first response would not alter to the logically low level even if any inputs are applied as a second-time input.

As mentioned above, the function of the DNA circuit depends on the initial concentrations. As a result, after the response to the first-time input, as the concentrations of DNA strands change to a different distribution of concentrations, there is no guarantee that the circuit will operate normally for the second-time input. Therefore, regarding the DNA circuits with memory and learning capabilities, the fundamental requirement to respond appropriately to inputs that are repeatedly applied to the circuit for learning is necessary. To this end, a mechanism is needed to initialize the concentrations changed after input to the initial concentrations at an appropriate timing prior to the next input. In this section, this requirement is referred to as "renewability," and such a circuit is called a "renewable circuit."

Revisiting the Definition of Renewable DNA Circuit

We consider a DNA circuit described by differential equations based on chemical kinetics:

$$\dot{x} = f(x), \quad x(0) = x_0, \tag{22.1}$$

where $x = [x_1 \ x_2 \ \cdots \ x_n]^T \in R^n$ (R is the set of real numbers) is the state vector whose element x_i ($i = 1, \ldots, n$) is the concentration of DNA strands, $x_0 \in R^n$ is a vector of initial concentrations, and $f : R^n \to R^n$ is a function giving the reaction rates. In this section, for the sake of simplicity $\dot{x}$ denotes the time derivative of x. According to [9], the following concept has been introduced to define the renewability of DNA circuits.

Definition 1 (*Total fuel quantity and total waste quantity* [9]) For the DNA circuit (22.1), let I_{fuel} and I_{waste} be index sets such that

$$I_{\text{fuel}} = \{i \in N \mid x_i(0) > 0, \ i = 1, \ldots, n\}, \tag{22.2}$$
$$I_{\text{waste}} = \{i \in N \mid x_i(0) = 0, \ i = 1, \ldots, n\}, \tag{22.3}$$

where N denotes the set of natural numbers, $I_{\text{fuel}} \bigcup I_{\text{waste}} = \{1, \ldots, n\}$ and $I_{\text{fuel}} \bigcap I_{\text{waste}} = \phi$. Then, "total fuel quantity" $T_{\text{fuel}} : R \to R$ and "total waste quantity" $T_{\text{waste}} : R \to R$ are defined as follows:

$$T_{\text{fuel}}(t) = \sum_{i \in I_{\text{fuel}}} x_i(t), \tag{22.4}$$

$$T_{\text{waste}}(t) = \sum_{i \in I_{\text{waste}}} x_i(t). \tag{22.5}$$

□

Then, the renewable DNA circuit is defined as follows.

Definition 2 (*Renewable DNA circuit* [9]) For the DNA circuit (22.1), let $x(t; x_0)$ be a solution of (22.1) at time t starting from x_0. Say for any given $t_r > 0$, there exists renewable reactions $g : R^n \to R^n$ such that for $t \geq t_r$,

$$\dot{\bar{x}} = g(\bar{x}), \ \bar{x}(t_r) = x(t_r; x_0), \tag{22.6}$$

and the time $t_1 > t_r$ such that

$$\bar{T}_{\text{fuel}}(t) > T_{\text{fuel}}(t_r), \ \forall t \geq t_1, \tag{22.7}$$

where $\bar{T}_{\text{fuel}} : R \to R$ is the total fuel quantity of (22.6), calculated using the following equation:

$$\bar{T}_{\text{fuel}}(t) = \sum_{i \in I_{\text{fuel}}} \bar{x}_i(t), \ t \geq t_r. \tag{22.8}$$

Then, the DNA circuit is referred to as "renewable" with respect to x_0. Moreover, for a given time $t_e > t_1$, which is typically the end time of the renewable reaction (22.6), the efficiency of renewability is defined as follows:

$$r_e = 1 - \frac{T_{\text{fuel}}(0) - \bar{T}_{\text{fuel}}(t_e)}{T_{\text{fuel}}(0) - T_{\text{fuel}}(t_r)}. \tag{22.9}$$

□

In an ideal renewable mechanism, the efficiency r_e approaches 1.

22.3 How to Design a Renewable DNA Circuit

Strategies for Designing Renewable Circuits

When the input strands are applied to a DNA circuit under an initial condition, the DNA reaction system transitions to a new steady state in which the free energy is minimized. This means that it is impossible for the concentrations of the DNA strands to return autonomously to their initial concentrations after the response, and therefore, a "renewable" mechanism is required. The design methodologies of renewable DNA circuits can be categorized into different strategies.

The first strategy is to change the (steady state) concentrations by additionally applying "auxiliary" strands to the circuit at time t_r (see Definition 2) to direct the reaction flows in the initialization direction. The second strategy is to irradiate the DNA circuits with light at a specific wavelength at time t_r to change the reaction rate constants in order to direct the reaction flows in the initialization direction. This method is known as photoresponsive control. The third strategy, which is separate from Definition 2, is to make the circuit function correspond to transient or quasi-steady-state characteristics rather than to the steady state of the response. Although this strategy includes a variety of mechanisms to realize renewable DNA circuits, we introduce a renewable design with degradative enzymes.

Renewable Mechanism by Auxiliary Strands

The method of initializing circuit functions by additionally administering an auxiliary/external strand into a DNA circuit after responses dates back to 2000 [2], where the transition of open and closed structures of the DNA tweezer was controlled by the addition of auxiliary strands named "fuel DNA." Although concepts such as "renewable" were not explicitly used in this study, it can be said that it pioneered the study of renewable DNA circuits. Another example includes the renewable design of the seesaw gate [5] by introducing the "extractor" strands for initialization, where an elaborate mechanism with a hairpin-based motif was proposed to restore the concentrations of double-stranded DNA gates for driving the seesaw architecture by additionally administering the extractor strand after the response [10]. Similarly, a renewable design of the seesaw architecture was proposed in [11], which successfully initialized some logic gates, such as AND, OR, and a combinational circuit with one AND and two OR gates by additionally administering "pullback" strands for initialization. It should be noted that although the initialization by auxiliary strands at the appropriate timing t_r is an effective method, the "waste" strands will accumulate while repeating the administration of the inputs for the main response and the auxiliary strands for initialization.

Renewable Mechanism by Photo-Responsive Control

It is possible to initialize the DNA circuits by utilizing "photoisomerization" of photoresponsive molecules to control the reaction flow of DNA strand displacement reactions [12], and various photoresponsive molecules have been developed [13, 14]. In the following, a renewable design using a photoresponsive molecule, azobenzene, is introduced as a typical example.

Azobenzene has two structures that can be switched by photoisomerization: trans and cis forms under blue light (BL) and ultraviolet (UV) irradiation, respectively (Fig. 22.1a). As the cis forms destabilize the double-helix structure of double-stranded DNA (Fig. 22.1b), the binding affinity can be modulated by UV irradiation by applying azobenzene modification within DNA base sequences [13]. For example, in Fig. 22.1c, the azobenzene modification is applied to the 3' end toehold domains of the upper and lower strands, denoted by t^* and s, respectively. Under BL irradiation,

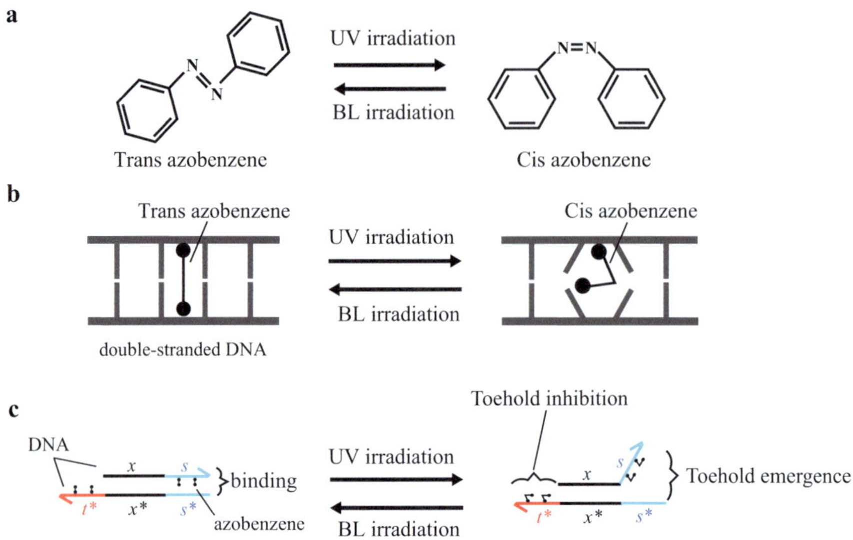

Fig. 22.1 Photoisomerization of azobenzene and its effect. **a** Photoisomerization of azobenzene; **b** conformation change of azobenzene between trans (left) and cis (right) structures in double-stranded DNA; and **c** effect of azobenzene modification on the DNA strand. The black line, x, indicates a long/recognition domain (typically 20 $\sim$ nt), while the non-black-colored lines, t and s, indicate short/toehold domains (typically 3–6 nt). In addition, x and x^* and s and s^* each represent complementary base sequences

as the trans-azobenzene does not affect the stability of the double-helix structure, domains x and s bind with affinity determined by sequence complementarity. In addition, the toehold-mediated strand displacement reactions starting from the toehold domain t^* are accessible. Meanwhile, under UV irradiation, the conformation change to the cis form by photoisomerization weakens the binding affinity of domain s, making the toehold-mediated strand displacement starting from s^* more accessible (referred to as "toehold emergence" in [9]), while that starting from t^* is less accessible (referred to as "toehold inhibition").

Then, a renewable DNA strand displacement reaction can be designed as shown in Fig. 22.2 [9]: In the initial state (State 1), applying the input strand to the gate strand triggers the DNA strand displacement reaction starting from the toehold t^*, which is the main reaction generating the output and waste strands. After the output response in State 2, by irradiating UV light, azobenzene-modified toeholds provide the starting domain for renewable reactions (State 3), where toeholds t^* and s^* become toehold inhibition and toehold emergence, respectively. As a result, under UV irradiation, the concentrations of input and gate strands are recovered in State 4 because the DNA strand displacement reaction by the output and waste strands is much more dominant than that of the input and gate strands. Finally, the BL irradiation can complete the initialization of the reaction system.

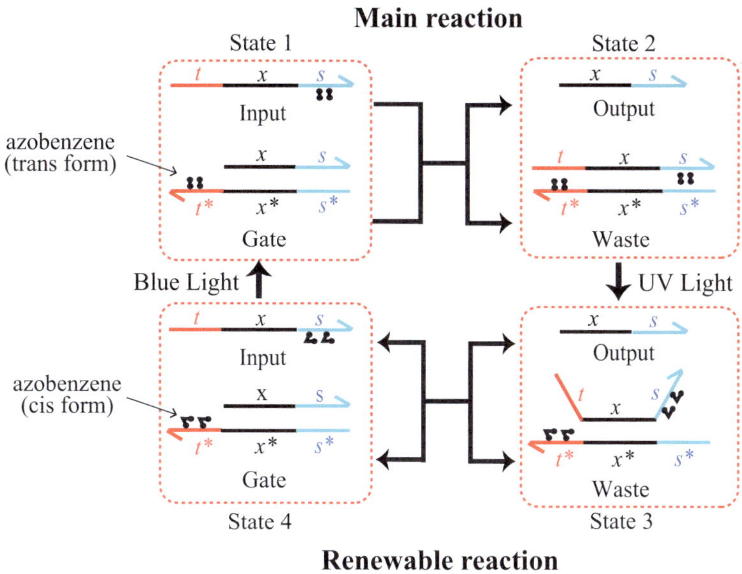

Fig. 22.2 Renewable DNA strand displacement reaction by photoresponsive control. The reaction in which the input strand given in State 1 is replaced by the output strand in State 2 via binding with the gate strand is called the DNA strand replacement reaction. By using the photoresponsive control, State 2 returns to State1 via State 3 and State 4

Renewable Mechanism by Degradative Enzyme

As the administration of the input strand to the DNA circuit triggers a series of DNA reactions, in principle, it is expected that the concentrations of the DNA strands in the circuit tend to recover to their initial concentrations gradually by removing the input strand from the reaction field. For example, in Ref. [15], an RNA strand was employed as an input strand, and an autonomous renewable mechanism was proposed using the degradative enzyme RNase H endoribonuclease (Fig. 22.3a), where the output amplitude, not steady-state concentration, was regarded as the output of the circuit because the output response became transient in the presence of the degradative enzyme. It should be noted that a renewable mechanism using degradative enzymes can also be accomplished by the degradation of the input strand. For example, in Fig. 22.3b, the input strand can be either RNA or DNA by selecting the appropriate degradative enzyme.

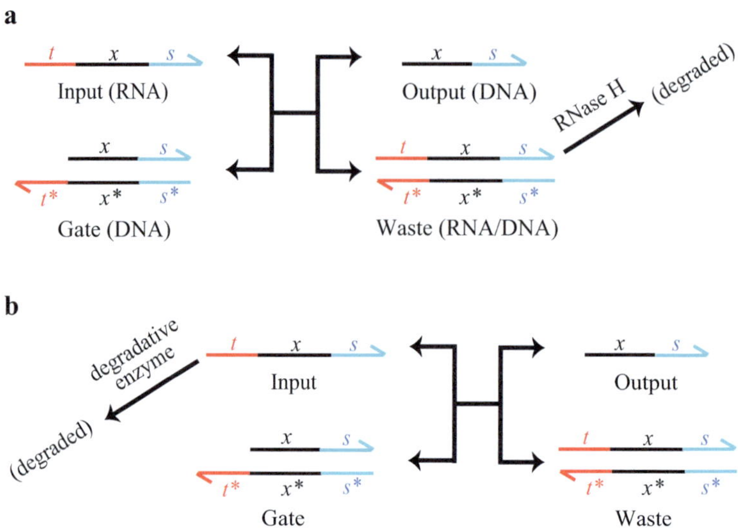

Fig. 22.3 Renewable mechanism by dissipative control with degradative enzyme for the waste (**a**) and input (**b**) strands

22.4 DNA Circuits that Acquire Classical Conditioning

In this section, we introduce the "classical conditioning circuit" as an example of a DNA circuit with memory and autonomous learning capabilities [16].

Consider the two-input and one-output classical conditioning circuit named "Pavlovian circuit" in Fig. 22.4a, where the inputs I_1 and I_2 conceptually correspond to "feed-related" and "bell-related" inputs in the context of Pavlovian conditioning, respectively. Under the pre-learning condition, only input I_1 can induce the "saliva" output O, as shown in Fig. 22.4b, which is so-called YES logic in the field of computer science. However, according to a conceptual viewpoint obtained from the Pavlovian dog experiment [17], in the classical conditioning mechanism, the input I_2 also can induce the output response when training for learning takes place, in which both inputs I_1 and I_2 are simultaneously applied to the circuit repeatedly. Namely, under the post-learning condition, the circuit function is altered to OR logic, as shown in Fig. 22.4c. Moreover, the circuit should be plastic; if the only input I_2 is applied to the circuit repeatedly, the acquired OR logic should be lost gradually so that finally the circuit function returns to the YES logic. In [16], the circuit with such learning acquisition and forgetting was defined as "classical conditioning circuit."

Figure 22.5 shows a schematic view of the classical conditioning circuit designed based on the DNA strand displacement reactions, where the single-stranded DNAs I_1 and I_2 are "feed-related" and "bell-related" inputs, respectively, and the single-stranded DNA O is the "saliva" output. The double-stranded DNAs M_1 and M_2 are defined as "memory" gates and have an important role in connecting pathways

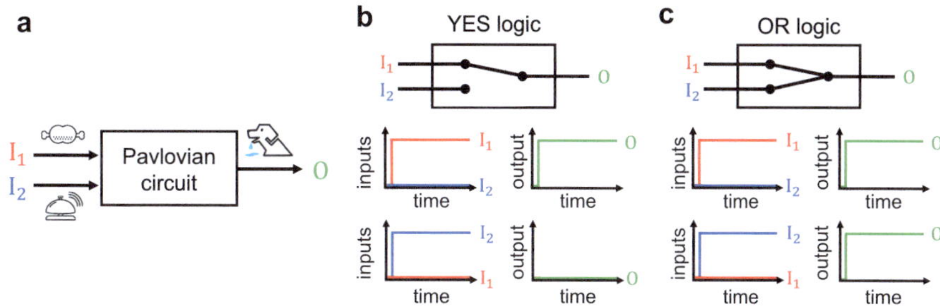

Fig. 22.4 Classical conditioning circuit [16]. **a** The classical conditioning circuit is a logical circuit with two inputs, I_1 and I_2, and an output, O. **b** Under the pre-learning condition, the circuit functions as a YES logic, where the output O responds to only I_1, but not to I_2. **c** Under the post-learning condition, the circuit functions as an OR logic, where the output O responds not only to I_1 but also to I_2

between inputs and outputs. Under the pre-learning condition, because the memory gate M_1 associated with the "feed-related" input is allocated an initial concentration, the output O is induced upon I_1 administration while generating the intermediate strands W_1 and Y_1. Meanwhile, because the memory gate M_2 associated with the "bell-related" input has no initial concentration, no output O is induced upon I_2 administration, whereas the intermediate strands W_2 and Y_{2p} are generated by strand displacement reaction with gate M_{2p}. Thus, what if both I_1 and I_2 are simultaneously given to the circuit? The output O induced by the input I_1 can associate with the "reward" strand R_2; as a result, the intermediate strand Y_2 is generated by the strand displacement reaction. In this simultaneous input case, as the intermediate strand W_2 is also generated by input I_2, the strand displacement reaction between Y_2 and W_2 can generate the memory gate M_2, implying that the memory gate is accumulated. Here, we call the concentration accumulation of the memory gate "updated."

Figure 22.6 shows the simulation results, demonstrating how the responses of the circuit alter by learning, where the responses (M_1, M_2, and O) to three consecutive inputs (I_1 and I_2) are indicated. It should be noted that because the circuit incorporates the renewable mechanism using a degradative enzyme, the concentrations of inputs and output strands return to their initial concentrations after responses. As the bell I_2 as the first input and meat I_1 as the second input do not satisfy the learning condition (simultaneous inputs), the bell I_2 as the third input does not induce the output (upper and bottom panels in Fig. 22.6a). This is because no update (concentration accumulation) of memory gate M_2 occurs (middle panel in Fig. 22.6a). Meanwhile, as demonstrated in Fig. 22.6b, when the simultaneous inputs I_1 and I_2, which satisfy the learning condition, are given as the second input, the bell I_2 as the third input induces the output response because an update of memory gate M_2 occurs. It should be noted that repeatedly providing simultaneous inputs that satisfy the learning condition increases the concentration of memory gate M_2 more in stages. It was also reported that after acquiring the OR logic after learning, repeated application of only

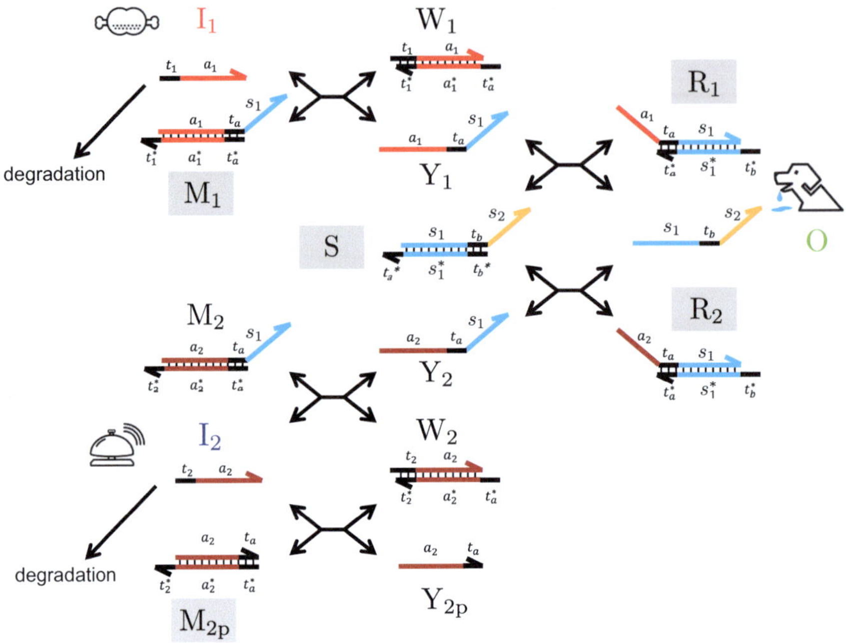

Fig. 22.5 Schematic view of the classical conditioning circuit [16]. The reactions are described with six single-stranded DNAs (Y_1, Y_2, Y_1, Y_2, Y_{2p}, and O) and eight double-stranded DNAs (M_1, M_2, M_{2p}, W_1, W_2, S, R_1, and R_2). The labels t_1, t_2, t_a, and t_b are the toehold domains, and the labels a_1, a_2, s_1, and s_2 are the recognition domains. The bi-directional arrows connecting the DNA strands denote reversible DNA strand displacement reactions. The double-stranded DNAs that have initial concentrations under the pre-learning condition are denoted by shadowed, square boxes

the I_2 input caused forgetting of the OR logic and a return to the YES logic. Additionally, by introducing the renewable mechanism by photoresponsive control with azobenzene, further performance improvement of the circuit can be achieved.

22.5 Summary and Further Readings

This section provided an overview of the design of DNA circuits with memory and learning capabilities. In particular, we focused on DNA circuits that possess an autonomous learning mechanism, rather than learning circuits that assume offline weight updating. As DNA circuits must respond to repeated inputs in a time-responsive manner to perform the learning task, how to incorporate the renewable mechanism is a fundamental design issue. Then, we outlined three representative methods as basic strategies for the renewable mechanism. Although the DNA circuit that acquires conditional reflections was introduced as an example of the circuit design, plastic adaptation, which is the essence of memory and learning capabilities, has been extensively studied in the field of DNA computing [18–21].

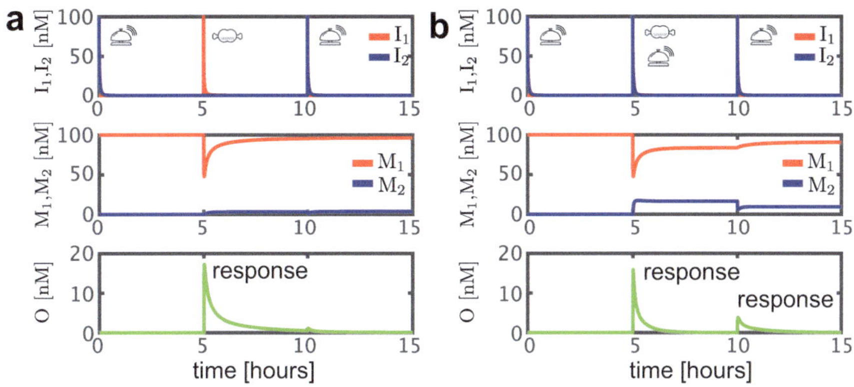

Fig. 22.6 Acquiring a conditioned reflex depending on input patterns [16]. The upper, middle, and lower panels indicate the time-course data of inputs, memory gates, and outputs, respectively. The three repetitive input patterns are applied to the circuit

References

1. Srinivas, N., Parkin, J., Seelig, G., Winfree, E., Soloveichik, D.: Enzyme-free nucleic acid dynamical systems. Science **358**, eaal2052 (2017)
2. Yurke, B., Turberfield, A., Mills, A., Simmel, F., Neumann, J.: A DNA-fuelled molecular machine made of DNA. Nature **406**, 605–608 (2000)
3. Stojanovic, M., Mitchell, T., Stefanovic, D.: Deoxyribozyme-based logic gates. J. Am. Chem. Soc. **124**, 3555–3561 (2002)
4. Seelig, G., Soloveichik, D., Zhang, D., Winfree, E.: Enzyme-free nucleic acid logic circuits. Science **314**, 1585–1588 (2006)
5. Qian, L., Winfree, E.: Scaling up digital circuit computation with DNA strand displacement cascades. Science **332**, 1196–1201 (2011)
6. Song, T., Garg, S., Mokhtar, R., Bui, H., Reif, J.: Analog computation by DNA strand displacement circuits. ACS Synt. Biol. **5**, 898–912 (2016)
7. Chatterjee, G., Dalchau, N., Muscat, R.A., Phillips, A., Seelig, G.: A spatially localized architecture for fast and modular DNA computing. Nat. Nanotechnol. **12**(9), 920–927 (2017)
8. Okumura, S., Gines, G., Lobato-Dauzier, N., Baccouche, A., Deteix, R., Fujii, T., Rondelez, Y., Genot, A.: Nonlinear decision-making with enzymatic neural networks. Nature **610**, 496–501 (2022)
9. Tamba, M., Murayama, K., Asanuma, H., Nakakuki, T.: Renewable DNA proportional-integral controller with photoresponsive molecules. Micromachines **13** (2022)
10. Eshra, A., Shah, S., Song, T., Reif, J.: Renewable DNA hairpin-based logic circuits. IEEE Trans. Nanotechnol. **18**, 252–259 (2019)
11. Garg, S., Shah, S., Bui. H., Song, T., Mokhtar, R., Reif, J.: Renewable time-responsive DNA circuits. Small e1801470 (2018)
12. Song, X., Eshra, A., Dwyer, C., Reif, J.: Renewable DNA seesaw logic circuits enabled by photoregulation of toehold-mediated strand displacement. RSC Adv. **7**, 28130–28144 (2017)
13. Asanuma, H., Liang, X., Nishioka, H., Matsunaga, D., Liu, M., Komiyama, M.: Synthesis of azobenzene-tethered DNA for reversible photo-regulation of DNA functions: hybridization and transcription. Nat. Protoc. **2**, 203–212 (2007)
14. Qiu, Z., Nakamura, S., Fujimoto, K.: Reversible photo-cross-linking of the GCN4 peptide containing 3-cyanovinylcarbazole amino acid to double-stranded DNA Org. Biomol. Chem. **17**, 6277–6283 (2019)

15. Del Grosso, E, Irmisch, P., Gentile, S., Prins, L.J., Seidel, R., Ricci, F.: Dissipative control over the toehold-mediated DNA strand displacement reaction. Angew. Chem., Int. Ed. **61** (2022)
16. Nakakuki, T., Toyonari, M., Aso, K., Murayama, K., Asanuma, H., de Greef, T.: DNA reaction system that acquires classical conditioning. ACS Synth. Biol. **13** (2024)
17. Pavlov, I.P.: Conditioned reflexes: an investigation of the physiological activity of the cerebral cortex. Ann. Neurosci. **17** (2010)
18. Lakin, M.R., Stefanovic, D.: Supervised learning in adaptive DNA strand displacement networks. ACS Synth. Biol. **5**, 885–897 (2016)
19. Cherry, K.M., Qian, L.: Scaling up molecular pattern recognition with DNA-based winner-take-all neural networks. Nature **559**, 370–376 (2018)
20. Zou, C., Wei, X., Zhang, Q., Zhou, C.A.: Novel adaptive linear neuron based on DNA strand displacement reaction network. IEEE/ACM Trans. Comput. Biol. Bioinf. **19**, 1424–1434 (2020)
21. Zou, C., Zhang, Q., Zhou, C., Cao, W.: A nonlinear neural network based on an analog DNA toehold mediated strand displacement reaction circuit. Nanoscale **14**, 6585–6599 (2022)

Chapter 23
Development of Artificial Reaction–Diffusion System Toward Chemical Computation for Molecular Cybernetics from Theoretical and Experimental Aspects

Ibuki Kawamata and Keita Abe

23.1 Theory of Reaction–Diffusion-Based Computation

23.1.1 Introduction

Pattern formation is an interesting yet important phenomenon in Biology for cell development, animal coloration, colony growth, and so on [1–3]. The physics behind such pattern formations is called reaction–diffusion, where molecules can regulate the production and degradation of other molecules while diffusing over given space with molecule-specific speeds [4]. Although Biology has their own purpose for pattern formation, mimicking such reaction–diffusion-based phenomena with artificially synthesized molecules has been tested [5–7]. For example, 2D and 3D patterns such as dots, stripes, or spirals have been demonstrated using chemical oscillations [8–10]. Most of these processes are explained by Turing patterns, where two types of chemical species, activator and inhibitor, are main factors to form the patterns [11] Chemically formed patterns are not just beautiful, but it would be possible to be used as a scaffold for engineering a smart material that can spontaneously shape its structure.

For that direction, it is required to control the fundamental parameters of molecules such as reaction and diffusion speeds in more detail, which can show Turing

I. Kawamata (✉)
Kyoto University, Kyoto, Japan
e-mail: kawamata.ibuki.8p@kyoto-u.ac.jp

K. Abe
Tohoku University, Miyagi, Japan
e-mail: keita.abe.e1@tohoku.ac.jp

S. Murata et al. (eds.), *Molecular Robotics II*,
https://doi.org/10.1007/978-981-95-2394-8_23

patterns and more. One of the methodologies to control the parameters is employing programmable materials such as DNA that can regulate their interactions and molecular weights by defining a proper base sequence [12, 13]. For instance, spiral waves, ring pattern, stable gradient of DNA concentration have been demonstrated using well-defined initial boundary shapes [14–16]. Functions that cannot be achieved by only DNA could be also experimentally realized using non-DNA factors such as enzymes, microbeads, hydrogel, and light [17, 18].

Along with the technique for the experimental realization of DNA reaction–diffusion systems, theoretical support to program the patterns to be formed is necessary. In this section, DNA reaction–diffusion is discussed in both theoretical and experimental aspects.

23.1.2 Cellular Automata

To theoretically explain the pattern formation, although a continuous model such as Turing dynamics is one of the common methods, it is sometimes convenient to use a discrete model for the sake of simplicity. Among such discrete models, cellular automaton is a computational model capable of representing systems that develop in both space and time [19]. The concept of cellular automata has been successfully exploited for various applications such as animal skin patterns, reaction–diffusion systems, and molecular computation [20–22]. To represent a system of interest, cellular automata use discretized time and space where space is divided into cells that follow simplified state transition rules [23] (Fig. 23.1).

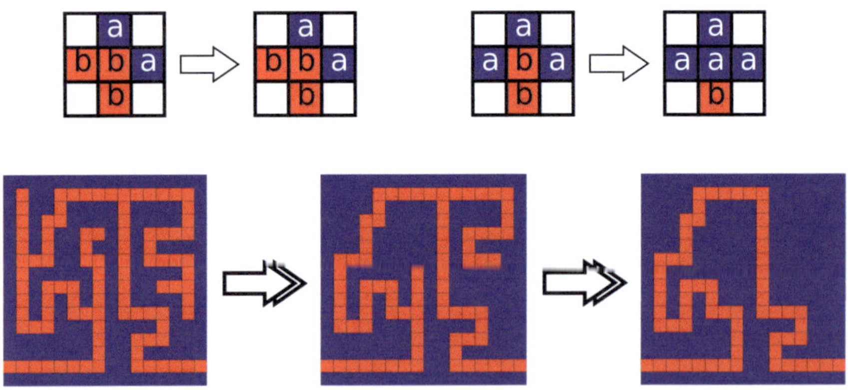

Fig. 23.1 Example of cellular automaton that solves maze. The state 'a' (blue) represents walls, and the state 'b' (red) represents a possible path. (Top) As a transition rule, a cell in the center counts the number of neighboring cells that are in state 'a' and changes its state into 'a' if the number is bigger or equal to three. (Bottom) From a given initial configuration, the state transition occurs until the maze is solved. Image from reference [23]

In standard cellular automata, the state transitions of cells are all synchronized, where cells can change their state all at once. Moreover, a cell can distinguish the neighboring cells in north, south, east, and west directions, so that combinatorially different state transition rules can be defined. Thanks to such strict mathematical definition, one of the advantages of cellular automata is that computational capabilities of represented models can be determined upon the discussion of theoretical complexity. For example, a 1D cellular automaton (also known as elementary cellular automaton) with a simple transition rule specified as rule-110) is proved to be computationally universal [24]. Because cellular automaton is a simple but powerful tool for representation and analysis, it has been used in the fields of molecular robotics and molecular cybernetics [25].

23.1.3 Concept of Gellular Automata

As a molecular realization of cellular automata, we have introduced a concept of gellular automata (Fig. 23.2) [26, 27]. Theoretical framework of the gellular automata imposes several restrictions to cellular automata in consideration of experimental implementation by means of molecules, molecular diffusion, and chemical interaction.

In gellular automata, a solution in a container made of hydrogel is regarded as a cell of cellular automata. Having multiple containers in a regular pattern can define a grid of cells. Each cell contains a large molecule (such as polymers) to represent the state. A large molecule cannot diffuse from a cell so that a state of the cell remains unchanged until the molecules diffuse from the neighboring cells. Unlike the large molecules, the diffusive small molecules (such as DNA) can move from a cell to neighboring cells, corresponding to cell–cell communication. After the communication, chemical reaction between large and small molecules corresponds

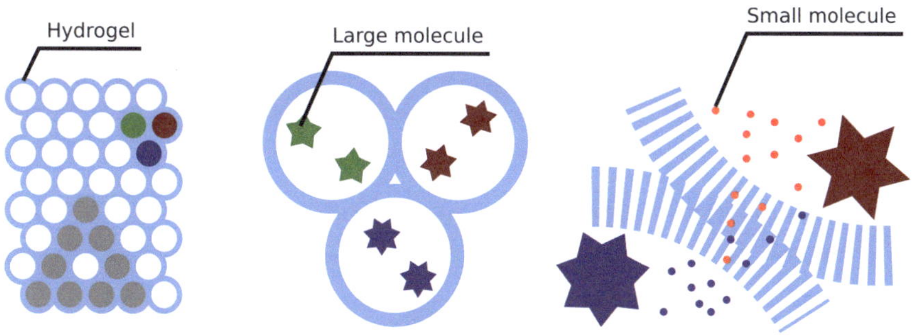

Fig. 23.2 Concept image if gellular automata. Each circle (light blue) represents a cell encapsulated with a shell of hydrogel. The state of each cell (colored circle) is recognized by large molecules (colored stars) inside the cell. The cell–cell communications are done by small molecules that move from one cell to neighboring cells across the hydrogel. Image from reference [27]

to the state transition in a cell. As a result of experimental implementation, the gellular automaton is subject to strong physical limitations, which can be modeled as non-synchronized update, totalistic rule, and non-camouflage property [28], as explained one by one in the following paragraph.

Unlike the standard cellular automata, the state transitions of gellular automata will not take place in a synchronized manner. In addition to that, a cell cannot distinguish neighboring cells because the same small molecules can come from different directions, which is called totalistic property. Moreover, the non-camouflage property means that a cell cannot distinguish whether small molecules come from the neighboring cells or the cell itself because the location information is lost by the diffusion of molecules. Even in such strong limitations, we have proved that gellular automaton with non-camouflage property is computational universal by constructing a computational model called packet system [28].

23.1.4 Pattern Formation as a Result of Chemical Reaction

What kind of pattern can we form using the framework of gellular automata? For example, consider the following chemical reactions [27].

$$A + B + S \xrightarrow{k_1} Z + M$$

$$B + C + S \xrightarrow{k_1} X + M$$

$$C + A + S \xrightarrow{k_1} Y + M$$

$$X + Y \xrightarrow{k_2} \phi$$

$$Y + Z \xrightarrow{k_2} \phi$$

$$Z + X \xrightarrow{k_2} \phi$$

$$M + T \xrightarrow{k_3} \phi$$

$$M + U \xrightarrow{k_4} M + G$$

$$G + S \xrightarrow{k_5} G$$

$$X + G \xrightarrow{k_6} X + A + P$$

$$Y + G \xrightarrow{k_6} Y + B + Q$$

$$Z + G \xrightarrow{k_6} Z + C + R,$$

where letters A, B, … correspond to molecular species and ϕ denotes an empty set. In each line, molecules at the left-hand side react to each other according to the rate constant (denoted by k with subscript number) next to the arrow and change to the new molecules in the right-hand side. Here, we employ the values of rate constants as k1 = k4 = k6 = 10, k2 = 1.0 × 105, k3 = 1.0 × 103, k5 = 1.0 in arbitrary units.

To regard the chemical reactions as gellular automaton, we assume that space is discretized in a hexagonal lattice. To represent the state of each cell in the lattice, concentration of all the molecules are used. In this specific model, molecules A, B, and C are small ones so that they can pass through the gel by diffusion. In particular, the diffusion coefficient of them is 1.0 × 10–4 in arbitrary units. Other molecules are large molecules that cannot diffuse and are stopped inside a cell. As an initial state of gellular automaton, consider that all the cells, except for two special ones, possess molecules S, T, and U with concentrations 1/6, 1/20, and 1 in arbitrary units, respectively. The special cells trigger whole reactions, where X or Y molecules with concentration of 1/6 are placed with 1-cell gap. The special cells also contain 1/6 of M and 1 of U. The concentrations of other unspecified molecules are zero at the beginning.

The setup above will impose the cell to have a different state from neighboring cells, resulting in a checkerboard-like pattern with 3 colors in hexagonal lattice when the concentration of P, Q, and R is visualized in red, green, and blue, respectively. To demonstrate the pattern, we solved the differential equations and obtained a spontaneous regular pattern (Fig. 23.3).

23.1.5 Gellular Automaton Modeled Using DNA

The above example is based on ideal chemical species and reactions with unrealistic kinetics and chemical reactions. To make the framework of gellular automata more realistic, we introduce another concrete example using DNA as chemical species (Fig. 23.4) [23]. In this system, we employed the mathematical model of a physically implemented DNA bistable system [29]. In the bistable system, the state of a dynamic system eventually becomes either A or B depending on both the initial state and external stimuli. We regard the external stimuli as a diffusion of molecules from neighboring cells.

Like the example in the previously mentioned cellular automaton, we regard the state A as a wall, and B as a possible path, where transition from path to wall

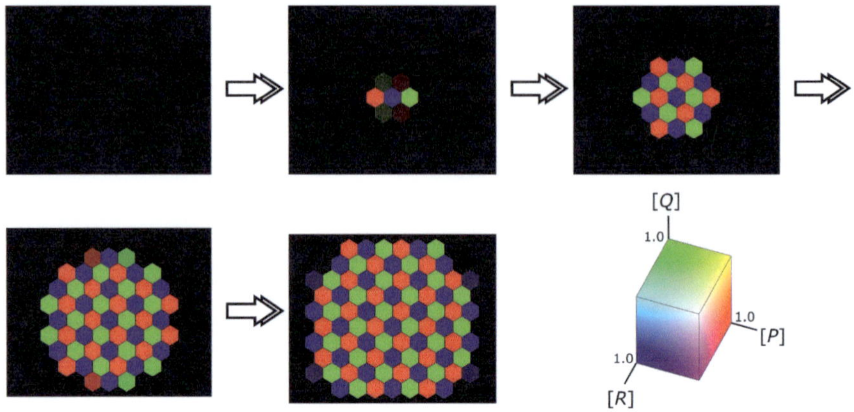

Fig. 23.3 Possible implementation of gellular automata by chemical reaction network. Following the equations, a regular pattern appears as a result of reaction–diffusion among cells. The color represents the concentration of chemical species P, Q, and R. Image from reference [27]

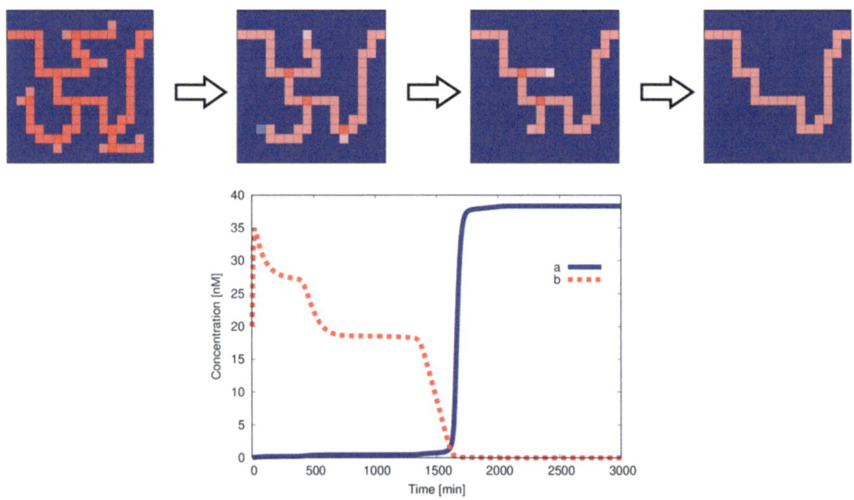

Fig. 23.4 Possible implementation for maze solving gellular automaton using DNA chemical reaction network. From the initial state, reaction–diffusion reactions take place to leave the answer of the given maze. The blue and red colors represent the concentration of chemical species A and B, respectively. The graph on the right represents the continuous concentration development of a specific cell. Image from reference [23]

takes place to solve the maze. The chemical reactions of this system result in partial differential equations with 26 variables and 19 parameters, details of which can be found in the literature [23]. As the actual system is driven by enzymatic reactions, we use Michaelis–Menten formulas for equations. As shown in the figure, although the diffusion and reaction take place in continuous space, time, and concentration, the resultant phenomenon is discretized like a cellular automaton as expected.

The powerful aspect of the maze-solving gellular automaton is not just the conceptually feasible implementation by DNA, but also the extension capability such as employing hexagonal lattice or conducting a computation based on simple state transitions, which are more computationally meaningful than solving a maze.

23.2 Experiment of DNA-Based Artificial Reaction–Diffusion System

23.2.1 DNA-Based Cellular Automata

The concept of linking gel automata with DNA computing was demonstrated in experimental systems using hydrogels [30]. A matrix built by hydrogel is used as cells, and the state transitions are represented by reactions in the prepared DNA solution within those cells. The state of a cell is propagated to neighboring cells as a result of the diffusion of DNA, serving as the trigger for the next state transition. Additionally, DNA bound to a polymer as a molecule that remains within the cell (Anchor DNA) is introduced. The polymer part of the Anchor DNA cannot diffuse to the other cells because of the hydrogel matrix constituting the walls (Fig. 23.5a).

To experimentally demonstrate the implementation of gellular automaton, we fabricated honeycomb-shaped hydrogel and prepared the DNA logic gate solutions in each hexagonal cell (Fig. 23.5). Since DNA propagation relies on diffusion, the concentration decreases during the transmission to adjacent cells. Therefore, an AND gate, which outputs a signal only when two types of specific DNAs are inputted, is combined with an amplifier (AMP) gate to ensure the intensity of the signal. The AND gate is designed to react with two types of input DNA and output catalyst DNA. When this catalyst is input into the AMP gate, it catalytically amplifies signal DNA. The combined gate is referred to COM gate. Furthermore, incorporating a pair of fluorophore and quencher, the concentration of signal is measured as fluorescence intensity by designing the signal to dissociate the quencher through strand displacement reactions.

In Fig. 23.5b, DNA solutions were placed in thirteen hexagonal cells, and the temporal changes in fluorescence were observed. In eight of these cells, the gates appropriately worked in each and fluorescence intensity gradually increased from the bottom to the top cells and resulted in the arrow-shaped pattern formation.

When the COM gate and the pair of fluorophore and quencher are anchored, inputs for the AND gate diffuse to the neighboring cells and spatiotemporal pattern formation is achieved in the hydrogel medium.

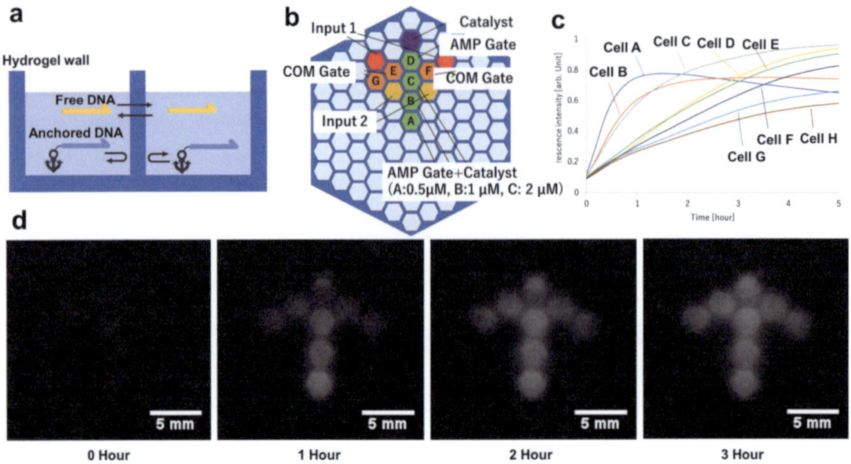

Fig. 23.5 Gellar automaton generating arrow-shaped pattern. **a** Hydrogel wall separates the cells and anchored DNA cannot move to the neighboring cells in contrast to free DNA, which can diffuse to neighboring cells. **b** Gate and catalyst DNA solutions are prepared in A-H cells. **c** In the cells A, B, and C, the amplification speed of the AMP gate has a gradient due to the concentration of the catalyst, with the intensity increasing in the order. Cell D undergoes amplification due to the catalyst diffused from the cell above, while cells E, F, G, and H begin amplification through the COM gate upon receiving input DNA from neighbor cells. **d** An arrow-shaped pattern formed in the hydrogel

23.2.2 Weighted Voronoi Pattern Formation by DNA Reaction–Diffusion System

Spatiotemporal information processing using DNA can be applied not only to digital patterns like cellular automata but also to the formation of analogue patterns [15–17, 31, 32]. As one example, here, we introduce a system that applies the DNA anchoring method mentioned in the previous section to form weighted Voronoi patterns [33].

The target pattern formation here is a bisector pattern line between two sources. First, two types of molecules are designed to diffuse equal speed and react to each other to stop their diffusion. Placing the molecules at two points in a reaction field separately, the molecules diffuse, and reactions occur where the diffusion fronts overlap. As a result, diffusion of molecules stops there, and the reaction products are localized along the bisector line. We call the pattern obtained in this way as a bisector pattern.

As a mechanism behind the system, we materialized the reaction–diffusion system by combination of the anchoring method and a DNA logic gate. This DNA logic gate consists of four DNA strands, input A, B, gate, and output, where the logic gate is designed to release an output strand when two input strands react to each other. In the process, the inputs bind to the gate DNA. Making use of the reaction, we anchored the gate DNA, placed the input A and B at the left and right sources separately, and used the anchored gate to capture the input strands when they react to each other.

In Fig. 23.6a, input A (green) and B (magenta) are prepared at the left and right sources, diffused into the hydrogel medium where the gate DNA is anchored. Several hours later, white bisector representing the colocalization of input A and B appeared and grew in time. This mechanism can be used for generating Voronoi diagrams, classifying the space into the regions referring to the nearest given points.

Additionally, the diffusion speed of input A and B can be tuned through diffusion modulation [34]. This method uses two DNAs: Trap and Competitor for a single target DNA. The Trap is anchored and can be hybridized with the target to stop the diffusion, and the Competitor releases the target from the Trap by strand displacement reaction. Since the Competitor is designed to hybridize with the Trap using the same number of bases, the process of hybridization, branch migration, and releasing is reversible. Based on the chemical equilibrium, the time of the binding between the target and Trap depends on the concentration of the Competitor, so the net diffusion speed of the target can be controlled by the Competitor concentration. By adjusting the diffusion speed of input A, the position of the bisector shifts closer to the point of input A source. This method is applied for imposing weight for the points to generate a weighted Voronoi diagram (Fig. 23.6b and c).

23.2.3 Parallel and Cascaded Pattern Formation

By paralleling and cascading basic reactions, it is possible to achieve more advanced reaction systems. While there are other reported materials for realizing artificial reaction–diffusion systems [34–37], the high programmability suitable for such usage is a unique feature of DNA. By appropriately designing the base sequences, DNA can provide multiple complementary strands that have the same number of bases, equivalent stability, and drastically different sequences that do not interfere with each other. These base sequences are called orthonormal sequences [38] and by effectively incorporating them into reaction domains of DNA, parallelization and cascading of reactions become possible.

Utilizing this property, DNA-based reaction–diffusion systems can realize parallel and cascaded pattern formation [39]. For parallelization and cascading, an improvement in spatial resolution in the observation area for each pattern is required. Taking the bisector pattern as an example, it is necessary to form two or three bisectors between two diffusion sources. Therefore, it is essential to be able to draw sharp bisectors in relation to the width of the two diffusion sources.

In the following example, a sharp bisector pattern is achieved by adopting a hybridization process that is faster than strand displacement reactions in the program for the bisector pattern. In this design, two DNA strands are used to form a single bisector. They have two reaction domains, alternatively hybridize with each other, and form double-stranded polymers (Fig. 23.7a). Since large molecules have a smaller diffusion coefficient according to Stokes–Einstein–Sutherland equation, as the polymerization progresses, the diffusion coefficient of the DNAs decreases. When the

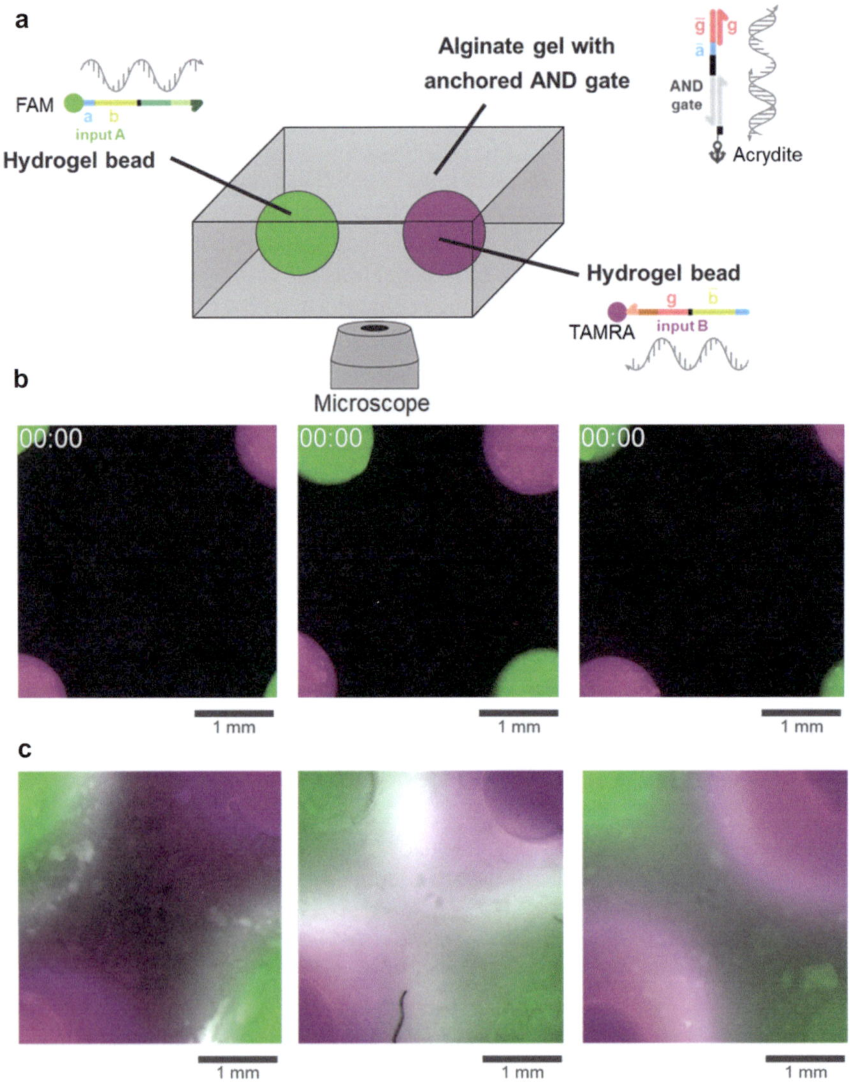

Fig. 23.6 Voronoi patterns formed in DNA-based reaction–diffusion systems. **a** The input A (green) and B (magenta) form bisector patterns, resulting from the process of the AND gate in the field. The positions of the bisectors were tuned using the diffusion modulation method. **b** In the initial state of the reaction–diffusion system, Trap and Competitor are used for input A (left) and for input B (right). **c** Patterns formed after 6 h. The diffusion modulation method leads to the formation of differently weighted Voronoi diagrams from the same initial condition. Image from reference [33]

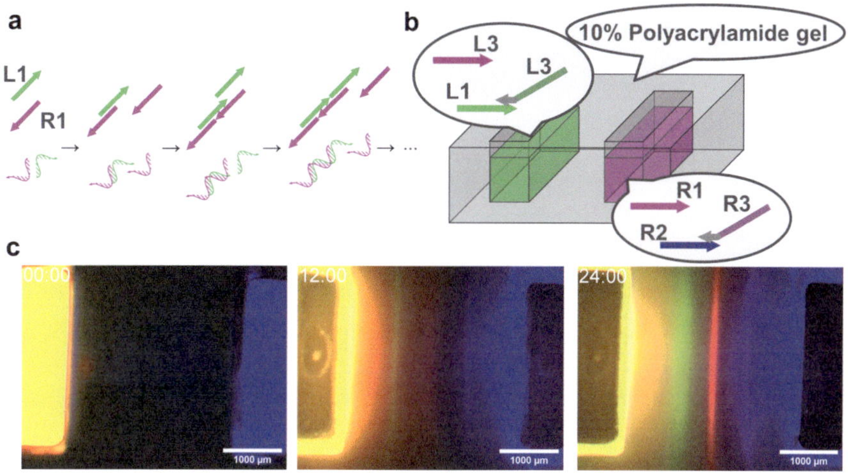

Fig. 23.7 Cascaded pattern formation in a DNA-based reaction–diffusion system. **a** DNA L1 and R1 hybridize each other alternatively and form polymers. **b** Three pairs of DNA (L1 and R1, L2 and R2, and L3 and R3) are prepared in the pockets fabricated in the polyacrylamide gel and diffuse from there. **c** Pairs 1 and 2 form green and blue bisectors at the left and right, and after that, pair 3 forms an additional red bisector. Image from reference [39]

DNAs diffuse from the separated points, they react on the bisector, forming polymers there, and a bisector pattern appears as a result.

This design is applied for parallelized and cascaded pattern formations. We used two pairs of DNA, which have orthonormal nucleotide base sequences and adjuster DNAs to form two bisectors between a pair of sources. Additionally, they applied other orthonormal sequences for polymerization to the adjusters. Pairs 1 and 2 diffuse from the sides, and pair 3 works as adjusters of them and is released in the process of the polymerization of pairs 1 and 2. When we prepare the pairs as shown in Fig. 23.7b, green and blue bisectors form at the left and right by the polymerization of pairs 1 and 2, respectively, and then, additional bisector (red) forms by the pair 3 (Fig. 23.7c).

23.3 Summary and Future Perspective

In this section, we introduced theories of DNA-based reaction–diffusion systems and experimental demonstrations. These are a promising technology applicable in molecular cybernetics, where spatiotemporal collaboration of multiple components is required. Furthermore, the perspective of utilizing molecular design for artificial reaction–diffusion systems holds the potential to advance the fabrication of large structures beyond the conventional milli/microscale in the field of molecular robotics.

In that context, this approach is believed to have significant implications, expanding the applicability of molecular cybernetics [40].

References

1. Fritsch, C., et al.: Drosophila under the lens: imaging from chromosomes to whole embryos. Chromosome Res. **14**, 451–461 (2006)
2. Torii, K.U.: Two-dimensional spatial patterning in developmental systems. Trends Cell Biol. **22**, 438–446 (2012)
3. Jacob, E.B., et al.: Bacterial linguistic communication and social intelligence. Trends Microbiol. **12**, 366–372 (2004)
4. Kondo, S., et al.: Reaction-diffusion model as a framework for understanding biological pattern formation. Science **329**, 1616–1620 (2010)
5. Luo, N., et al.: Synthetic pattern formation. Biochem. **58**, 1478–1483 (2019)
6. Lee, K.-J., et al.: Experimental observation of self-replicating spots in a reaction–diffusion system. Nature **369**, 215–218 (1994)
7. Ouyang, Q., et al.: Transition from a uniform state to hexagonal and striped Turing patterns. Nature **352**, 610–612 (1991)
8. Pearson, J.E.: Complex patterns in a simple system. Science **261**, 189–192 (1993)
9. Bánsági, T., et al.: Tomography of reaction-diffusion microemulsions reveals three-dimensional turing patterns. Science **331**, 1309–1312 (2011)
10. Vanag, V.K., et al.: Segmented spiral waves in a reaction-diffusion system. Proc. Natl. Acad. Sci. U.S.A. **100**, 14635–14638 (2003)
11. Turing, A.M.: The chemical basis of morphogenesis. Philos. Trans. R. Soc. B **237**, 37–72 (1952)
12. Zhang, D.Y., et al.: Control of DNA strand displacement kinetics using toehold exchange. J. Am. Chem. Soc. **131**, 17303–17314 (2009)
13. Smith, D.E., et al.: Dynamical scaling of DNA diffusion coefficients. Macromolecules **29**, 1372–1373 (1996)
14. Padirac, A., et al.: Spatial waves in synthetic biochemical networks. J. Am. Chem. Soc. **135**, 14586–14592 (2013)
15. Chen, S., et al.: Programmable patterns in a DNA-based reaction–diffusion system. Soft Matter **16**, 3555–3563 (2020)
16. Zenk, J., et al.: Stable DNA-based reaction–diffusion patterns. RSC Adv. **7**, 18032–18040 (2017)
17. Zadorin, A.S., et al.: Synthesis and materialization of a reaction–diffusion French flag pattern. Nat. Chem. **9**, 990–996 (2017)
18. Chirieleison, S.M., et al.: Pattern transformation with DNA circuits. Nat. Chem. **5**, 1000–1005 (2013)
19. Wolfram, S.: Cellular automata as models of complexity. Nature **311**, 419–424 (1984)
20. Manukyan, L., et al.: A living mesoscopic cellular automaton made of skin scales. Nature **544**, 173–179 (2017)
21. Adamatzky, A.: Reaction–diffusion path planning in a hybrid chemical and cellular-automaton processor. Chaos Solitons Fractals **16**, 727–736 (2003)
22. Scalise, D., et al.: Emulating cellular automata in chemical reaction–diffusion networks. Nat. Comput. **15**, 197–214 (2015)
23. Kawamata, I., et al.: Discrete DNA reaction-diffusion model for implementing simple cellular automaton. Lect. Notes Comput. Sci. **9726**, 168–181 (2016)
24. Cook, M.: Universality in elementary cellular automata. Complex Syst **15**, 1–40 (2004)
25. Isokawa, T., et al.: Universal totalistic asynchonous cellular automaton and its possible implementation by DNA. Lect. Notes Comput. Sci. **9726**, 182–195 (2016)

26. Hagiya, M., et al.: On DNA-based gellular automata. Lect. Notes Comput. Sci. **8553**, 177–189 (2014)
27. Kawamata, I. et al.: Pattern formation and computation by autonomous chemical reaction diffusion model inspired by cellular automata. 2016 *Fourth International Symposium on Computing and Networking*, pp. 215–221 (2016)
28. Yamashita, T., et al.: Turing-completeness of asynchronous non-camouflage cellular automata. Inf. Comput. **274**, 104539 (2020)
29. Padirac, A., et al.: Bottom-up construction of in vitro switchable memories. Proc. Natl. Acad. Sci. U.S.A. **109**, E3212–E3220 (2012)
30. Hosoya, T., et al.: Pattern formation on discrete gel matrix based on DNA computing. New Gener. Comput. **37**, 97–111 (2019)
31. Urtel, G., et al.: DNA-based long-lived reaction–diffusion patterning in a host hydrogel. Soft Matter **15**, 9343–9351 (2019)
32. Dorsey, P.J. et al.: DNA reaction–diffusion attractor patterns. Angew. Chem. **133**, 342–348 (2021)
33. Abe, K., et al.: Programmable reactions and diffusion using DNA for pattern formation in hydrogel medium. Mol. Syst. Des. Eng. **4**, 639–643 (2019)
34. Belousov, B.P.: An oscillating reaction and its mechanism. Sborn Ref. Radiat Med. **145**, (1959)
35. Zhabotinsky, A.M.: Periodical oxidation of malonic acid in solution (a study of the Belousov reaction kinetics). Biofizika **9**, 306–311 (1964)
36. Zaikin, A.N., Zhabotinsky, A.M.: Concentration wave propagation in two-dimensional liquid-phase self-oscillating system. Nature **225**, 535–537 (1970)
37. Lovrak, M., et al.: Free-standing supramolecular hydrogel objects by reaction-diffusion. Nat. Commun. **8**, 15317 (2017)
38. Kitajima, T., et al.: Design of code words for DNA computers and nanostructures with consideration of hybridization kinetics. Lect Notes Comput Sci **4848** (2008)
39. Abe, K., et al.: Cascaded pattern formation in hydrogel medium using the polymerisation approach. Soft Matter **17**, 6160–6167 (2021)
40. Murata, S., et al.: Molecular cybernetics: challenges toward cellular chemical artificial intelligence. Adv. Func. Mater. **32**(37), 2201866 (2022)

Chapter 24
Control Theory for Diffusion-Based Molecular Communication Systems

Taishi Kotsuka and Yutaka Hori

Abstract Molecular communication (MC) is an emerging field that aims to establish chemically driven information exchange mechanisms between molecular robots and cells. This technology enables the complete chemical implementation of multicellular molecular systems capable of achieving complicated tasks by coordination, leading to applications such as targeted drug delivery, biocomputing, and tissue engineering. Control theory provides useful concepts and tools that provide insights for better understanding and designing the dynamics of MC systems. In this section, we provide an overview of synthetic MC systems from the perspective of systems and control. Specifically, this section focuses on control-theoretic models and analysis methods for understanding the fundamental characteristics of diffusion-based MC systems.

24.1 Introduction

Molecular communication (MC) is a biochemical process where cells transmit and receive information using molecules as carriers. This form of communication is fundamental to life, enabling cells to coordinate their activities, respond to their environment, and maintain homeostasis in a robust and efficient manner based on chemically driven information exchange.

These natural biochemical processes of cellular communication have inspired engineering-oriented research that aims to build coordinated motion and behavior of synthetic biomolecular circuits and molecular robots [9, 17]. The ability to rationally design MC systems will open up many applications, including medical engineering [8, 13], biocomputing [3, 23], tissue engineering [4], and the internet of bio-nano things [1]. Currently, many efforts are underway to develop theoretical and experi-

T. Kotsuka
University of California, Santa Barbara, CA, USA
e-mail: tkotsuka@ucsb.edu

Y. Hori (✉)
Keio University, Yokohama, Kanagawa, Japan
e-mail: yhori@appi.keio.ac.jp

S. Murata et al. (eds.), *Molecular Robotics II*,
https://doi.org/10.1007/978-981-95-2394-8_24

mental techniques for overcoming challenges such as the robustness and scalability of MC systems.

This section presents an introductory overview of synthetic MC systems from the perspective of systems and control. Due to the page limitations, this section specifically focuses on diffusion-based MC systems. Interested readers are also referred to other review articles [5, 7, 12].

The rest of this section is organized as follows. In Sect. 24.2, we describe a basic mechanism of MC and its engineering applications. Section 24.3 introduces system-theoretic models of a simple diffusion-based MC channel between two molecular robots. We then extend the model for multicellular MC systems and show the stability analysis of multicellular MC systems in Sect. 24.4.

24.2 Basic of Molecular Communication

In MC, information is transmitted by a small molecule that acts as an effector to regulate reactions within the receiver host. The small molecule is transmitted either by (i) passive diffusion or (ii) active transport mechanisms. (i) Passive diffusion-based molecular communication uses the passive movement of a small molecule, an information carrier, that moves down the concentration gradient. This type of communication is seen in the quorum sensing of bacterial cell populations, exemplified by bioluminescent marine bacteria *Vibrio fischeri* [24]. In quorum sensing, bacteria produce and release acyl-homoserine lactone (AHL) autoinducers, information carrier molecules that freely diffuse in and out of the cell. This molecule is then sensed by the neighboring cells using a receptor protein, LuxR, to trigger downstream reactions, allowing for decentralized sensing of population density and population-level coordination through MC. On the other hand, (ii) active transport-based molecular communication uses external energy stored in the cells to move information carrier molecules against the concentration gradient [11]. Examples of active transport include the mass movement of fluids such as blood and lymph, which enables communication between tissues and organs in multicellular organisms, and neuronal signal transmission based on chemical messengers released from synaptic vesicles in response to an action potential [22].

Among them, diffusion-based molecular communication receives particular attention as an engineering machinery to build synthetic molecular communication systems (MC systems) that enable coordination of compartmentalized reaction systems. To be more specific, MC systems are broadly defined as systems consisting of molecular robots and MC channels as illustrated in Fig. 24.1. Molecular robots are compartmentalized biomolecular reaction systems such as genetic circuits and DNA circuits enclosed in bacterial cells, vesicles, or gels. MC channels refer to the communication medium consisting of a signal molecule and a solution between the compartments.

Experimental efforts have been made to build libraries of circuit parts, compartments, and testbeds to facilitate the engineering process of MC systems. Examples of

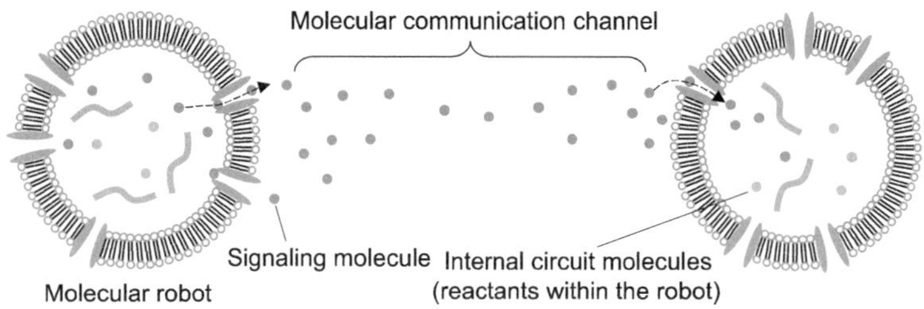

Fig. 24.1 Molecular robots communicating by diffusion-based MC

experimentally demonstrated MC systems include drug delivery systems [10], logic gates and adders for biocomputing [3], and spatial organization [17]. Theoretical studies, on the other hand, have revealed many information and system-theoretic properties of MC systems, such as channel capacity [2], frequency response [7, 16, 20], stability and robustness of the diffusion-based communication [6, 19, 21].

These efforts facilitate the rational design of multicellular MC systems that harness the collective power of individual molecular robots to create sophisticated behaviors. In the next section, we will see a basic system-theoretic modeling framework for MC systems.

24.3 System-Theoretic Models of MC Channels

The governing equation of the diffusion-based MC systems is the diffusion equation of the signal molecules coupled with the local reaction terms representing the reaction kinetics within the molecular robots. The specific form of the equation varies depending on the density of the molecular robots. When the molecular robots are densely dispersed in space, the MC system is approximately modeled by a reaction-diffusion equation, where the spatially homogeneous reaction term represents the local reaction kinetics at each point in the space [14, 15]. This type of model is relatively simple to handle, but it is hard to capture density-dependent behaviors, especially when the density of the molecular robots is low. This drawback is addressed by recently proposed models that assume that the local reaction occurs only at the boundary of the diffusion domain. This results in a diffusion equation with a dynamic boundary condition, which is harder to analyze but captures the spatio-temporal kinetics at a higher resolution [20]. In the rest of this section, we will see the details of the diffusion equation and how the boundary conditions are determined.

We begin with a simple case where the MC system consists of two molecular robots located at the ends of one-dimensional fluidic environment $\Omega = [0, L]$ as shown in Fig. 24.2a. Let $c(t, r)$ denote the concentration of the signal molecule at time t and position r. In the MC channel, the signal molecules move in a fluidic

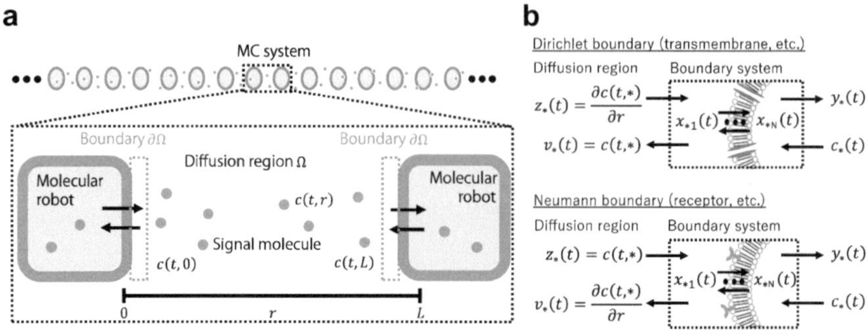

Fig. 24.2 **a** A multicellular MC system and a bidirectional MC channel between two molecular robots. **b** Input-output relation at the boundary between the diffusion region and the molecular robot for different membrane signal transfer mechanisms. Adapted with permission from [20]

medium according to Fick's law, and thus, the dynamics of the concentration $c(t, r)$ of the signal molecules can be modeled by the diffusion equation

$$\frac{\partial c(t, r)}{\partial t} = \mu \frac{\partial^2 c(t, r)}{\partial r^2}, \tag{24.1}$$

where μ is the diffusion coefficient.

The boundary condition depends on the membrane signal transfer mechanisms of the molecular robots since the molecular concentration at the boundaries changes dynamically based on the membrane signal transfer and the inflow/outflow of the signal molecules from/to the diffusion region. Specifically, the boundary becomes a Dirichlet-type condition when the membrane signal transfer mechanism involves the secretion and the absorption of molecules, while it becomes a dynamic Neumann boundary condition in the case of ligand-receptor reception. Figure 24.2b illustrates the input-output relation at the boundary of the diffusion domain Ω (boundary system), where

$$\begin{cases} v_*(t) = c(t, *), \ z_*(t) = \dfrac{\partial c(t, *)}{\partial r} & (\partial\Omega \text{ is Dirichlet}) \\ v_*(t) = \dfrac{\partial c(t, *)}{\partial r}, \ z_*(t) = c(t, *) & (\partial\Omega \text{ is Neumann}) \end{cases} \tag{24.2}$$

at position $*$ $(* = 0, L)$. The boundary system captures the dynamics of membrane transfer or signal reception of molecules. Thus, the characteristics of the entire MC channel are determined by the combination of the membrane signal transfer mechanism and the diffusion coefficient μ.

As is the case with classic feedback control and communication theory, frequency-domain analysis of the MC channel facilitates the understanding of fundamental properties of the systems, such as stability, robustness, and frequency response. Let $\mathcal{G}^\sharp(s)$ denote the multi-input multi-output (MIMO) transfer function of the MC chan-

Table 24.1 Transfer functions $\mathcal{G}^\sharp(s)$ for various boundary conditions. $\hat{s}$ is defined by $\hat{s} = \sqrt{s}/\sqrt{\mu}$

From	To	
	Dirichlet (d)	Neumann (n)
Dirichlet (d)	$\begin{bmatrix} -\hat{s}\dfrac{1}{\tanh(L\hat{s})} & \hat{s}\dfrac{1}{\sinh(L\hat{s})} \\ -\hat{s}\dfrac{1}{\sinh(L\hat{s})} & \hat{s}\dfrac{1}{\tanh(L\hat{s})} \end{bmatrix}$	$\begin{bmatrix} -\hat{s}\tanh(L\hat{s}) & \dfrac{1}{\cosh(L\hat{s})} \\ \dfrac{1}{\cosh(L\hat{s})} & \dfrac{1}{\hat{s}}\tanh(L\hat{s}) \end{bmatrix}$
Neumann (n)	$\begin{bmatrix} \dfrac{1}{\hat{s}}\tanh(L\hat{s}) & \dfrac{1}{\cosh(L\hat{s})} \\ \dfrac{1}{\cosh(L\hat{s})} & -\hat{s}\tanh(L\hat{s}) \end{bmatrix}$	$\begin{bmatrix} -\dfrac{1}{\hat{s}}\dfrac{1}{\tanh(L\hat{s})} & \dfrac{1}{\hat{s}}\dfrac{1}{\sinh(L\hat{s})} \\ -\dfrac{1}{\hat{s}}\dfrac{1}{\sinh(L\hat{s})} & \dfrac{1}{\hat{s}}\dfrac{1}{\tanh(L\hat{s})} \end{bmatrix}$

nel, where the input and the output are defined by $[v_0(s), v_L(s)]^{\mathrm{T}}$ and $[z_0(s), z_L(s)]^{\mathrm{T}}$, respectively, and $v_*(s) = \mathcal{L}[v_*(t)]$ and $z_*(s) = \mathcal{L}[z_*(t)]$ with the Laplace transform operator $\mathcal{L}$. Table 24.1 shows a list of transfer functions $\mathcal{G}^\sharp(s)$ for different combinations of boundary systems, where $\sharp =$ dn, dd, nd, nn represents the combinations of the boundary mechanisms at both ends of the region Ω, with d denoting Dirichlet boundary and n denoting Neumann boundary (see Ref. [20] for the derivation). For example, $\mathcal{G}^{\mathrm{dn}}(s)$ corresponds to the (1, 2) entry in Table 24.1.

As shown in Table 24.1, the transfer functions for the MC channel contain hyperbolic functions, unlike the standard real rational transfer functions for lumped parameter systems. These transfer functions are obtained by the combination of the simple transfer function

$$g(s, r) = \exp\left(-\sqrt{\frac{s}{\mu}}r\right), \tag{24.3}$$

representing the passive diffusion for distance r in one-dimensional space $[0, \infty)$. For example, Fig. 24.3 illustrates the block diagram of the transfer function from $c(s, 0)$ to $\partial c(s, L)/\partial r$, which is the (2, 1)th entry of $\mathcal{G}^{\mathrm{dn}}(s)$. As illustrated in the figure, the input-output relation of the MC channel is fully characterized by the transfer function $g(s, r)$.

The transfer function enables the frequency response analysis of the MC channel and provides design guidelines for the channel length and diffusion coefficient for various boundary mechanisms. For further details, readers are referred to reference [20]. In the next section, we further combine the system model for the two communicating molecular robots to introduce a model of multi-agent molecular robots, or multicellular systems, that exhibit cooperative behaviors using MC.

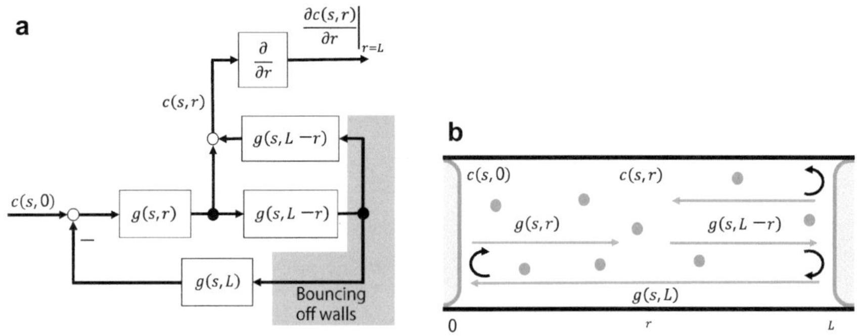

Fig. 24.3 **a** Block diagram of the diffusion process. **b** Physical interpretation of the block diagram

24.4 System-Theoretic Model and Analysis for Multicellular MC Systems

Multicellular MC systems refer to a swarm of molecular robots and cells that interact with each other by MC. As discussed in Sect. 24.2, bacterial quorum sensing is an example of such systems, which enables cells to *sense* the population density and generate collective dynamical behaviors such as state synchronization [9] and pattern formation [17]. In this section, we consider a simple model of multicellular MC systems and introduce a method for analyzing the stability of an equilibrium point, which is a fundamental and important property of the system related to state synchronization.

Suppose a swarm of molecular robots is aligned in the one-dimensional spatial domain as shown in the top of Fig. 24.2a. Let n denote the total number of molecular robots, and let the $i-1$-th MC channel Ω_i be defined as the channel between the $i-1$-th and i-th molecular robots. Assuming that the number of molecular robots is sufficiently large, we approximately model the system using a periodic boundary, or periodic channel $\Omega_1 = \Omega_n$. In other words, the left and the right ends of the space are connected. We denote the concentration of the signal molecules at position $r \in [0, L]$ in the i-th MC channel by $c_i(t, r)$, where L is the communication length of each channel. In practice, the communication length between molecular robots may not be uniform, but we here consider the average length L to approximately model the collective behavior of many molecular robots.

The dynamics of the concentration $c_i(t, r)$ in the MC channel can be modeled by the diffusion equation (24.1). When the molecular robots have a transmembrane-type signal transfer mechanism, the boundary conditions at $r = 0$ and $r = L$ in each MC channel are Dirichlet boundary conditions. Thus, the transfer function between any two molecular robots can be modeled by $\mathcal{G}^{\mathrm{dd}}(s)$ shown in Table 24.1. We define $w_i(t)$ as the input flux to the ith molecular robot using the output $z_L(t) = \partial c_i(t, L)/\partial r$ of the ith MC channel Ω_i and the output $z_0(t) = \partial c_{i+1}(t, 0)/\partial r$ of the $i+1$-th MC channel Ω_{i+1}. Specifically,

$$w_i(t) = \frac{\mu}{\Delta r}\left(\frac{\partial c_{i+1}(t,0)}{\partial r} - \frac{\partial c_i(t,L)}{\partial r}\right), \tag{24.4}$$

where Δr is the size of the molecular robot. The output $y_i(t)$ is the concentration of the signal molecule emitted from the $i-1$-th molecular robot. The conversion from $w_i(t)$ to $y_i(t)$ is due to the local reaction within the molecular robot, which can be modeled by the law of mass action. Thus, $y_i(t) = \mathcal{H}[w_i(t)]$, where $\mathcal{H}$ is an operator that encapsulates the (nonlinear) dynamics of the molecular robot.

The local properties of the multicellular MC system can be analyzed using the transfer functions. In particular, the local stability is related to the homeostasis of the multicellular MC system around a certain equilibrium. To this end, we hereafter consider a linearized system around an equilibrium point. Let $Y_i(s) := \mathcal{L}[y_i(t)]$ and $W_i(s) := \mathcal{L}[w_i(t)]$ denote the Laplace transform of the signals $y_i(t)$ and $w_i(t)$, respectively. Then, the multicellular MC system can be represented by the feedback of the molecular robots and the MC channels,

$$\boldsymbol{Y}(s) = H(s)\boldsymbol{W}(s),\ \ \boldsymbol{W}(s) = G(s)\boldsymbol{Y}(s), \tag{24.5}$$

where $\boldsymbol{Y}(s) = [Y_1(s), Y_2(s), \ldots, Y_n(s)]^{\mathrm{T}}$ and $\boldsymbol{W}(s) = [W_1(s), W_2(s), \ldots, W_n(s)]^{\mathrm{T}}$. The transfer function matrix $H(s)$ is the diagonal matrix of the form $H(s) = h(s)I_n$, where $h(s)$ is obtained from the linearized model of the local reaction $\mathcal{H}$. The transfer function matrix $G(s)$ becomes the circulant tridiagonal matrix of the form

$$G_{ij}(s) = \begin{cases} 2\mathcal{G}^{\mathrm{dd}}_{11}(s) & \text{if } i = j \\ -\mathcal{G}^{\mathrm{dd}}_{21}(s) & \text{if } i = j \pm 1 \text{ or } (i = 1 \text{ and } j = n) \text{ or } (i = n \text{ and } j = 1) \\ 0 & \text{otherwise,} \end{cases} \tag{24.6}$$

where $G_{ij}(s)$ is the (i, j) entry of $G(s)$.

The characteristic structure of the matrix $G(s)$ enables to reduce the stability analysis of the MIMO system (24.5) into simpler ones. The following theorem shows a necessary and sufficient stability condition for the closed-loop system (24.5).

Theorem 1 [21] *The linearized multicellular MC system (24.5) is asymptotically stable if and only if all roots of the characteristic equations*

$$p_i(s) = 1 - \lambda_i(s)h(s) = 0 \tag{24.7}$$

lie in the open left half of the complex plane for all i, where

$$\lambda_i(s) = 2\mathcal{G}^{\mathrm{dd}}_{11}(s) - 2\mathcal{G}^{\mathrm{dd}}_{21}(s)\cos\left(2\pi\frac{i-1}{n}\right). \tag{24.8}$$

and

$$i = \begin{cases} 1, 2, \ldots, \frac{n}{2} + 1 & \text{when } n \text{ is even} \\ 1, 2, \ldots, \frac{n+1}{2} & \text{when } n \text{ is odd.} \end{cases} \tag{24.9}$$

Theorem 1 shows that the stability analysis of the MIMO system (24.5) can be decomposed into that of scalar subsystems, facilitating the analysis using simpler methods in control engineering.

Example. Consider the multicellular MC system with $n = 100$ molecular robots, where each molecular robot has an activator-repressor-diffuser (ARD) genetic circuit [14]. In what follows, we analyze the stability of the system around a homogeneous equilibrium point of this system for different values of the communication length L and the diffusion coefficient μ.

We consider the mathematical model of the MC system with the ARD genetic circuit presented in Ref. [21]. Linearizing the system around the equilibrium point, the dynamics of the local reaction is given by

$$h(s) = \frac{10s^3 + 6.1469s^2 + 0.128s + 0.0003}{s^4 + 1.1147s^3 + 0.0701s^2 + 0.0003s}, \tag{24.10}$$

which has two poles in the right-half complex plane (unstable). Using Theorem 1, the stability of the closed-loop system can be characterized by the roots of the characteristic equation (24.7) for $i = 1, 2, \cdots, 51$. Figure 24.4a (top) shows examples of the Nyquist plot for $i = 1$ (black) and 10 (gray) for $L = 70\,\mu\text{m}$, $\mu = 10\,\mu\text{m}^2 \cdot \text{min}^{-1}$, which encircles $-1 + j0$ twice. Checking the Nyquist plot for all $i = 1, 2, \cdots 51$, we conclude that the subsystems are Hurwitz. As shown in Fig. 24.4b (right), the states, or the concentrations, of the repressor molecule of the molecular robots converge to the same value, achieving homeostasis when the system is stable. On the other hand, the Nyquist plot for $i = 1$ (black) does not encircle $-1 + j0$ when $L = 10\,\mu\text{m}$ and $\mu = 10\,\mu\text{m}^2 \cdot \text{min}^{-1}$ indicating that the system becomes unstable. The corresponding time-course simulation of the concentrations does not converge to the original homogeneous equilibrium as shown in Fig. 24.4b (left).

Figure 24.4c shows the parameter map for the stability for various communication lengths L and diffusion coefficients μ. The result implies that the multicellular MC system becomes stable as the communication length L is greater, which means that the MC system with a lower density of molecular robots can more easily transition to another state. This characteristic is consistent with the quorum-sensing mechanism [24].

24.5 Summary and Further Readings

This section has provided an overview of synthetic MC systems from the perspective of systems and control. Specifically, this section has focused on a system-theoretic tool for understanding fundamental characteristics of diffusion-based MC systems.

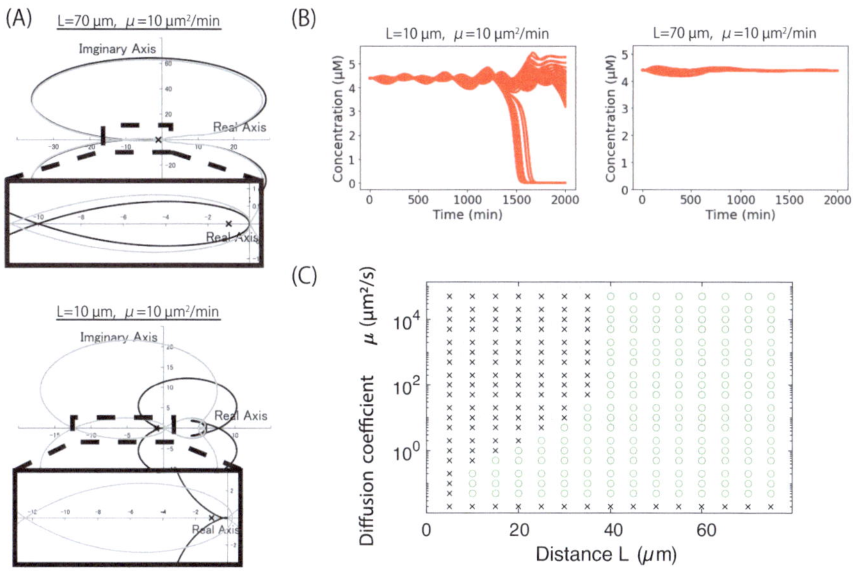

Fig. 24.4 **a** The Nyquist plot of the open-loop transfer function $-\lambda_i(j\omega)h(j\omega)$ for $i = 1$ (black) and $i = 10$ (gray). The cross mark represents $-1 + j0$. **b** The behavior of the concentration of the repressor molecule in the 100 molecular robots around the spatially homogeneous equilibrium point **c**. The parameter map for the stability of the multicellular MC system for different communication lengths L and diffusion coefficients μ. Adapted with permission from [21]

With the establishment of control theory for MC systems, similar to conventional mechanical and electrical feedback systems, we expect to be able to design molecular robot populations that achieve the desired functionality, which would be useful for many applications in the medical field, synthetic biology, and industry.

In addition to the stability analysis of multicellular MC systems introduced in Sect. 24.4, many system-theoretic analyses exist such as the frequency response analysis of MC channels [16, 18, 20], the stability analysis of connected MC systems [6], and the analysis of the parameter and structure conditions for pattern formation [14, 15]. Interested readers are referred to these literatures for further details.

References

1. Akyildiz, I.F., Pierobon, M., Balasubramaniam, S., Koucheryavy, Y.: The internet of bio-nano things. IEEE Commun. Mag. **53**(3), 32–40 (2015)
2. Atakan, B., Akan, O.B.: On channel capacity and error compensation in molecular communication. In: Proceedings of the Transactions on Computational Systems Biology, vol. X, pp. 59–80. Springer, Berlin, Heidelberg (2008)
3. Ausländer, D., Ausländer, S., Pierrat, X., Hellmann, L., Rachid, L., Fussenegger, M.: Programmable full-adder computations in communicating three-dimensional cell cultures. Nat. Methods **15**(1), 57–60 (2018)
4. Basu, S., Gerchman, Y., Collins, C.H., Arnold, F.H., Weiss, R.: A synthetic multicellular system for programmed pattern formation. Nature **434**(7037), 1130–1134 (2005)

5. Bi, D., Almpanis, A., Noel, A., Deng, Y., Schober, R.: A survey of molecular communication in cell biology: establishing a new hierarchy for interdisciplinary applications. IEEE Commun. Surv. Tutor. **23**(3), 1494–1545 (2021)
6. Casten, R.G., Holland, C.J.: Stability properties of solutions to systems of reaction-diffusion equations. SIAM J. Appl. Math. **33**(2), 353–364 (1977)
7. Chude-Okonkwo, U.A.K., Malekian, R., Maharaj, B.T.: Diffusion-controlled interface kinetics-inclusive system-theoretic propagation models for molecular communication systems. EURASIP J. Adv. Signal Process. **2015**(1), 1–23 (2015)
8. Chude-Okonkwo, U.A.K., Malekian, R., Maharaj, B.T., Vasilakos, A.V.: Molecular communication and nanonetwork for targeted drug delivery: a survey. IEEE Commun. Surv. Tutor. **19**(4), 3046–3096 (2017)
9. Danino, T., Mondragón-Palomino, O., Tsimring, L., Hasty, J.: A synchronized quorum of genetic clocks. Nature **463**(7279), 326 (2010)
10. Omar Din, M., Danino, T., Prindle, A., Skalak, M., Selimkhanov, J., Allen, K., Julio, E., Atolia, E., Tsimring, L.S., Bhatia, S.N., et al.: Synchronized cycles of bacterial lysis for in vivo delivery. Nature **536**(1), 81–85 (2016)
11. Enomoto, A., Moore, M.J., Suda, T., Oiwa, K.: Design of self-organizing microtubule networks for molecular communication. Nano Commun. Netw. **2**(1), 16–24 (2011)
12. Farsad, N., Birkan Yilmaz, H., Eckford, A., Chae, C.B., Guo, W.: A comprehensive survey of recent advancements in molecular communication. IEEE Commun. Surv. Tutor. **18**(3), 1887–1919 (2016)
13. Femminella, M., Reali, G., Vasilakos, A.V.: Molecular communications model for drug delivery. IEEE Trans. Nanobiosci. **14**(7), 935–945 (2015)
14. Hori, Y., Miyazako, H., Kumagai, S., Hara, S.: Coordinated spatial pattern formation in biomolecular communication networks. IEEE Trans. Mol. Biol. Multi-scale Commun. **6**(2), 111–121 (2015)
15. Hsia, J., Holtz, W.J., Huang, D.C., Arcak, M., Maharbiz, M.M.: A feedback quenched oscillator produces turing patterning with one diffuser. PLOS Comput. Biol. **8**(1), e1002331 (2012)
16. Huang, Yu., Ji, F., Wei, Z., Wen, M., Chen, X., Tang, Y., Guo, W.: Frequency domain analysis and equalization for molecular communication. IEEE Trans. Signal Process. **69**, 1952–1967 (2021)
17. Karig, D., Michael Martini, K., Lu, T., DeLateur, N.A., Goldenfeld, N., Weiss, R.: Stochastic turing patterns in a synthetic bacterial population. In: Proceedings of the National Academy of Sciences of the United States of America, vol. 115, no. 26, pp. 6572–6577 (2018)
18. Kitada, S., Kotsuka, T., Hori, Y.: Analysis of signal distortion in molecular communication channels using frequency response. New Gen. Comput. **42**, 253–270 (2024)
19. Kotsuka, T., Hori, Y.: Spatial frequency-based characterization of disturbance rejection in molecular communication systems. IEEE Trans. Mol. Biol. Multi-scale Commun. **8**(1), 36–43 (2022)
20. Kotsuka, T., Hori, Y.: A control-theoretic model for bidirectional molecular communication systems. IEEE Trans. Mol. Biol. Multi-scale Commun. **9**(3), 274–285 (2023)
21. Kotsuka, T., Hori, Y.: Stability analysis for large-scale multi-agent molecular communication systems. IEEE Trans. Nanobiosci. **23**(3), 507–517 (2024)
22. Malak, D., Akan, O.B.: Molecular communication nanonetworks inside human body. Nano Commun. Netw. **3**(1), 19–35 (2012)
23. Regot, S., Macia, J., Conde, N., Furukawa, K., Kjellén, J., Peeters, T., Hohmann, S., de Nadal, E., Posas, F., Solé, R.: Distributed biological computation with multicellular engineered networks. Nature **469**(7329), 207–211 (2011)
24. Waters, C.M., Bassler, B.L.: Quorum sensing: cell-to-cell communication in bacteria. Ann. Rev. Cell Dev. Biol. **21**, 319–346 (2005)

Chapter 25
Networks of Classical Conditioning Gates

Shun-ichi Azuma

25.1 Introduction

A key technology for realizing molecular robots is chemically synthesized artificial intelligence (chemical AI), which has the ability to process information from sensors, operate actuators, and learn from sensor data. Motivated by this, a research project to develop chemical AI, called the *Molecular Cybernetics Project*, was launched in 2021 [1] and has been working toward the creation of chemical AI.

One of the prototypes of chemical AI that the project aims to develop is a liposome-based molecular machine that can learn a kind of conditioned reflex through the process called classical conditioning (see, e.g., [2] for classical conditioning). Once such a machine is realized, the next step is to configure a network of such machines to create more advanced functionality. On the other hand, there are many technical challenges to achieve this. In particular, how to train such a network is one of the fairly basic.

Following this line of research, a mathematical model of classical conditioning and a training method based on it have been recently developed in [3]. This section introduces the result.

Note that this section deals with two logical functions, a projection with respect to the first input and a logical OR, whose truth tables are shown in Table 25.1. The logical OR operation of the binary variables x and y is denoted by $x \bigvee y$.

S. Azuma (✉)
Kyoto University, Kyoto, Japan
e-mail: sazuma@i.kyoto-u.ac.jp

S. Murata et al. (eds.), *Molecular Robotics II*,
https://doi.org/10.1007/978-981-95-2394-8_25

Table 25.1 Truth tables of projection and logical OR

Input 1	Input 2	Output	
		Projection w.r.t. input 1	Logical OR
0	0	0	0
0	1	0	1
1	0	1	1
1	1	1	1

25.2 Classical Conditioning and Its Model

Classical conditioning is a learning process of acquiring a conditioned reflex by presenting a conditioned stimulus with an unconditioned stimulus. "Pavlov's dog" is the most well-known experiment on classical conditioning (see, e.g., [2]), which can be summarized as follows. If one feeds a dog repeatedly while ringing a bell, the dog will eventually start salivating at the sound of the bell. On the other hand, if one simply rings a bell repeatedly without doing anything else for the dog that salivates at the sound of the bell, the dog will stop salivating at the sound of the bell. That is, if a particular stimulus was presented in the dog's environment at the time of feeding, the stimulus becomes associated with the food and causes salivation on its own.

A model of Pavlov's classical conditioning, developed in [3], is shown in Fig. 25.1. The model is a finite state machine in the discrete time domain $t = 0,1,\ldots$. It has two inputs, indicated as Inputs 1 and 2 in the figure, and one output. They take a binary value at each time instant. The state takes one of the values PRJ and OR and switches between the two values according to the two inputs. When the state is PRJ, the gate operates as the projection with respect to Input 1, whose output is equal to Input 1. When the state is OR, the gate operates as a logical OR gate, whose output is equal to 1 if and only if at least one of the two inputs is equal to 1. See Table 25.1 for the input–output relations of the projection and logical OR.

The state changes when two special inputs, which are (1,1) and (0,1), are applied to it several times in a row. If the current state is PRJ, the state is changed to OR by entering the value (1,1) to the pair of Inputs 1 and 2 several times in a row. On the other hand, if the state is OR, the state is changed to PRJ by entering the value

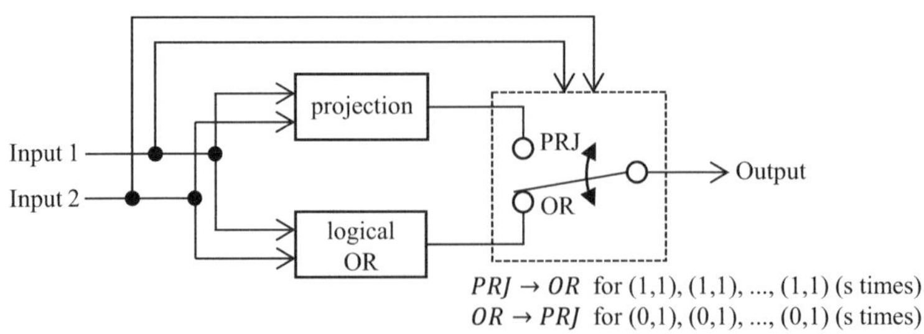

Fig. 25.1 Model of classical conditioning (classical conditioning gate)

(0,1) several times in a row. The input values that cause the state transition are called *training inputs*. We use s to represent the times that enter the same input value for changing the state. The number s is called the *gate training time*.

This model is called a *classical conditioning gate*.

The correspondence between this model and Pavlov's dog experiment is as follows:

- Input 1: Feeding the dog or not (0: not feeding, 1: feeding).
- Input 2: Ringing a bell or not (0: not ringing, 1: ringing).
- State PRJ: Responding (start salivating) only when the dog is being fed.
- State OR: Responding when the dog is being fed or hears the bell.

Note that, as described above, the input–output relation is equivalent to either projection or logical OR at any time instant, depending on the state. In other words, the model is a logic gate whose function type is switched by a toggle.

25.3 A Network of Classical Conditioning Gates and Its Training Problem

By itself, a classical conditioning gate is just a single logic gate whose function is switched, but when networked, it becomes an advanced learning machine.

Figure 25.2 shows a network model in a binary tree structure, where a box represents a classical conditioning gate. We consider the network to consist of m layers. The layers are indexed by $1,2,\ldots,m$ from the input side as shown in the figure. Since the i-th layer has 2^{m-i} gates, the network has $\sum_{i=1}^{m} 2^{m-i}$ gates. The network has 2^m inputs, denoted by $x = \left(x_{11}, x_{12}, x_{21}, x_{22}, \ldots, x_{(2^{m-1})1}, x_{(2^{m-1})2}\right)$, and one output, denoted by y. Let the gates be indexed as $1,2,\ldots,\sum_{i=1}^{m} 2^{m-i}$ from the top of layer 1 to the gate of layer m. Figure 25.3 illustrates a network of $m = 3$.

The state of the network is the collection of the states of the gates. It is called the *state pattern* and expressed as the tuple of the state values of gates $1,2,\ldots,m$. Figure 25.3 shows a network in the state pattern $(PRJ, PRJ, PRJ, PRJ, OR, PRJ, PRJ)$, where the state of each gate is indicated inside the box.

The input–output relation of the network at time t is determined by the state pattern, i.e., the states of the 2^{m-i} gates, at time t. For example, the input–output relation is $y = x_{11} \bigvee x_{21}$ for the network in Fig. 25.3.

Note here that there may be several state patterns which correspond to a certain input–output relation. For example, consider the network in Fig. 25.4. The state pattern differs from that in Fig. 25.3, while the input–output relation is $y = x_{11} \bigvee x_{21}$, which is the same as Fig. 25.3.

Note also that, although it appears that any input–output relation in the form of a logical disjunction can be realized in the network, it is not true. For example, $y = x_{11} \bigvee x_{22}$ cannot be realized in the network of $m = 3$ because gate 2 outputs

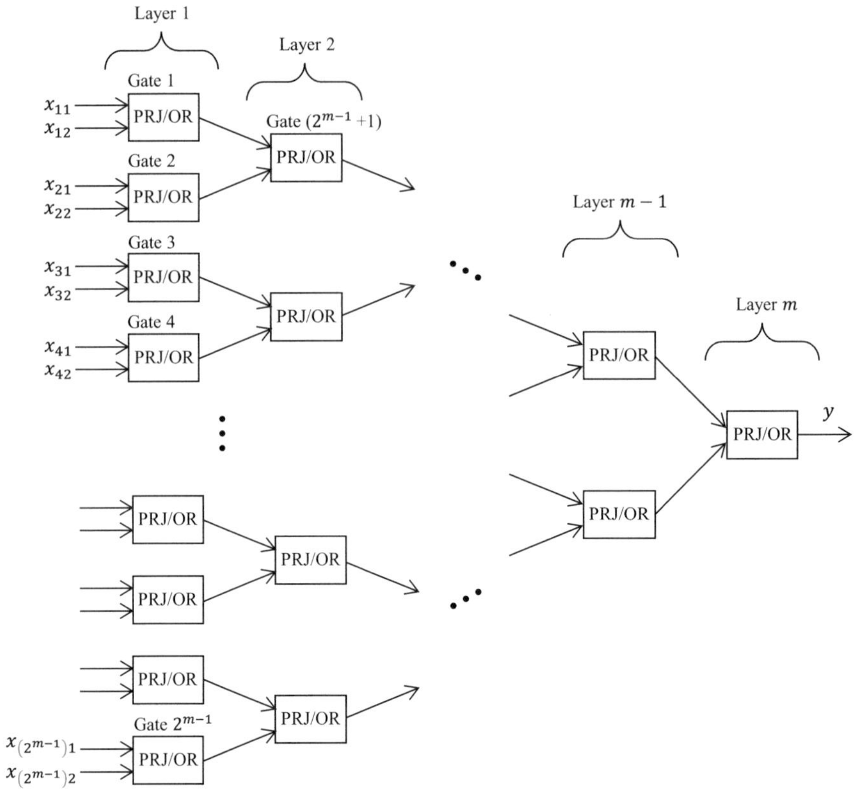

Fig. 25.2 Network of classical conditioning gates

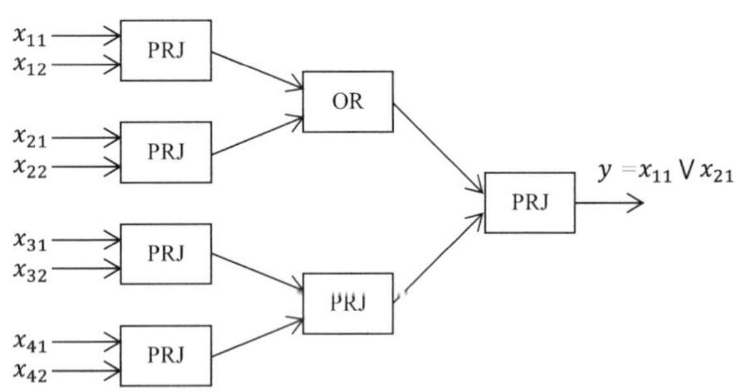

Fig. 25.3 Network with m = 3 and the state pattern (PRJ, PRJ, PRJ, PRJ, OR, PRJ, PRJ)

either x_{21} (when the state is PRJ) or $x_{21} \bigvee x_{22}$(when the state is OR) and the resulting output y contains x_{21}, $x_{21} \bigvee x_{22}$, or neither (if the output y contains x_{22}, then it will appear along with x_{21}).

Now, we state about the training of a network. By noting the definition of classical conditioning gates, the state pattern of the network can change by some input sequence (thus, the input–output relation can change). For example, consider the

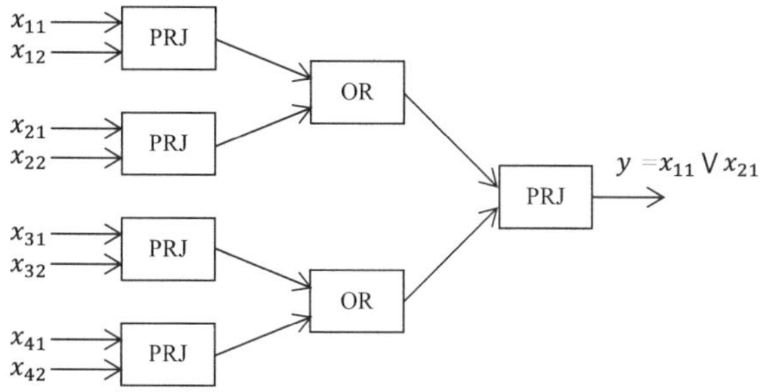

Fig. 25.4 Network with m = 3 and the state pattern (PRJ, PRJ, PRJ, PRJ, OR, OR, PRJ)

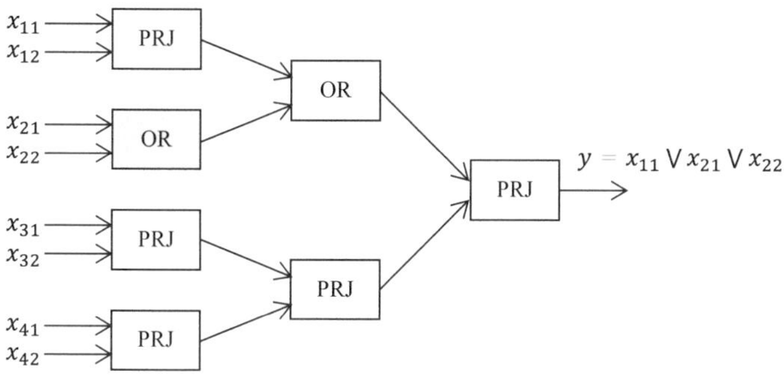

Fig. 25.5 Network with m = 3 and the state pattern (PRJ, OR, PRJ, PRJ, OR, PRJ, PRJ)

network in Fig. 25.3, whose input–output relation is $y = x_{11} \bigvee x_{21}$ as stated above. Then if the input $(x_{11}, x_{12}, x_{21}, x_{22}, x_{31}, x_{32}, x_{41}, x_{42}) = (1{,}0,\ 1{,}1,\ 0{,}0,\ 0{,}0)$ is applied to the network s times in a row, then the networks become as shown in Fig. 25.5 and the input–output relation becomes $y = x_{11} \bigvee x_{21} \bigvee x_{22}$. In this case, gates 1, 2, 3, and 4 have the inputs (1,0), (1,1), (0,0), and (0,0), respectively, gates 5 and 6 have the inputs (1,1) and (0,0), respectively, and gate 7 has the input (1,0). As the result, the state of gate 2 changes from PRJ to OR and the states of the other gates do not change. In this way, the input–output relation of the network can be changed by applying some input sequence. Realizing a desired input–output relation in such a way is called *training*, and finding an input sequence for training is the *training problem*.

25.4 Training Method

Next, a training method for networks is presented. A key technique of training is the *flipping principle*, which is shown as follows.

Consider a single classical conditioning gate, whose state is assumed to be arbitrarily given. By definition, if either (0,0) or (1,0) is applied to the gate, then the state does not change and the output becomes 0 for the input (0,0) and 1 for the input (1,0). This means that the gate outputs any binary value while preserving the state.

This fact provides the following property, called the *flipping principle*, for the networks in Fig. 25.2: for each gate, there exists an input sequence such that the state of the gate is flipped while the state of the other gates is preserved.

The flipping principle is illustrated as follows. Consider the network in Fig. 25.3 with the state pattern (PRJ, PRJ, PRJ, PRJ, OR, PRJ, PRJ). Suppose that we want the state of gate 2 to flip. Then if the input (1,0, 1,1, 0,0, 0,0) is applied to the network s times in a row, then the networks become as shown in Fig. 25.5, where the state of gate 2 becomes OR and the state of the other gates are the same as Fig. 25.3.

The flipping principle allows us to realize the desired input–output relation by flipping the state of gates one by one. This establishes the following algorithm for training.

Algorithm 1 (Training a network with gate training time S)

Input: The initial state pattern $(S_1, S_2, \dots, S_n)$ of the network, where n is the number of gates in the network, and a desired input–output relation $y = f(x)$.

Step 1: Derive the Target State Patterns

Find one of the state patterns corresponding to the desired input–output relation and it is denoted by $(D_1, D_2, \dots, D_n)$. If there does not exist such a state pattern, then halt and return "infeasible" (which means that the desired input–output relation cannot be realized in the network).

Step 2: List the Gates for Which the State Has to Be Flipped

Make the list of the gates whose states have to be flipped, i.e., the list of the indexes $i \in \{1,2, \dots, n\}$ such that $S_i \neq D_i$.

Step 3: Apply the Input Sequence to Flip the State of a Gate in the List

If the list is empty, then halt; otherwise, pick an element i from the list and determine the input value to flip the state of gate i while preserving the state of the other gates based on the flipping principle. Apply the input value to the network in s times in a row. (As the result, the state of the gate is flipped at this point).

Step 4: Exclude the Gate for Which the State Has Been Flipped from the List

Delete i picked in Step 3 from the list. Go to Step 3.

The following example demonstrates how the algorithm works.

Consider the network in Fig. 25.3 with $s = 3$ and the state pattern.

$$(S_1, S_2, S_3, S_4, S_5, S_6, S_7) = (PRJ, PRJ, PRJ, PRJ, OR, PRJ, PRJ)$$

Suppose that the desired input–output relation is given as $y = x_{11} \bigvee x_{21} \bigvee x_{22} \bigvee x_{31}$.

Step 1 derives the state pattern

$$(D_1, D_2, D_3, D_4, D_5, D_6, D_7) = (PRJ, OR, PRJ, PRJ, OR, PRJ, OR)$$

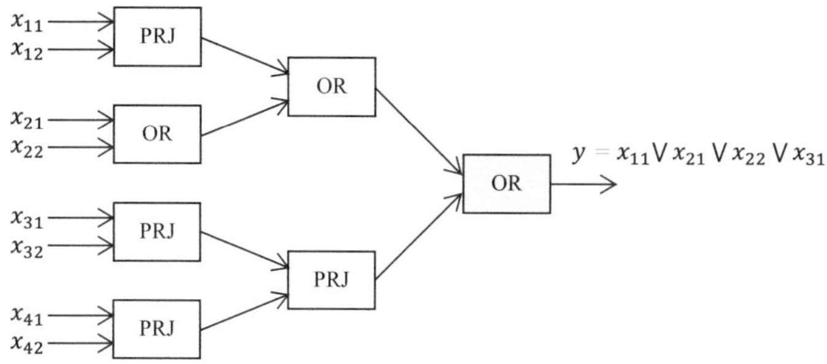

Fig. 25.6 Network with m = 3 and the state pattern (PRJ, OR, PRJ, PRJ, OR, PRJ, OR)

that corresponds to the desired input–output relation. In Step 2, the list {2,7} is obtained, because$S_2 \neq D_2$, $S_7 \neq D_7$, and $S_i = D_i$ for $i \in \{1, 3, 4, 5, 6\}$, and thus the states of gates 2 and 7 have to be flipped. Since the list is not empty, Step 3 picks an element from the list. Here, we assume that 2 is picked. Then the input value to flip the state of gate 2 is generated as (1,0, 1,1, 0,0, 0,0) and applied it to the network three times in a row. As the result, the state of gate 2 changes from PRJ to OR as shown in Fig. 25.5. Step 4 removes the element 2 from the list, and the resulting list is {7}. The execution point then moves to Step 3. In a similar manner, Step 3 picks 7 and the input value to flip the state of gate 7 is generated as $(1, 0, 0, 0, 1, 0, 0, 0)$. The input is applied to the network three times in a row, which changes the state of gate 7 from PRJ to OR as shown in Fig. 25.6. As the result, the desired input–output relation is obtained. Step 4 removes the element 7 from the list, and the list becomes empty. The algorithm terminates on the next run of Step 3.

References

1. Murata, S., Toyota, T., Nomura, S.I.M., Nakakuki, T., Kuzuya, A.: Molecular cybernetics: challenges toward cellular chemical AI. Adv. Func. Mater. **32**, 2201866 (2022)
2. Hart-Davis, A.: Pavlov's Dog: Groundbreaking Experiments in Psychology. Elwin Street Limited (2015)
3. Azuma, S., Takakura. D., Ariizumi, R., Asai, T.: Networks of Classical Conditioning Gates and Their Learning. New Generation Computing (2024)

Chapter 26
Computation by Collective Behavior of Distributed Elements

Teijiro Isokawa

Abstract Developing molecular-scale devices for processing information, i.e., molecular-scale computers, is crucial for creating robotic systems at this scale. In this section, we present a framework for performing computation using molecular-scale components. We base our approach on a Brownian Circuit, which enables asynchronous (clock-free) operations with sufficient computational power. We demonstrate two types of implementations of the Brownian Circuit: asynchronous Cellular Automaton-based and Swarm Agents-based (robot-based) approaches. Both implementations operate in a collective manner. Moreover, each element in the system is simple and uniform, and the interactions among these elements can produce complex operations within the Brownian Circuit.

26.1 Introduction

Molecular-scale autonomous robotic systems, or molecular robots, are expected to have various applications, such as in vivo drug delivery and artificial organ formation [5, 6]. To complete these molecular robots, several challenges must be addressed, including integrating robot components, supplying energy to these components, and making them work together.

One key challenge is controlling molecular-scale components to achieve the desired behavior within a molecular-scale system. Thus, developing molecular-scale "computers" capable of accepting various sensory inputs, processing these inputs with internal memories, and generating output signals to control actuators is essential. Numerous molecular implementations of the computational elements are possible. In this section, we focus on the framework for building molecular-scale computing devices, rather than on the specific implementations of these computational elements.

We begin by introducing an asynchronous token-based computational model called the Brownian circuit [4, 7]. This circuit is computationally universal, meaning

T. Isokawa (✉)
Graduate School of Engineering, University of Hyogo, Himeji, Japan
e-mail: isokawa@eng.u-hyogo.ac.jp

S. Murata et al. (eds.), *Molecular Robotics II*,
https://doi.org/10.1007/978-981-95-2394-8_26

that various components such as arithmetic and logic operations, and memory, can be constructed using it. The Brownian circuit has the advantage of not requiring clock signals, which are used in conventional computers to synchronize component states and can cause heat dissipation.

We then describe two types of implementations for performing operations within a Brownian circuit: a cellular automaton (CA) [1] and a robot-based implementation called Swarm Agents [2]. Both implementations are characterized by their use of several simple elements, along with their state transitions or structural changes, can facilitate the operations required within the Brownian circuit.

The remainder of this section is organized as follows. Section 26.2 describes the fundamentals of the Brownian circuit and provides an example of its components. Section 26.3 presents the asynchronous CA-based approach for constructing a Brownian circuit. Section 26.4 presents the robot-based (Swarm Agent-based) approach. Finally, Sect. 26.5 concludes this section.

26.2 Brownian Circuit

Brownian circuits consist of several types of elements: signals, logic elements that process these signals, and signal wires that connect the logic elements. A signal in Brownian circuits is an indivisible token, and also signals never fuse. Signals travel along signal wires, and upon reaching the input terminal of a logic element, are processed and then exit from the output terminal. Signals can travel bidirectionally, except in the case of the *ratchet* element (described later). Thus, signals on a wire can move both forward and backward. Additionally, with the exception of the ratchet element, signals can travel from output terminals back to input terminals of logic elements. Signal wires, which route signals between logic elements, can be bent and crossed. We used the set of elements defined in [3] as examples in Brownian circuits.

The first element in this set is the *Hub* element, depicted in Fig. 26.1a. This element features three terminals, which can function as either inputs or outputs. When a signal (represented by the black blob in the figure) arrives at one terminal, it can exit from any of the three terminals, including the terminal where it arrived.

Next, we present a logical element called the *Conservative Join* (*CJoin*), depicted in Fig. 26.1b. The CJoin has two pairs of terminals, which can serve as either input or output terminals. Its primary function is to synchronize two signals arriving at the input terminals. When two signals are simultaneously present at the input terminals—one at each terminal—the CJoin transfers each signal to one of the output terminals. If only one signal is present at an input terminal, the CJoin does not process it; instead, the signal either remains at the input terminal or travels back along the signal wire to return to the input terminal.

An element called the *ratchet*, shown in Fig. 26.1c is also described. This element features input and output terminals. Signals entering the input terminals are transferred to the output terminals, but the movement of signals in the reverse direction is restricted. In other words, the ratchet allows signals to flow in only one direc-

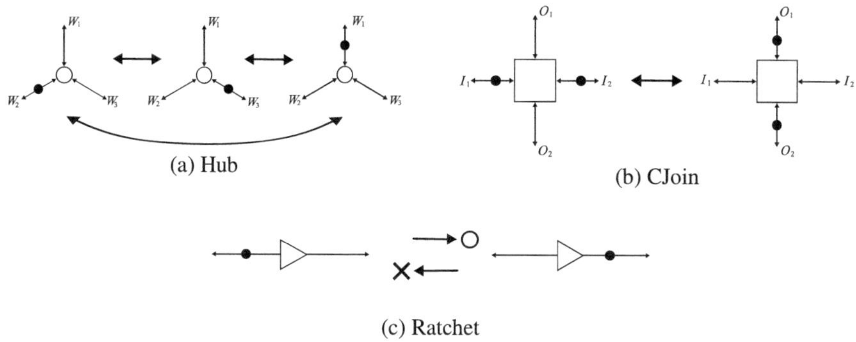

Fig. 26.1 Elements in the Brownian circuit: **a** Hub for routing signals. **b** CJoin for synchronizing signals. **c** Ratchet for restricting signal direction. Signals are represented as blobs in these figures

tion, similar to a diode in electronic circuits. Although not essential for ensuring the computational power of the Brownian circuit, the ratchet enhances the circuit's computational speed. Without the ratchet, unrestricted signal return on signal lines, hubs, and CJoin elements would significantly reduce computational efficiency. The ratchet helps mitigate this effect by preventing such unrestricted signal return.

Various types of computing components can be constructed using these circuit elements. For example, a circuit that acts as a half adder is presented using these elements. A half adder built with the Brownian circuit elements is shown in Fig. 26.2a. This circuit computes an additive output value and carry from two binary input values using dual-rail encoding. In this encoding scheme, two signal wires, labeled 0 and 1, represent the binary states. A signal on the 0 wire indicates an input of 0, while a signal on the 1 wire indicates an input of 1. No signal on either wire means no input is present, and signals on both wires do not ensure proper component operation.

Figure 26.2b demonstrates the operation of the half adder with inputs A and B set to 1. In this case, the additive output should be 0 and the carry 1. The signals travel through the circuit along the signal wires to find the appropriate CJoin from the four available, corresponding to the combinations of the two input values. The gray areas in Fig. 26.2b show the possible signal wire paths. If both input signals find a suitable CJoin, the signals are processed at that CJoin to produce the half adder's output.

To construct computational components using the Brownian circuit elements in [3] on other systems, the following elements need to be included: (a) signals, (b) signal wires that can be bent and crossed, and (c) logic elements (Hub, Conservative Join, and ratchet. It is also necessary to ensure that these logic elements can be connected by signal wires.

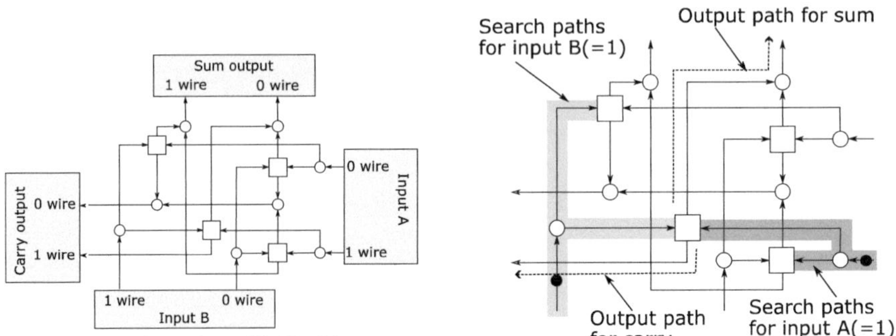

(a) Configuration of half adder

(b) Operation of half adder with inputs 1 and 1

Fig. 26.2 Implementation of a half adder in the Brownian circuit: **a** configuration of the half adder using dual rail encoding on signal wires, **b** operation of the half adder with inputs of 1 and 1

26.3 Cellular Automaton-Based Approach

26.3.1 Triangular Brownian Cellular Automaton: TBCA

In this section, we describe the implementation of a Brownian circuit based on an asynchronous CA. A CA is a system consisting of identical finite automata, known as cells, arranged in a regular spatial grid. Each cell has one of several possible states and is connected to its neighboring cells, allowing state exchanges between them. The states of the cells are updated according to predefined transition rules (transition functions). Unlike typical CAs, where all cells are updated simultaneously, the CA cells in this section operate asynchronously, meaning that not all cells are updated at the same time.

The CA discussed in this section is called a Triangular Brownian Cellular Automaton (TBCA), characterized by a two-dimensional array of identically shaped triangular cells (Fig. 26.3). Each cell has a finite number of states and can read its own state as well as the states of its three neighboring cells. If the states of these four cells conform to a specific pattern (as shown on the left-hand side of one of the transition rules), all four cells are updated simultaneously. The representation of the transition rule used in this section is illustrated in Fig. 26.4, and the following notation is employed:

$$(p_c, p_n, p_e, p_w) \rightarrow (p'_c, p'_n, p'_e, p'_w), \tag{26.1}$$

where p_c, p_n, p_e, amd p_w are the states of a cell and its adjacent cells before transition, and p'_c, p'_n, p'_e, and p'_w are their states after transition. This means that if the state of a cell is p_c, the state of its neighbor to the north is p_n, the state of its neighbor to the east is p_e, and the state of its neighbor to the west is p_w, then these cell states are set to p'_c, p'_n, p'_e, and p'_w, respectively. We assume that the transition rules have

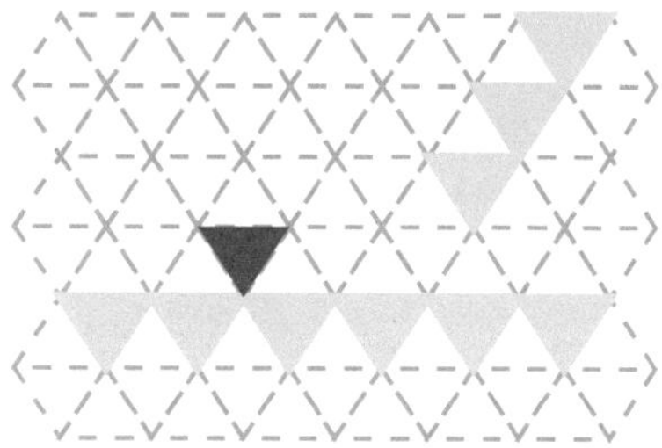

Fig. 26.3 Cellular space of TBCA

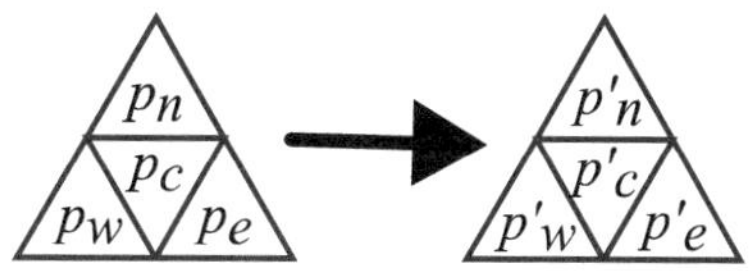

Fig. 26.4 Transition function in TBCA

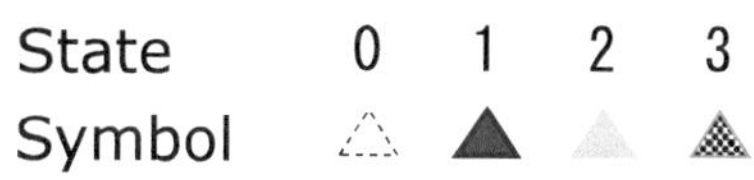

Fig. 26.5 Four states in TBCA

the rotation and reflection symmetric properties. If we define one transition rule as in Eq. (26.1), the following rules were also defined:

$$(p_c, p_n, p_w, p_e) \rightarrow (p'_c, p'_n, p'_w, p'_e), \quad (p_c, p_w, p_n, p_e) \rightarrow (p'_c, p'_w, p'_n, p'_e),$$
$$(p_c, p_w, p_e, p_n) \rightarrow (p'_c, p'_w, p'_e, p'_n), \quad (p_c, p_e, p_w, p_n) \rightarrow (p'_c, p'_e, p'_w, p'_n),$$
$$(p_c, p_e, p_n, p_w) \rightarrow (p'_c, p'_e, p'_n, p'_w).$$

In the TBCA, cell transitions do not need to be updated synchronously; cells undergo random transitions independently of each other. However, to prevent a cell from being updated in two different states simultaneously, updates are constrained so that neighboring cells never undergo transition at the same time.

26.3.2 Building Brownian Circuit Elements on TBCA

In this section, we show how to build the elements of a Brownian circuit using the TBCA and how these elements can be interconnected with signal wires built on the TBCA. There are four cell states, with symbols corresponding to these states shown in Fig. 26.5. The five defined transition rules are shown in Fig. 26.6.

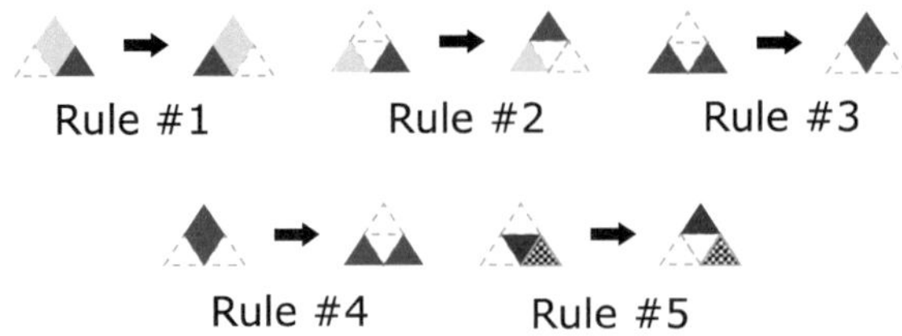

Fig. 26.6 Five transition rules in TBCA

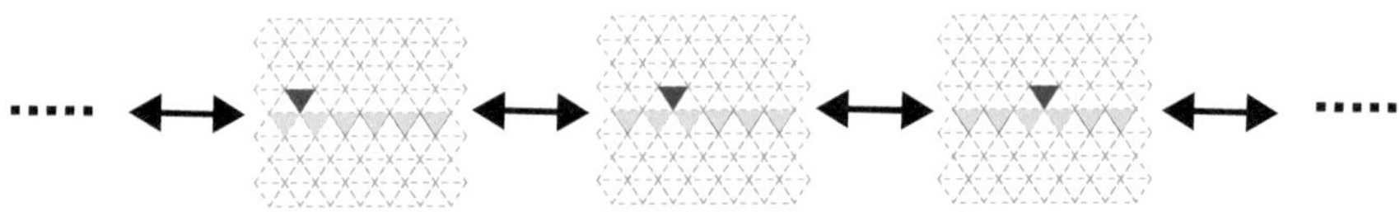

Fig. 26.7 Movement of a signal in TBCA applying the transition rule #2

26.3.2.1 Signal and Signal Wire

Initially, we present the construction of the signals and signal wires. Figure 26.7 shows the configuration of a signal on a straight-signal wire and its transitions according to the transition rule. The signals on the signal wires are represented by cells in state 1, and the signal wires are represented by an array of cells in state 2. By applying transition rule #2, along with the rules for rotation and inversion of this rule shown in Fig. 26.6, the signal can move along the signal wire. The direction in which transition rule #2 is applied determines the direction in which the signal moves, allowing the signal to move randomly along the signal wire.

26.3.2.2 Hub

The construction of the hub element is shown in Fig. 26.8a. In the TBCA, a hub can be constructed by arranging one signal line over another, allowing the signal to travel along either line by applying transition rule #2 for signal propagation. When a signal reaches a hub element junction, its output destination is determined by the updates of neighboring cells. In other words, the signals input to the hub element can be output from any terminal, including the one where the signal was originally input. The hub element allows the signal wire to bend at a 60° angle; even when bent, the signal can continue moving by applying transition rule #2.

The properties of the signal and signal wires are further described using another hub element, as shown in Fig. 26.8b. Due to the triangular shape of the cells in the TBCA, an alternative configuration for the signal, signal wire, or hub element can be defined with inversely symmetrical structures. For any cell with similar signal wires or hubs, signals can proceed according to transition rule #2; however, different types of signal wires cannot be directly connected, meaning signals cannot be transferred

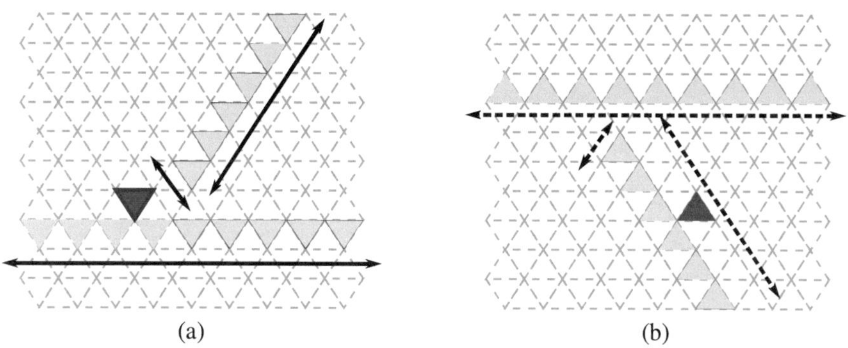

Fig. 26.8 Implementation of the Hub element in TBCA: **a** hub with a signal, **b** another type of hub. Due to the triangular shape of the cells, two types of signals and signal wires can be defined

between different them. A solution to this problem is the use of ratchet elements, which are described in the section on ratchet construction.

26.3.2.3 Crossing of Signal Wires

The elements for crossing signal wires are described below. Crossings between signal wires are necessary in most circuits. In ordinary electrical circuits, these crossings are typically achieved by arranging the signal wires in multiple layers. However, systems such as cellular automata do not have a layered structure in their cell space, making it necessary to implement an element specifically for crossing signal wires. In the TBCA, two different types of signal wires can be used as crossing wires. Figure 26.9 shows a crossing element with these two types of signal wires. The solid and dashed arrows in the figure indicate the areas where the two signals need to be crossed. Using transition rules #1 and #2, each signal in the crossed wires can move in the correct direction.

26.3.2.4 Conservative Join

In this section, the implementation of the CJoin element in the TBCA is described. The CJoin element is designed to transfer two signals to each of the two output terminals when two signals are present at both input terminals concurrently. The configuration of the CJoin element in the TBCA is shown in Fig. 26.10, where the solid and dashed arrows indicate the travel paths of the two signals. In this TBCA implementation, the positions of the input and output terminals differ from those in the original CJoin element (Fig. 26.10). However, this positional adjustment does not affect the operation, as the input and output terminal pairs are separately defined and function independently. The signal wires connected to these elements were specified according to their types, with the input and output wires being of different types.

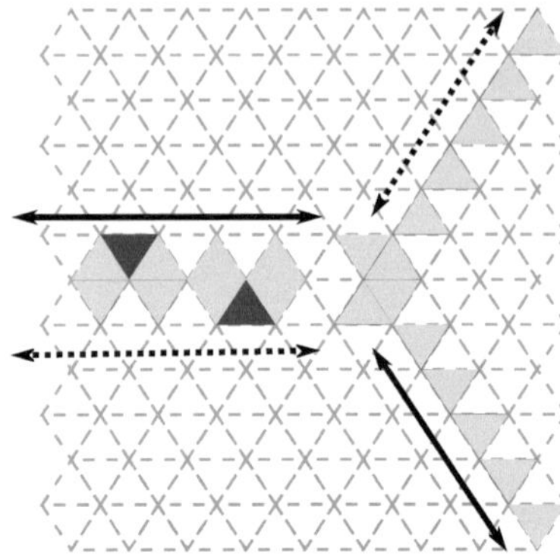

Fig. 26.9 Crossing element in TBCA. Different types of signal wires are connected to the input and output terminals of this element

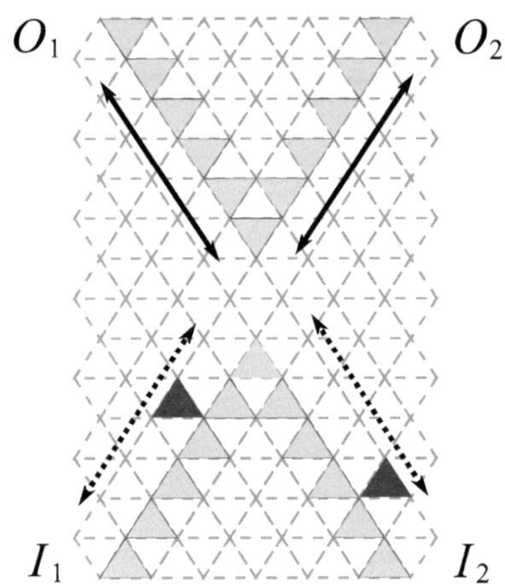

Fig. 26.10 CJoin element in TBCA. Different types of signal wires are connected to the input and output terminals of this element

When two signals reach the center of the CJoin element through the input wires simultaneously, the CJoin processes these signals and directs them to the output wires. The transfer of signals occurs in two steps, governed by transition rules #3 and #4.

26.3.2.5 Ratchet and Signal Wire Converter

The implementation of the ratchet element in the TBCA is also described. Figure 26.11 shows the configuration of the ratchet element on the TBCA, including the direction of the signal within this element. The ratchet element is constructed using state 3, with transition rule #5 governing its operation. When a signal enters the ratchet element, it can exit through the output terminal, but it is prevented from returning past the ratchet element.

The ratchet element is connected to different types of signal wires, causing the signal type to change as it passes through the element. This ability to convert signal

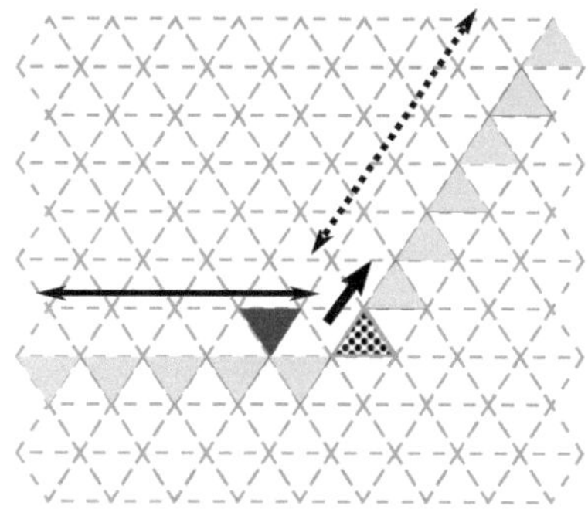

Fig. 26.11 Ratchet element in TBCA. Different types of signal wires are connected to the input and output terminals

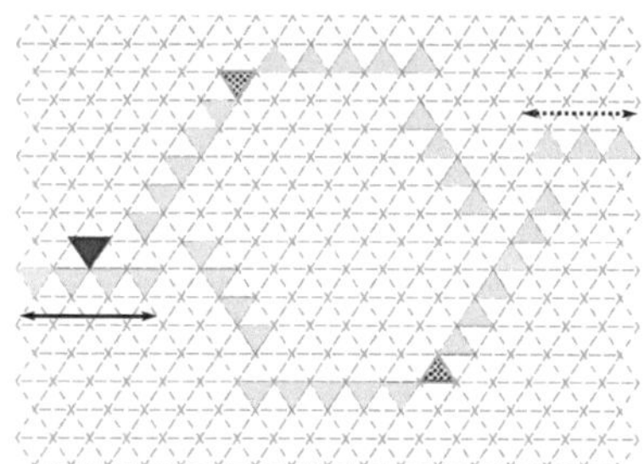

Fig. 26.12 Converter for changing signal wire types in TBCA by using ratchet elements

types is particularly useful, as different types of signal wires can be connected using ratchet elements. Figure 26.12 shows the configuration of a signal wire converter using two ratchets, where the input and output terminals of the two ratchets are connected. This setup allows one type of signal to be transformed into another. This conversion capability is beneficial for elements like the Crossing and CJoin, where the signal wire types at the input and output terminals are otherwise fixed.

26.3.2.6 Computational Universality on TBCA

The computational power of the TBCA is equivalent to that of the original Brownian circuit, provided the cellular space of the TBCA is sufficiently large. This equivalence is demonstrated by the ability to construct all elements of the Brownian circuit within the TBCA and connect them accordingly. As discussed in previous sections, each element can be implemented on the TBCA and placed anywhere within the cellular space. Some elements, such as the Crossing and CJoin, are restricted by the types of signal wires attached to their input and output terminals. However, these types can be modified using a signal wire converter with ratchets. Although some additional cell space is required to configure the circuit, the elements configured in the cellular space can still be connected. Consequently, the TBCA can replicate any structure of the Brownian circuit within the cellular space, performing the same operations as the original circuit. Therefore, the TBCA and the Brownian circuit possess the same computational capabilities, confirming the TBCA's computational universality.

26.4 Swarm Agents-Based Approach

26.4.1 *Swarm Agents*

In this section, we present an alternative implementation of the Brownian circuit called Swarm Agents. The system consists of several robots, each with internal states that define its specific functions. These robots are configured in two-dimensional space, similar to the TBCA; unlike TBCA, certain robots in this system can move within the space and interact with others. By connecting to each other, these robots can form groups that make up the various components of the Brownian circuit. Four distinct types of robots have been defined in this system, each playing a role in constructing the elements of the Brownian circuit. Figure 26.13 illustrates the robots and their respective states. In this configuration, a signal within the Brownian circuit is represented by robots in state 1, which moves through the space. The structural elements of the Brownian circuit, such as the signal wire, Hub, CJoin, and ratchet, are constructed using robots in state 2, which function as guide rails for the signals moving within the space. Additionally, robots in states 3 and 4 are utilized to build the Hub and CJoin functionalities, contributing to the overall operation of the system.

26.4.2 *Building Brownian Circuit Elements in Swarm Agents*

26.4.2.1 Signal and Signal Wire

In the Brownian circuit, a signal is represented by a robot in state 1, which moves through a two-dimensional space by exerting a forward-directed force. To control the direction of this signal, state 2 robots are employed. These robots are pinned and aligned in space, allowing the signal to move along them by colliding with the state 2 robots. Figure 26.14 shows this setup, where the signal is represented by a state 1 robot, and the signal wire is formed by state 2 robots. The collisions with the state 2 robots can randomly alter the signal's direction, making these robots function as walls or signal wires that control the signal's path. The behavior of these signals closely resembles that of a Brownian circuit, where signals can move forward or backward along the signal wire by interacting with the signal wire.

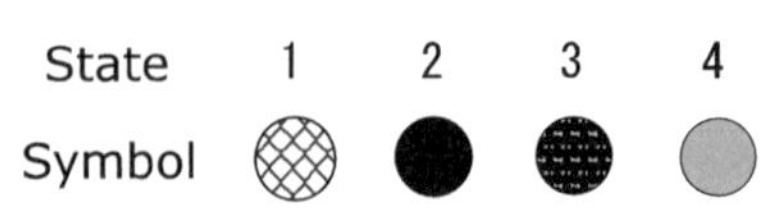

Fig. 26.13 Four states for robots in Swarm Agents

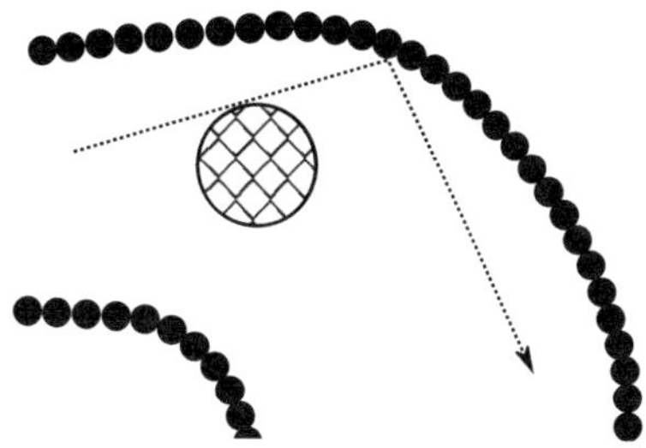

Fig. 26.14 Signal with signal wire. Robots in state 2 act as guide rails for signals

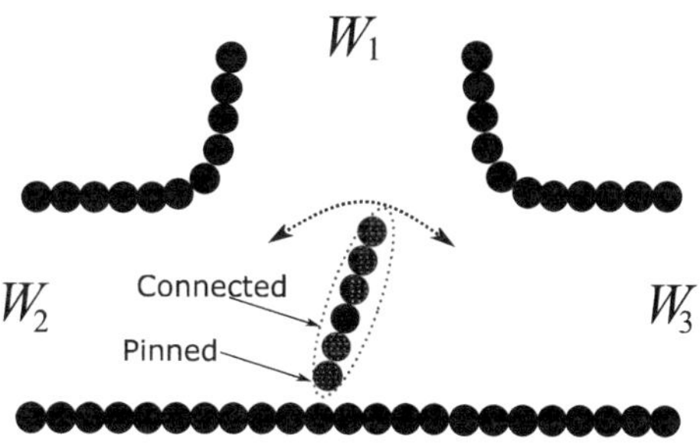

Fig. 26.15 Implementation of the hub element in Swarm Agents

26.4.2.2 Hub

Robots in state 2 are useful in constructing various mechanisms for controlling signal direction. Figure 26.15 shows the configuration of a hub element, where robots in state 2 form the primary structure of the hub, and robots in state 3 create a beam structure. One tip of the beam, composed of robots in state 3, is pinned in place, allowing it to rotate around this point, while the robot at the other tip exerts force in either the left or right direction, enabling the beam to rotate similarly to a windshield wiper. When a signal enters a terminal in this hub, the rotating beam and the hub's structure alter the signal's direction, eventually guiding it to exit through another terminal.

26.4.2.3 Conservative Join

The implementation of a CJoin is shown in Fig. 26.16, resembling a mechanical ratchet rather than the ratchet used in the Brownian circuit. This configuration includes a wheel gear that rotates counterclockwise and a pawl that prevents the wheel gear from rotating. The wheel gear is composed of robots in state 4, with the central robot pinned, allowing the gear to rotate in space. The pawl is also formed by a series of robots in state 4, where one tip is pinned, exerting a downward force. The configuration is fine-tuned to ensure that the downward force of the pawl is weaker than the force exerted by the rods in the wheel when pushed by two signals. Figure 26.17 shows the operation of the CJoin when two signals arrive simultaneously. These signals push the rods of the wheel, causing the pawl to release the rod,

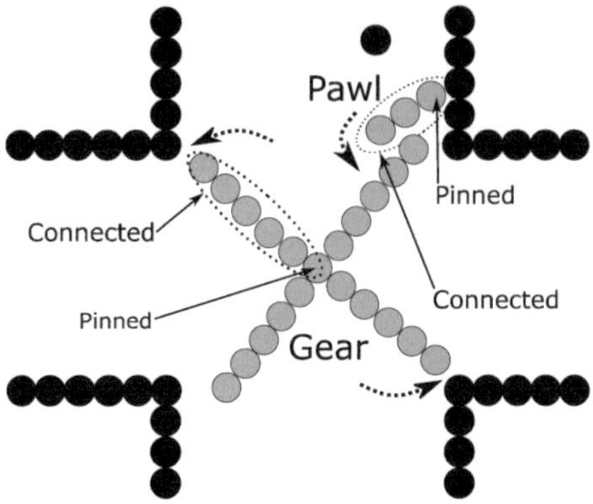

Fig. 26.16 Configuration of CJoin in Swarm Agents, including a gear wheel and pawl

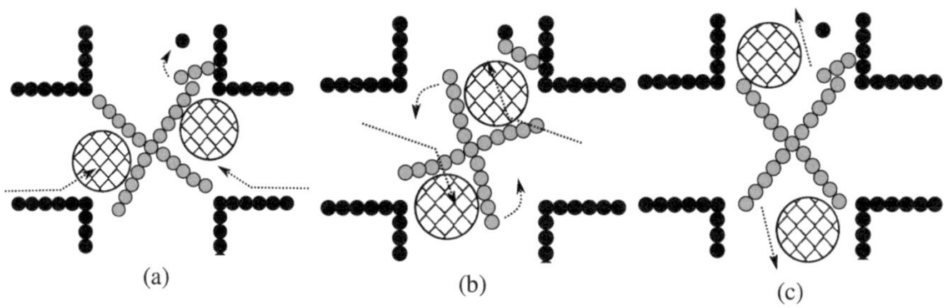

Fig. 26.17 Operations of the CJoin with two signals: **a** input signals can push the wheel, **b** each signal proceeds to its output terminal, and **c** each signal exits the CJoin element

allowing the wheel to rotate counterclockwise. Consequently, each signal is released from the output terminals, and the wheel is locked again by the pawl.

26.4.2.4 Ratchet

A ratchet in the Brownian circuit can be constructed using robots in state 2, similar to the construction of signal wires, as shown in Fig. 26.18. In this configuration, the input and output terminals of the ratchet are placed on the left and right sides, respectively. Figure 26.19a shows the operation of a ratchet when a signal is present at the input terminal. The signal flows along the path in the ratchet and exits through the output terminal. However, if a signal is at the output terminal and attempts to return to the input terminal, the ratchet's path prevents the signal from returning, as shown in Fig. 26.19b. This structure does not guarantee that signals will never pass from the output to the input; however, this does not critically impact the computational power of the system.

Fig. 26.18 Configuration of the ratchet element, with the input terminal on the left and the output terminal on the right

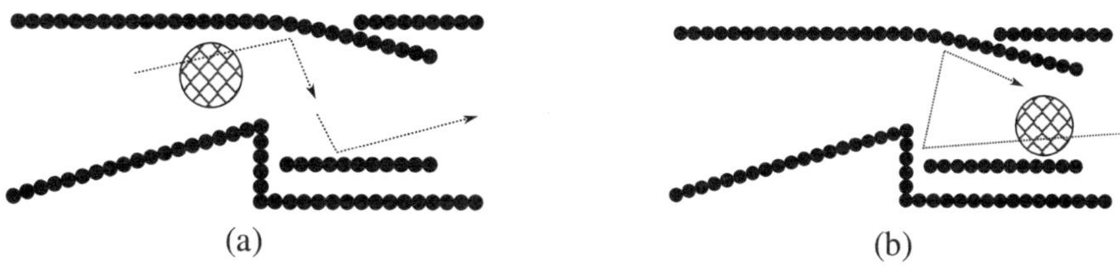

Fig. 26.19 Operations of the ratchet element: **a** a signal passes from the input to the output terminals and **b** a signal struggles to return from the output terminal

Fig. 26.20 Configuration of the crossing element used to cross two signal wires

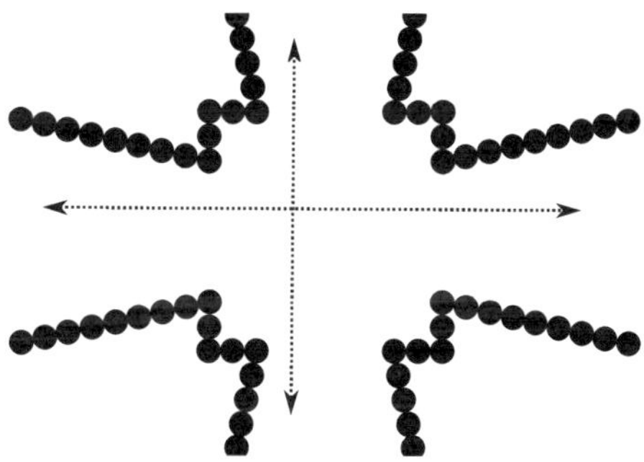

26.4.2.5 Crossing of Signal Wires

A ratchet structure can be used to adjust the positions of signals and construct the crossing of signal wires. Figure 26.20 shows the implementation of the crossing element for the signals, utilizing only robots in state 2. The operation of signal crossing is illustrated in Fig. 26.21, where one signal arrives at the north input and exists through the south output, while the other signal moves from the west input to the east output. This implementation does not prevent signal collisions within the element; hence, it is necessary to ensure that the frequency of the signals in the circuit is not high.

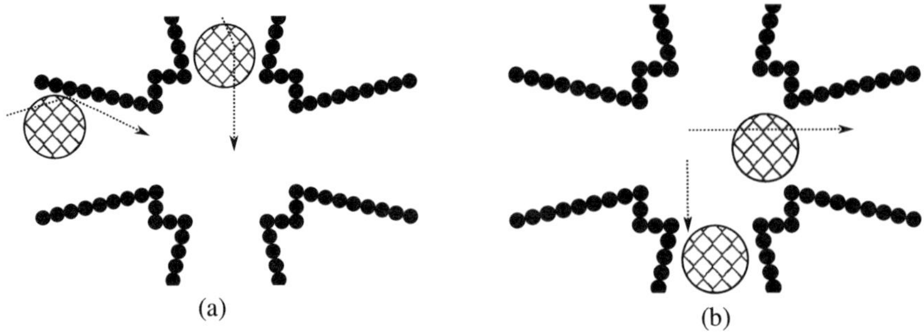

Fig. 26.21 Operations of the crossing element: **a** the first signal moves from north to south through the element, **b** the second signal moves from west to east

26.4.2.6 Computational Universality on Swarm Agents

The computational power of the Swarm Agents is equivalent to that of the original Brownian circuit. In this system, all elements can be constructed using a Swarm Agents, with these elements connected by signal wires. By employing the elements implemented by the Swarm Agents, the operations in the Brownian circuit can be replicated by the swarm robots, ensuring that the computational results of both the Brownian circuit and Swarm Agents are identical.

26.5 Summary

In this section, we present a computational scheme using distributed elements, potentially useful for implementation in molecular systems. The Brownian circuit, described in Sect. 26.2, serves as a foundational computational model due to its asynchronous operation, eliminating the need for clock signals to synchronize elements and allowing for signal fluctuations within the system. Two systems have been presented for implementing the Brownian circuit: the TBCA and Swarm Agents.

References

1. Isokawa, T., Peper, F., Lee, J.: On computatinal universality for a triangular Brownian cellular automaton with 4 states and 5 rules. In: Proceedings of 24th IFIP WG 1.5 International Workshop on Cellular Automata and Discrete Complex Systems (Local proceedings, AUTOMATA2018), pp. 33–44 (2018)
2. Kimura, M., Isokawa, T., Peper, F., Nakade, S., Lee, J., Kamiura, N.: On implementation of brownian circuits by swarm agents. In: Proceedings of 2023 11th International Symposium on Computing and Networking Workshops (CANDARW), pp. 114–120 (2023)

3. Lee, J., Peper, F.: On Brownian cellular automata. In: Proceedings of Automata 2008, pp. 278–291 (2008)
4. Lee, J., Peper, F., Cotofana, S., Naruse, M., Ohtsu, M., Kawazoe, T., Takahashi, Y., Shimokawa, T., Kish, L., Kubota, T.: Brownian circuits: designs. Int. J. Unconv. Comput. **12**(5–6), 341–362 (2016)
5. Murata, S., Konagaya, A., Kobayashi, S., Hagiya, M.: Molecular robotics: a new paradigm for artifacts. New Gen. Comput. **31**(1), 27–45 (2013)
6. Murata, S., Toyota, T., Nomura, S.I.M., Nakakuki, T., Kuzuya, A.: Molecular cybernetics: challenges toward cellular chemical artificial intelligence Advan. Funct. Mater. **32**(37), 2270207 (2022)
7. Peper, F., Lee, J., Carmona, J., Cortadella, J., Morita, K.: Brownian circuits: fundamentals. ACM J. Emerg. Technol. Comput. Syst. **9**(1), 3:1—24 (2013)

Chapter 27
The Critical Role of Resource Competition in Designing Synthetic Genetic Circuits

Naosato Takagi and Daisuke Kiga

Abstract Synthetic biology aims to engineer novel genetic circuits with predictable behavior for diverse applications in biotechnology, medicine, and research. However, inherent resource competition within biological systems poses significant challenges to achieve this predictability. This review highlights the importance to account resource competition in the design and analysis of synthetic genetic circuits. These studies revealed how competition for regulatory proteins, proteases, ribosomes, ATP, DNA-binding sites, and other cellular resources affects the functionality and stability of these circuits. Therefore, this discussion underscores the need to enhance the robustness and stability of molecular systems by harnessing their inherent features. It also offers valuable insights into unexplored design principles that have yet to be applied in the field of molecular robotics.

27.1 Competition in Signal Transduction System for Regulatory Proteins and Proteases

The concept of retroactivity, introduced by Domitilla Del Vecchio in 2008, is crucial for understanding how downstream components influence the upstream elements of transcriptional networks [1]. Retroactivity is similar to the output impedance in electrical systems, where feedback from downstream components affects upstream behavior. This is significant when the concentrations of transcription factors are comparable to or lower than those of the promoter-binding sites. This was also significant when the binding affinity was high. To mitigate retroactivity, researchers have proposed insulation mechanisms based on electronic amplifiers. This approach relies on feedback mechanisms using phosphorylation-dephosphorylation cycles, which can attenuate retroactivity due to their fast reaction times.

N. Takagi · D. Kiga (✉)
Waseda University, Tokyo, Japan
e-mail: kiga@waseda.jp

N. Takagi
e-mail: naosato-bsbl8kyd@ruri.waseda.jp

S. Murata et al. (eds.), *Molecular Robotics II*,
https://doi.org/10.1007/978-981-95-2394-8_27

Building on this, Jayanthi's 2013 study examined the temporal dynamics of gene transcription influenced by retroactivity [2]. The study showed that retroactivity could cause a sign-sensitive delay, with a delayed response to induction and a faster response to de-induction, compared to an isolated system. This delay is due to sequestration of transcription factors by downstream binding sites. This study provides mathematical tools for predicting the effect of retroactivity on gene circuits, aiding in the design and analysis of synthetic biological systems.

In 2012, Tek-Hyung Lee and Narendra Maheshri explored the regulatory role of repeated decoy transcription factor-binding sites on gene expression [3]. Using a synthetic system in yeast, researchers showed that repeated decoy sites decrease target gene expression by sequestering transcription factors and reducing their ability to activate target promoters. This study concluded that decoy-binding sites could significantly alter gene expression by sequestering transcription factors, offering a mechanism for fine-tuning gene expression in both natural and synthetic systems.

Moriya's 2014 study comprehensively examined how downstream genes, such as reporter genes, influence the dynamics of synthetic genetic circuits [4]. Researchers have identified two primary effects: competition for regulatory proteins and competition for proteases. The regulatory proteins expressed in synthetic genetic circuits are often limited in quantity. When downstream genes are introduced, their promoters compete with those of the regulatory genes for binding to the regulatory proteins (Fig. 27.1a). This significantly alters gene expression dynamics and the overall behavior of the genetic circuit. For instance, the introduction of a downstream green fluorescent protein (GFP) reporter gene alters the oscillation dynamics of a synthetic genetic circuit model. This competition is similar to the effect caused by an increase in the number of decoy sites in a genetic circuit, which decreases the expression levels by binding activator proteins.

The second significant effect is the competition of enzymes for protein degradation. Proteins within a cell often share the same degradation machinery, such as the ubiquitin–proteasome system. When downstream genes were introduced, the downstream reporter proteins competed with the regulatory proteins for the same degradation mechanisms (Fig. 27.1b). This influences the stability and levels of the

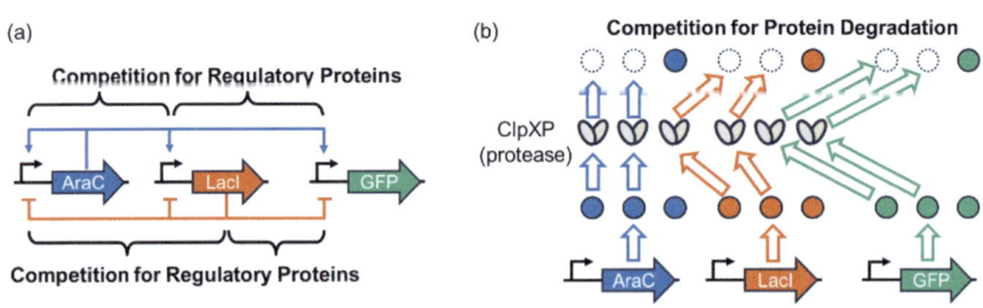

Fig. 27.1 Competition for Regulatory Proteins and Protein Degradation. **a** Competition for regulatory proteins (AraC, LacI) by GFP promoter. AraC activates transcription, while LacI represses transcription. **b** Competition for protein degradation (ClpXP) by GFP protein

regulatory proteins, thereby affecting the dynamics of the genetic circuit. In this study, we demonstrated that introducing a GFP reporter gene led to a longer oscillation period in a synthetic genetic circuit model. This was due to competition for the ClpXP protease, which is responsible for the degradation of SsrA-tagged proteins, including AraC, LacI, and GFP. This increased competition resulted in a longer falling sub-phase during the oscillation period, indicating reduced degradation efficiency of the regulatory proteins.

In 2019, Moriya et al. explored the effects of retroactivity and competition for proteases using in silico and in vivo experiments [5]. Using mathematical modeling, they demonstrated that a genetic regulatory circuit can differentiate the dynamics when it is connected to either of the two reporter genes, which differ only in their promoter sequences. The researchers confirmed these predictions by observing the oscillations in GFP reporter expression using a microscope. Further modeling showed that different perturbations in upstream regulatory circuits are caused by competition from target proteins for their tag-specific proteases, rather than from regulatory protein-binding sites on promoter DNA to regulatory proteins.

They also drew parallels between the molecular competition in genetic circuits and the competition for voltage in electronic circuits, proposing that the output impedance of the shared DC power is isomorphic to the amount of a shared cellular component, such as the tag-specific protease. This analogy helps in understanding the trade-offs between robustness and energy consumption in genetic circuits. These findings underscore the necessity of considering retroactivity and protease sharing when introducing downstream genes to construct comprehensive mathematical models that can accurately predict interactions within synthetic genetic circuits.

27.2 Competition in Translation System for Ribosome by mRNA

Chizzolini et al. [6] demonstrated that in the cell-free transcription-translation (PURE) system, genes positioned earlier in an operon are expressed at higher levels than those located further downstream. This positional effect was more pronounced than the influence of the T7 transcriptional promoter strength on protein expression [6]. This study highlights the importance of gene position within operons to optimize protein yields in synthetic biology applications, particularly in cell-free systems where transcription and translation are decoupled from the cellular context.

Siegal-Gaskins et al. [7] explored the dynamics of resource competition in a cell-free system using *Escherichia coli* extracts, focusing on the relationship between DNA concentration and resource competition [7]. Their findings indicated that as DNA concentrations increased, competition for transcriptional and translational machinery intensified, leading to reduced efficiency in gene expression. This study underscores the necessity of balancing DNA template concentrations to mitigate resource competition and ensure predictable performance of synthetic genetic circuits.

Senda et al. [8] investigated the intricacies of competition for ribosomes in cell-free systems, particularly when T7 promoter variants were used to tune the expression levels [8]. Researchers first measured the activity of each T7 promoter variant with one or two base substitutions at critical sequence positions in a single cell-free solution. By simultaneously expressing multiple fluorescent proteins, each at various expression levels, they further comprehensively analyzed how competition for ribosomes impacted protein synthesis. Their research revealed that at low total mRNA concentrations in a test tube, ribosomal competition was minimal, allowing for predictable protein expression based on the activity of each T7 promoter variant. However, as the mRNA concentration increases, the competition for ribosomes intensifies. As a result, the translation of each protein could not be directly predicted from each promoter activity alone, but was significantly influenced by the ribosome-binding site (RBS) strength. Messenger RNAs with stronger ribosome-binding sites (RBSs) were favored. Weaker RBSs are more adversely affected by competition, leading to lower protein yields and reduced predictability of protein expression. This study highlights the need to optimize promoter and RBS strength by considering the competition for ribosomes when designing multigene expression systems.

Enomoto et al. [9] improved the predictability of in vivo gene expression from in vitro gene expression by identifying and managing translational bottlenecks [9]. Achieving a clear correlation between in vitro and in vivo gene expression is crucial for the development of molecular breadboards, allowing rapid prototyping and testing of genetic circuits in cell-free systems before transferring them to living cells. Their study revealed that high mRNA levels could overwhelm the translational machinery, creating a bottleneck that disrupts the correlation between mRNA and protein levels. This issue is particularly evident when using strong promoters, such as the T7 promoter, which is widely employed in genetic engineering. Significant differences in mRNA levels among the promoter variants resulted in minimal differences in protein levels. This is because excess mRNA generated by high concentrations of T7 RNA polymerase is not being processed efficiently by ribosomes, creating a translational bottleneck. To address this issue, researchers have adjusted experimental conditions to avoid translational bottlenecks. By reducing the concentration of the T7 RNA polymerase inducer, mRNA levels were lowered to within the capacity of the translational machinery. Under these optimized conditions, the in vitro and in vivo protein expression levels from a set of T7 promoter variants correlated well. This finding underscores the importance of fine-tuning transcriptional activity to match the translational capacity of the cell, thereby ensuring that the observed protein levels accurately reflect promoter strength. This study's approach of modulating the inducer concentration to manage mRNA levels provides a practical strategy for circumventing translational bottlenecks, which can be applied to other genetic systems and organisms.

27.3 Competition in Metabolic System for ATP Between Glycolysis and Translation

Sato et al. explored the intricate competition for ATP between glycolysis and protein synthesis [10]. Purified glycolytic enzymes from *E. coli* were combined with a commercial reconstituted transcription-translation system known as the PURE system. The creatine phosphate-creatine kinase (CP-CK) system was excluded from the PURE system. The reconstituted coupling system was named G-PURE. The initial substrate for glycolysis was glucose-6-phosphate (G6P). G-PURE allowed a detailed examination of how shared elements, particularly ATP and phosphate (Pi) influenced the success of system coupling.

This study revealed that the kinetic properties of the systems determined the fate of their coupling. Glycolysis regenerates ATP from phosphate (Pi) and ADP, whereas protein synthesis consumes ATP and supplies both Pi and ADP. The success of protein synthesis depends on the initial glycolytic substrate. When fructose-1,6-bisphosphate (FBP) was used, the system successfully synthesized sfGFP from the reporter DNA. However, when G6P was used, the system failed because of early ATP consumption by glycolysis, which reduced the ATP available for the PURE system and decreased the supply of Pi and ADP necessary for glycolysis.

To further understand these dynamics, researchers developed a kinetic model called the G-PURE simulator. This model uses ordinary differential equations to describe the behavior of glycolysis and protein synthesis, allowing for simulations that align well with experimental results. The simulations revealed that the initial concentrations of common elements such as Pi and ADP are critical for the coupling of glycolysis and the PURE system. Without sufficient Pi, ATP is converted to ADP because of its consumption by the upstream glycolysis pathway. Additionally, while initial ADP is not essential for glycolytic flux, it contributes to the final ATP concentration and affects the efficiency of protein expression.

Experimental confirmation of these findings showed that higher glycolysis levels increased the Pi demand, which was consistent with the simulation results. Increasing initial ADP concentrations decreased Pi demand monotonically, regardless of the glycolysis level.

We also explored the role of exogenous ATPases in facilitating system coupling. The addition of *E. coli* glutamine synthase (GlnA) as an ATPase to the PURE system inhibited protein synthesis. However, addition of GlnA to the G-PURE system restored the coupling. The researchers demonstrated that exogenous ATPase could restore coupling in a concentration-dependent manner, confirming that it could modulate the availability of shared elements such as ATP and Pi.

In conclusion, this study highlights the importance of understanding the demand for shared elements, such as ATP and Pi, to determine the success of coupling glycolysis and protein synthesis. Additionally, it is necessary to consider that the demand for these elements varies with the balance of the constituent systems. These findings provide new insights into the metabolic tug-of-war between glycolysis and ATP utilization systems, which is crucial for creating a synthetic metabolism and cells capable of self-reproduction by integrating reconstituted biological pathways.

27.4 Cell Loading and Competition in Transcription System for DNA-Binding Sites by Transcriptional Regulators

Ceroni's 2015 study underscores the importance of mitigating the burden of synthetic constructs imposed on host cells due to competition for limited resources, such as transcriptional and translational machinery [11]. This study introduced a capacity monitor, a fluorescence-based method for the real-time measurement of bacterial gene expression capacity, using a synthetic constitutive GFP expression cassette. Quantifying these resources, the researchers highlighted that constructs with weak RBS and high-copy plasmids are more efficient, reducing the cellular burden and leading to greater efficiency, including both construct output and cell capacity.

In a 2018 study, researchers combined RNA-seq with an in vivo assay to identify the major transcriptional changes that occur in *E. coli* when inducible synthetic constructs are expressed [12]. RNA-seq data revealed that gene overexpression led to resource reallocation and significantly affected transcription. All tested constructs triggered the upregulation of σ32-regulated promoters, indicating that the heat-shock response is activated when native genes are expressed in non-native contexts. This finding suggests that the burden involves not only polymerases and ribosomes but also limited pools of post-translational regulation machinery, such as chaperones and proteases. By leveraging these promoters, researchers have developed a dCas9-based feedback regulation system that automatically adjusts the expression of synthetic constructs in response to the cellular burden. Cells equipped with this system demonstrated robust growth and enhanced protein yield, highlighting the effectiveness of this approach in mitigating the effects of synthetic gene expression.

Building on these findings, De Marchi et al. [13] explored the effects of resource competition and regulator crosstalk in more complex synthetic circuits using Staphylococcus aureus dead-Cas9 (sadCas9) as a programmable transcriptional regulator [13]. Researchers constructed and analyzed approximately 20 circuits, including static logic gates and dynamic incoherent feedforward loops (IFFLs), to investigate the impact of different designs on cellular resources and circuit behaviors. Findings revealed that the expression of reporter genes, such as Red Fluorescent Protein (RFP) can impose a substantial load on cellular resources, leading to nonlinear and unpredictable behavior in circuits. For instance, in NAND gate circuits, high expression levels of reporter genes deplete cellular resources, affecting the expression of other genes and altering the dynamic range, switch point, and slope of the input–output response curves.

Moreover, researchers have highlighted the competition for DNA-binding sites by transcriptional regulators as a significant factor influencing circuit performance.

Binding of sadCas9 to its target sites can interfere with other regulatory elements, leading to unexpected outcomes. This crosstalk is particularly evident in incoherent feedforward loops (iFFLs) circuits, where the proximity of the binding sites for LuxR and sadCas9 complexes results in multiphasic output behaviors. The Monod–Wyman–Changeux (MWC) mathematical model used in this study successfully captured these interactions, suggesting that the non-independent binding of the two regulators to DNA could generate complex output behaviors. This phenomenon underscores the importance of considering regulatory crosstalk and resource competition in the design and analysis of synthetic genetic circuits.

The orthogonality and specificity of the sadCas9 system were also demonstrated, showing that it could independently operate within the same cell alongside the traditional Streptococcus pyogenes-derived dCas9 (spdCas9) without crosstalk. This finding is significant for designing large-scale synthetic circuits, as it allows the simultaneous use of multiple CRISPRi systems to target different genes without interference. Additionally, researchers have demonstrated the utility of the sadCas9 system in fine-tuning synthetic circuits to overcome challenges associated with the transcriptional burden. By adjusting the expression of sadCas9, researchers were able to restore the desired input-dependent activation trend of a transcriptional activator cascade involving T7 RNA polymerase and the PT7 promoter.

In conclusion, these studies collectively emphasized the critical roles of resource competition and regulatory crosstalk in the design and functionality of synthetic genetic circuits. Understanding and mitigating these factors is essential for achieving predictable and stable behavior in engineered biological systems.

27.5 Conclusion

Resource competition within biological systems profoundly influences the design and functionality of synthetic genetic circuits. Studies have shown that the competition for regulatory proteins, degradation pathways, ribosomes, ATP, and DNA-binding sites can significantly affect gene expression dynamics and circuit stability. Addressing these challenges requires comprehensive mathematical models and strategic design approaches to mitigate resource competition and ensure predictable behavior. By understanding and managing these competitive interactions, synthetic biology can advance toward more reliable and efficient applications in biotechnology, medicine, and research.

References

1. Del Vecchio, D., et al.: Modular cell biology: retroactivity and insulation. Mol. Syst. Biol. **4**, 161 (2008)
2. Jayanthi, S., et al.: Retroactivity controls the temporal dynamics of gene transcription. ACS Synth. Biol. **2**, 431–441 (2013)
3. Lee, T.-H., Maheshri, N.: A regulatory role for repeated decoy transcription factor binding sites in target gene expression. Mol. Syst. Biol. **8**, 576 (2012)
4. Moriya, T., et al.: Effects of downstream genes on synthetic genetic circuits. BMC Syst. Biol. **8**, S4 (2014)
5. Moriya, T., et al.: Comparison between effects of retroactivity and resource competition upon change in downstream reporter genes of synthetic genetic circuits. Life **9**, 30 (2019).
6. Chizzolini, F., et al.: Gene position more strongly influences cell-free protein expression from operons than T7 transcriptional promoter strength. ACS Synth. Biol. **3**, 363–371 (2014)
7. Siegal-Gaskins, D., et al.: Gene circuit performance characterization and resource usage in a cell-free "breadboard." ACS Synth. Biol. **3**, 416–425 (2014)
8. Senda, N. et al.: Development of an expression-tunable multiple protein synthesis system in cell-free reactions using T7-promoter-variant series. Synth. Biol. 7, ysac029 (2022)
9. Enomoto, T. et al.: Correlation between in vitro and in vivo gene-expression strengths is dependent on bottleneck process. New Generat. Comput. (2024)
10. Sato, G., et al.: Metabolic tug-of-war between glycolysis and translation revealed by biochemical reconstitution. ACS Synth. Biol. **13**, 1572–1581 (2024)
11. Ceroni, F., et al.: Quantifying cellular capacity identifies gene expression designs with reduced burden. Nat. Methods **12**, 415–418 (2015)
12. Ceroni, F., et al.: Burden-driven feedback control of gene expression. Nat. Methods **15**, 387–393 (2018)
13. De Marchi, D., et al.: Design and model-driven analysis of synthetic circuits with the Staphylococcus aureus Dead-Cas9 (sadCas9) as a programmable transcriptional regulator in bacteria. ACS Synth. Biol. **13**, 763–780 (2024)

Part V
Journalist-in-Residence for Collaborative Engagement Between Journalists and Scientists

Chapter 28
Beyond Merely Reporting Science: A Reflection of the Journalist-in-Residence Approach in the "Molecular Cybernetics" Project

Kanako Takae and Mikihito Tanaka

28.1 Introduction

In the modern world, science communication involving researchers, government, and citizens has become a normative and essential practice to democratically guide the development of science and technology. Among various forms of science communication, the media plays a crucial role through science writing, serving as a bridge between the scientific community and the public.

As Tanaka points out, science writing broadly encompasses two distinct but interrelated functions: *science reporting* and *science journalism* [1]. The former focuses on disseminating scientific knowledge in an accessible and accurate manner, aiming to raise public scientific literacy and support informed decision-making. The latter, by contrast, emphasizes placing scientific knowledge within broader social and political contexts, and fosters critical dialogue around science-related public issues. While science reporting highlights the transmission of expert knowledge from scientists to the public, science journalism regards both experts and laypeople as participants in a shared communicative space, frequently addressing controversial or value-laden aspects of science. This conceptual distinction helps clarify the multiple roles the media can play in shaping public understanding and engagement with science.

K. Takae
Institute for Advanced Social Sciences, Waseda University, Nishiwaseda, Shinjyuku-Ku, Tokyo, Japan

M. Tanaka (✉)
Faculty of Political Science and Economics, Waseda University, Nishiwaseda, Shinjyuku-Ku, Tokyo, Japan
e-mail: mikihito.tanaka@gmail.com

S. Murata et al. (eds.), *Molecular Robotics II*,
https://doi.org/10.1007/978-981-95-2394-8_28

The different norms and functions of science reporting and science journalism, however, create an inherent tension between these two forms of science communication. As Tanaka explains, science reporting typically presupposes trust in the scientific community, granting scientists a form of "prior loyalty," meaning an assumed allegiance to scientific authorities as primary and legitimate sources of knowledge [1]. Science journalism, on the other hand, critically engages with science as a socially embedded activity. This critical stance often involves questioning or challenging mainstream scientific views, especially when scientific institutions face dysfunctions such as misconduct, conflicts of interest, or system-wide failures. Such a stance can clash sharply with expectations held by scientists, who may perceive journalistic critiques as inappropriate or even "anti-science." Thus, the normative differences between scientists and journalists, who operate under fundamentally different professional standards, often lead to friction in the context of science communication [2–4].

Nevertheless, the tension between science reporting and science journalism should not necessarily be viewed negatively. Indeed, a certain degree of critical tension can be considered essential for maintaining a healthy, independent journalistic practice. Rather than fostering cozy or overly harmonious relationships, journalism must function as an independent watchdog, capable of critically monitoring the scientific community, which holds significant epistemic authority and power. Robust science journalism thus requires a critical distance to effectively hold the scientific establishment accountable, especially given the influential role scientists and their knowledge play in shaping modern society.

Yet, framing science journalism solely as the exclusive domain of journalists—and depicting it dualistically as a conflict between journalists and scientists—can obscure the importance of scientists' own engagement with society. Such a perspective risks overlooking scientists' roles as active participants who share responsibility for addressing the societal implications of science. Indeed, scientists themselves have often been called upon to leave their "ivory towers" and communicate more directly with the public, thereby becoming active agents in science communication rather than mere passive sources of knowledge. Recognizing scientists as engaged communicators emphasizes that science journalism is collaborative, multi-directional process that relies on constructive dialogue between scientists and journalists, rather than on confrontation or isolation between the two.

To foster a better relationship between journalists and science, several initiatives have emerged to promote mutual understanding. One notable example is the Journalist-in-Residence (JIR) program, in which journalists spend extended periods embedded within research institutions, gaining firsthand insight and producing science-related content. Inspired by Artist-in-Residence (AIR) initiatives, where artists develop creative work while residing at specific locations, JIR programs similarly aim to enhance mutual understanding between journalists and scientists, thereby enriching science communication. Such initiatives are increasingly implemented globally (e.g., the Max Planck Institute) and have also emerged in Japan, including at institutions like Kyoto University's Mathematical Society.

Given this context, we experimented with our own JIR initiative to communicate and discuss the emerging field of molecular cybernetics. Our JIR trial took place within a large-scale project titled "Molecular Cybernetics: Development of the Minimal Artificial Brain by the Power of Chemistry," launched in 2020 with support from the Japan Society for the Promotion of Science (JSPS), aiming to explore chemical AI.

This chapter begins by situating our Journalist-in-Residence (JIR) program within the broader context of science journalism and Ethical, Legal, and Social Issues (ELSI), highlighting current challenges posed by PR-driven science communication. We then turn to the practical unfolding of our JIR program, examining early challenges and collaborative efforts undertaken by journalists and scientists. Finally, we discuss key issues and lessons learned from the project, with particular focus on fostering ELSI engagement and developing collaborative relationships between journalists and scientists—while preserving the critical and independent nature of journalism. Furthermore, this chapter is followed by contributions from three JIR participants—a science magazine journalist, a science fiction author, and a newspaper journalist—each offering distinct perspectives from their fields.

28.2 The Journalist-in-Residence Approach in Science Communication

This subsection explores the background of our JIR program, focusing on its relationship to Ethical, Legal, and Social Issues (ELSI) and the current challenges facing science journalism in a PR-driven media environment.

28.2.1 Rationale and Early Attempts at JIR

Journalist-in-Residence (hereafter JIR) program now plays a significant role in science communication. Previous studies suggest that JIR programs effectively promote both the dissemination of knowledge and public understanding about complicated issues, including science [5, 6]. This approach has particular social significance in emerging fields of science and technology. By involving journalists who represent the public in close cooperation with scientists, decisions regarding the direction of technological development are not left solely to scientists but are open for broader societal discussion. Consequently, JIR can serve as an important first step toward achieving genuine science governance within civil society [7–9].

However, JIR occupies a unique position within science communication because it relies on maintaining an equilibrium or productive tension between the different professional cultures of science and journalism. Unlike other science communication activities, such as science cafés or public symposia, JIR is not yet widely practiced nor broadly institutionalized.

Our specific JIR initiative originated when researchers in molecular cybernetics recognized the need for dialogue with citizens regarding the social implications of their research. Given the complexity of molecular cybernetics and its distance from practical applications, it was challenging for non-specialists to grasp its relevance through standard public-engagement methods. Traditional dialogue formats often restrict discussions to organizer-defined topics, reducing the real impact of citizen engagement.

To address these shortcomings, we proposed implementing a first JIR program a decade ago, emphasizing the journalist's function as a bridge between science and society. Although our initial plan was to establish a full-scale JIR in Japan, multiple obstacles arose. Previous research on scientist-media relationships in Japan, including the 2009 "Mass Media Literacy of Researchers" survey (conducted under a JST-RISTEX project), revealed a persistent lack of mutual understanding between the scientific community and the media—undermining the quality of science writing. While that survey recommended JIR as a potential solution for universities and research institutes, the proposal was not implemented at the time. One major impediment was deep-seated mistrust among Japanese scientists, who often viewed journalists as seeking to uncover hidden issues or scandals. This sentiment was epitomized by a 2009 public relations comment likening JIR to being "investigated for something that doesn't hurt," reflecting scientists' discomfort with perceived media intrusion.

Building on these past experiences and challenges, we decided to reintroduce the JIR model in the field of molecular cybernetics. This research area focuses on creating chemical artificial cells and exploring biological functions through synthetic molecules with physical properties. Although primarily a domain of basic research, molecular cybernetics has far-reaching implications for future social applications, making proactive public dialogue and Ethical, Legal, and Social Issues (ELSI) considerations essential. Because of the conceptual distance between the lab bench and potential societal impact, more effective communication was needed. As a result, our proposal to adopt JIR as the program's practical core earned support from the project's representative, Dr. Murata, paving the way for collaborative bottom-up engagement among scientists, journalists, and the broader community.

28.2.2 ELSI's Role in Science Journalism and the Necessity of Scientists' Engagement

Ethical, Legal, and Social Issues (ELSI) is central to science journalism. By examining the broader societal impacts of emerging technologies, ELSI research brings forward ethical and legal considerations that often necessitate public engagement and open debate. Science journalism plays a key role in fostering these discussions.

Yet, scientists themselves must also participate in this dialogue; it cannot be left solely to journalists and the public in a democratic society. Because public opinion indirectly shapes both research practices and policy decisions, scientists' direct involvement and transparency are essential. Moreover, their specialized expertise allows them to anticipate and address potential risks at early stages.

A notable example of scientists taking the lead in managing such risks was the 1975 Asilomar Conference, where guidelines were established for recombinant-DNA research. These included graded biological/physical containment tied to risk classifications and prohibitions on work with highly pathogenic agents. Crucially, Asilomar lifted a pre-meeting voluntary moratorium only after articulating community norms of responsibility, thereby linking experimental design to public governance and oversight rather than treating risk management as a purely technical matter. In this sense, the conference broadened scientific practice to encompass self-regulation, transparency, and societal accountability wider social considerations, such as self-regulation, transparency, and accountability to society.

Another watershed moment came in the 1990s with the Human Genome Project, which explicitly integrated ELSI into a large-scale research agenda. Acting on James Watson's proposal—he was a key figure in the project and co-discoverer of DNA's double helix—3–5% of the project's budget was allocated to ELSI research. This provision addressed a range of issues, including privacy and fairness in the use of genetic information; clinical integration such as informed consent and genetic counseling; and research ethics and governance around biobanking and data ownership. By embedding normative analysis within the project—across these key domains in which genomic risks arise—this allocation enabled a more comprehensive assessment of scientific progress in light of ethical and social concerns.

Significantly, these initiatives were instigated by scientists themselves, reflecting a growing awareness of—and responsibility for—the societal impacts of their work. Viewed in this light, science is no longer seen as purely objective or politically neutral. Rather, the development of science and technology, and the research that underpins them, are inherently value-laden endeavors: they do not merely describe the world but actively shape what it should become. As such, the ways in which science is conducted and its results are interpreted and applied reflect underlying value systems. This, in turn, poses a fundamental question: *What kind of world should we aim to build?* ELSI in science journalism invites both scientists and the public to grapple with this question together.

28.2.3 The Rise of PR-Driven Science and the Challenges It Creates

Despite the acknowledged importance of science journalism and ELSI, both now face increasing difficulties. Over the past few decades, science communication has shifted from traditional journalism to a more PR-driven model for the sake of the public understanding of science (PUS), a trend Martin W. Bauer describes as "PUS Inc." [10]. Although public relations (PR) was originally envisioned as a form of two-way communication aimed at helping organizations fulfill their public responsibilities, in practice it often functions as one-way promotion or marketing. Consequently, many science institutions treat the public sphere as a "market" in which to "sell" scientific achievements, employing press releases optimized for news cycles and large-scale media campaigns to gain visibility, funding, and influence. This has fueled "PR wars," in which well-funded organizations hire ex-journalists and communication experts, marginalizing independent journalism and limiting space for open, critical debate.

This PR-centric dynamic also makes the science community more reluctant to subject their internal workings of research to genuine public scrutiny. Instead, the emphasis shifts toward controlling public perception to maintain intellectual authority and preserve positive image of science. Framing research as uniquely beneficial or ethically responsible then becomes a strategic move—presented as an act of "social responsibility," yet often functioning more as a preemptive defense against external criticism.

Such pressures make it increasingly difficult to sustain critical, independent science journalism. PR strategies can escalate into "spin" [1, 11, 12]—the deliberate shaping of information to secure favorable coverage. Skilled PR practitioners adapt content to journalistic formats, reinforcing a cycle of "churnalism," wherein press releases are reproduced with minimal scrutiny. This challenge is further intensified by today's fast-paced media environment constrained by limited time and resources. As a result, journalists may end up amplifying the benefits highlighted in press materials while paying only superficial attention to ethical or risk-related concerns, thereby becoming partially complicit in the narrowing of public discourse.

This tendency can be observed when ethical considerations are reduced to ritualized phrases, such as "Ethical concerns remain and will require further debate," with little deeper exploration of the issues. Indeed, our recent study [13] analyzing Japanese newspapers coverage found that ethical discussions in science writing were typically superficial and formulaic, dominated by stock expressions like "This is expected to raise ethical questions," rather than genuine engagement with underlying issues.

Ultimately, this competitive, PR-driven climate shapes not only science journalism but also the quality of science reporting, reframing scientific discourse into consumable packages geared toward marketing and publicity. As a result, authentic intellectual curiosity and deeper public understanding risk being eclipsed by the need for quick, attention-grabbing narratives—undermining science communication, including ELSI-focused discussions, and diminishing its capacity to inform meaningful public discourse and democratic deliberation.

Our JIR project is an exploratory attempt to address this predicament by fostering a new relationship between journalists and scientists. It aims to reimagine the future of science communication amidst this PR-driven media trend, exploring how the quality of science reporting can be sustained and how the critical functions of science journalism and substantive ELSI discussions can be maintained. To tackle these challenges effectively, scientists and journalists must collaboratively move beyond their distinct professional norms and practices to forge a new and productive partnership. The JIR framework offers a promising avenue for realizing this possibility.

28.3 Project Activities

This subsection outlines how our JIR program unfolded in practice, focusing on early challenges, the development of shared concerns, and collaborative events aimed at building mutual understanding and trust between journalists and scientists.

28.3.1 Recruiting Journalists: Process and Challenges

In March 2021, we began recruiting for the JIR program by distributing a call for applications via social media and email lists. After interviewing several candidates online, five project leaders selected three journalists: (1) a science reporter from a major newspaper, (2) a novelist and science writer, and (3) a freelance journalist contributing to international tech magazines. Notably, no female journalists applied.

During the selection process, some scientists expressed reservations about independent journalistic reporting, worried it might dwell on research challenges and setbacks, which could result in unduly negative or overly critical coverage. In response, we explained core journalism principles and drew on insights from science and technology studies, fostering scientists' understanding and encouraging them to welcome the JIR journalists into their research community.

Introducing JIR also revealed organizational and financial obstacles. For instance, Waseda University, which hosted our JIR program, pointed out difficulties in granting researcher status to journalists lacking formal academic credentials. This highlighted a broader systemic issue: the limited institutional flexibility to accommodate professionals whose expertise is rooted in practice rather than academia. As a workaround, we ultimately hired them as research assistants.

28.3.2 Activities: Year One

During the program's first year (2021), journalists mostly pursued self-directed coverage and engaged in online discussions with scientists, given that COVID-19 restrictions limited lab visits. With few opportunities for extended, on-site observation, the original plan for an in-depth residency to capture "field" perspectives was significantly constrained. Online outreach designed to spark public interest also received less engagement than anticipated, undermining our aim to position journalists as a bridge between science and society.

Even so, these early activities helped clarify mutual concerns—such as editorial processes and manuscript handling—and led to the development of practical guidelines (e.g., standardized credit notation). Still, the anticipated "chemical reaction" between these two distinct professions had not yet emerged in any substantial way.

28.3.3 Activities: Year Two and Beyond

As pandemic restrictions eased in 2022, we aimed to deepen collaboration and align values between journalists and scientists. To begin, we piloted a science-communication initiative in molecular cybernetics with one of our JIR journalists, the novelist, Shingo Fujisaki. Around this time, he launched a new web series with Kodansha Bluebacks, pairing interviews with molecular-cybernetics scientists and short fictional scenarios envisioning how advances in the field might shape society 30 years from now. Unfortunately, the project elicited limited public response, and the feedback we received did not translate into substantial science-communication impact. In short, without a compelling news hook or "buzz" factor, broader engagement proved difficult.

Building on these lessons, we organized a session entitled "What Lies Beyond Buzz Science," jointly planned by journalists and scientists, held on November 13, 2022, during the annual molecular cybernetics conference. This session explored effective ways to capture public interest in science. As part of the discussion, we invited *Riri Ayae*, a "virtual science writer" known for explaining latest scientific findings and other news on Twitter and YouTube.

Continuing this theme, we hosted another session, "When Science Meets Headline," examining how science and the media interact. By inviting scientists to reflect on how press coverage shapes public opinion and scientific practices, we highlighted the tension researchers face in an era that increasingly demands immediate, "useful" results. As universities and institutes compete globally through press releases and the pursuit of prominent media headlines, the session posed a central question: *Should scientists enter this "headline competition," and if so, under what guidelines?* Two guest speakers then led an open forum on responsible science communication and on how scientists should respond to contemporary media pressures.

Following the lectures, we held an interactive session in which participants and speakers discussed their visions for the future of media dissemination of scientific information (i.e., science reporting) and the evolving role of scientists. Participants proposed ways the science–media relationship should evolve to address emerging challenges in science communication.

Regarding science journalism issues, we staged a special event at the 2022 annual meeting of the Japanese Society for Molecular Cybernetics titled "Moral Panic Caused by Molecular Cybernetics?! Discussion through Ethical Role-Playing." This live-action role-playing (LARP) activity was designed to explore ethical questions surrounding "enhancement" in bioethics and to facilitate collaboration between scientists and journalists. Participants were immersed in a near-future scenario in which enhancement technologies—for cognition and moral emotions—had become widespread, prompting them to confront ethical dilemmas that scientists could realistically face.

In a LARP setting, participants take on specific roles within a structured narrative, enabling a more immersive and embodied engagement with complex issues. This format encourages deep reflection through decision-making and dialogue, offering a vivid alternative to conventional panel discussions.

The event challenged participants to consider what stance scientists should take on cognitive and moral enhancement. By fostering shared reflection on scientists' social responsibilities and the broader societal impact of technological progress, the LARP format helped illuminate how science journalism can move beyond information transmission and become a space for ethical imagination. For further perspectives on this approach, see Chap. 30 by Mr. Mori in this volume.

28.4 Challenges and Future Directions

Through our JIR program, journalists and scientists developed a shared awareness of how a PR-driven media environment can adversely affect science reporting. Both groups recognized that "headline competition" and the race to capture public attention often shape—or even dominate—the coverage of scientific topics. As Ryoma Komiyama, one of our JIR journalists, points out in Chap. 31, journalists must prioritize compelling stories that hook readers. Yet "attention-grabbing" does not necessarily mean "low-quality." Komiyama cites examples like black holes and the Ig Nobel Prize, which, despite their limited direct social utility, can both captivate audiences and educate, providing a pathway to science reporting that is engaging yet substantive.

However, integrating ELSI topics into the JIR program presents distinct challenges. Scientists are often unfamiliar with formal ethical reasoning, which makes it difficult to spark interest and sustain active engagement in ethical debate. Against this backdrop, our LARP effectively drew scientists into substantive discussion around enhancement-related ethical questions. Nevertheless, it also revealed a tendency among both scientists and journalists to treat "ethics" as little more than everyday moral intuition—that is, a personal opinion rather than a subject of reasoned public deliberation. Although the LARP scenario aimed to push participants beyond personal intuitions by presenting ethical dilemmas, structured normative claims and arguments remained underdeveloped.

Furthermore, journalists' limited capacity to facilitate deeper ELSI discussions reflects a broader pattern, discussed in Sect. 28.2.3, in which ethical issues are routinely handled in a superficial or ritualistic manner within science journalism. Strengthening the capacities of both journalists and scientists for more rigorous ethical inquiry is essential. Without such capacities, neither group can fully appreciate how science journalism and scientific practice itself are shaped by a fundamental normative question: *What kind of world should we aim to build?*

A further complication lies in defining the ideal relationship between journalists and scientists within JIR programs. Traditional science journalism maintains critical distance from the scientific community to preserve its role as an independent watchdog. By contrast, embedding journalists closely with researchers can blur this boundary, potentially compromising journalistic norms, such as not sharing prepublication drafts with sources. At the same time, JIR programs seek to foster co-creation and dialogue, encouraging collaboration between journalists and scientists to identify key science communication challenges. Balancing these competing demands has proven difficult. In practice, JIR activities often defaulted to scientists' comfort zones, reducing productive tension and constraining meaningful ELSI discussions.

These challenges are further compounded by practical institutional constraints on the journalists' roles within the university. As Fujisaki notes in Chap. 29, journalists received partial access to university resources (e.g., Slack channels and research meetings), but the lack of formal workspaces and reliance on remote participation severely limited informal interactions. Consequently, the JIR journalists were often perceived as "science translators" rather than fully integrated resident journalists. Labeling them as "contract journalists" or "exclusive writers" might mitigate logistical concerns but risks compromising their journalistic independence. Therefore, creating a supportive institutional environment in which journalists can operate effectively inside universities is essential—not only to avoid the drift toward purely promotional activities but also to realize JIR's full potential as a platform for deep engagement and mutual transformation between journalists and scientists.

28.5 Conclusion

Our Journalist-in-Residence (JIR) program aimed to create new pathways for dialogue between journalists and scientists, particularly in an era of PR-driven science communication. We acknowledge, however, that our overall outcome did not fully meet the ambitious goals we initially envisioned. Several factors contributed to this shortfall: the JIR model remains underdeveloped in Japan, institutional frameworks to support such collaborations are insufficient, and the COVID-19 pandemic severely restricted crucial in-person interactions. Consequently, we faced significant difficulties in establishing shared objectives, given the professional and institutional divides separating journalists and scientists.

Nevertheless, we made meaningful progress in identifying critical challenges, especially those related to fostering more robust discussions of Ethical, Legal, and Social Issues (ELSI). Both journalists and scientists tended to view ELSI as something best handled by specialists, defaulting to passive learning from dedicated ELSI experts rather than active engagement with ethical questions themselves. This attitude reflects an underlying assumption that ELSI falls outside their professional domains. Furthermore, ethics was often framed as mere personal intuition or individual "opinions," thereby constraining opportunities for reasoned public deliberation toward the collaborative development of ethical frameworks.

To bridge this gap, we need frameworks that enable both journalists and scientists to participate more fully in normative debate. A promising starting point is virtue epistemology, as developed by Miranda Fricker [14], which foregrounds "epistemic virtues"—qualities that guide and sustain our epistemic practices of knowing. Viewing science journalism in light of these virtues clarifies its normative dimension and its potential to foster deeper ethical deliberation. In a functioning democratic society, citizens rely on diverse "epistemic resources" to make informed judgments. This concept, one of the central ideas in Fricker's work, is vividly illustrated by the term "sexual harassment." Before the phrase was coined, people experienced harassment but lacked a shared concept, an epistemic resource, that could render their suffering intelligible as a social injustice; as result, it was often dismissed. Coining the term itself supplied a new resource that enabled a personal harm to be understood and addressed collectively.

This model maps directly onto the distinct roles of science reporting and science journalism. Science reporting provides foundational epistemic resources by naming and defining new phenomena—much like early reporting on "long COVID," which validated patients' experiences and framed the condition as a public-health issue. Science journalism, in turn, furnishes deeper interpretive resources for ethical deliberation: for example, coverage of CRISPR goes beyond gene-editing facts to introduce frameworks such as concerns about "designer babies" and the historical context of eugenics. There are not mere facts but conceptual tools that enable the public to debate a technology's societal implications.

Epistemic resources thus extend beyond the bare reporting of facts, situating scientific findings within broader social contexts so that their wider implications become intelligible to the public. Both science reporting and science journalism—each in its own way—provide vital epistemic tools for interpreting and critically evaluating scientific knowledge. Cultivating epistemic virtues is therefore crucial to maintaining and improving these resources.

From this perspective, the media space—the arena in which science is communicated—is fundamentally normative. Journalists bear responsibility for ensuring accuracy, fairness, and depth in the information they provide, remaining vigilant about any imprecision or bias. They are also called upon to expand interpretive resources—a task especially critical for emerging technologies—by shaping coverage that helps audiences navigate new ethical or societal dimensions.[1]

1: According to Fricker, ignoring or distorting these interpretive resources can lead to what she refers to as hermeneutical injustice, where certain voices or perspectives are systematically marginalized or misunderstood.

By adopting epistemic virtues as a guiding framework, we can better articulate the roles of our JIR journalists. Komiyama's approach to science reporting nurtures the virtue of intellectual curiosity directed toward basic science. Fujisaki's use of science fiction broadens readers' interpretive frameworks, enriching their epistemic resources. Similarly, Mori's LARP experiment enhances our collective interpretive capacities regarding science and technology through immersive ethical debate.

Scientists, for their part, also play a critical role in producing epistemic resources; thus, the corresponding epistemic virtues apply to them as well. While the specific virtues required of journalists and scientists may vary due to their different professional norms, both groups share certain core commitments, especially a shared responsibility to provide epistemic resources for the public. In this regard, the following survey findings are highly suggestive. One international study [15] indicates that, although scientists' abilities are not in question worldwide, their "intentions"—such as whether they genuinely serve the public interest—are often viewed skeptically by citizens. Our large-scale domestic survey of Japanese respondents likewise shows that scientists are perceived as lacking ethical and civic awareness and as hesitant to discuss their research publicly.

Viewed through the lens ofepistemic virtue, JIR projects can foster clearer shared objectives and deeper mutual understanding, counteracting the superficiality of PR-driven media. This approach also reframes ethics not as antagonistic to science, but as integral to it—both as *normative knowledge* and as *virtue*—thereby enabling journalists and scientists to jointly engage with a central question for science and science communication: *What kind of world should we aim to build?*

Looking ahead, future JIR programs can embed these epistemic virtues in a more comprehensive normative framework, facilitating more effective collaboration between journalists and scientists—particularly in the context of ELSI-focused science journalism. Such a framework offers a constructive path forward, empowering both journalists and scientists to navigate and to resist the constraints of today's PR-driven media landscape.

References

1. Tanaka, M.: "Kagaku chi, media, jānarisumu" ["Scientific knowledge, media, and journalism"]. *Kagaku shakaigaku [Sociology of Science]*. Matsumoto M (ed.). University of Tokyo Press, pp 201–231 (2021)
2. Burkett, W.: News Reporting: Science, Medicine, and High Technology. Iowa State University Press (1986)
3. Dunwoody, S.: Science journalism: prospects in the digital age. In: Bucchi, M., Trench, B. (eds.) Routledge Handbook of Public Communication of Science and Technology. Routledge (2014)
4. Fox, F.: Beyond the hype: Inside Science's Biggest Media Scandals from Climategate to Covid, Elliot and Thompson Ltd. (2022)
5. Grunig, J.E.: Theory and practice of interactive media relations. Public Opin. Q. **35**, 18–23 (1990)
6. Culbertson, H.M., Chen, N.: International Public Relations. Lawrence Erlbaum Associates, Inc, New Jersey (1996)
7. Valenti, J.M., Tavana, G.: Report: continuing science education for environmental journalists and science writers. In situ with the experts. Sci. Commun. **27**, 300–310 (2005)
8. Semu Phumisa, J., Kalera Mhango, R.N., Makwinja, R., et al.: Enhancing media communication of health research: an evaluation of the Journalist in Residence Programme at the Malawi Liverpool Wellcome Trust Clinical Research Programme. Wellcome Open Res (2023). [Preprint]

9. Valenti, J.M.: Report: continuing science education for environmental journalists and science writers. In situ with the experts. Sci. Commun. **27**, 300–310 (2005)
10. Bauer, M.W., Bucchi, M.: Journalism. Science and Society Science Communication between News and Public Relations, Routledge (2008)
11. Gaber, I.: Government by spin: an analysis of the process. Media Cult. Soc. **22** (2000)
12. Leighton, A.: Spin: from tactic to tabloid. J. Public Aff. **6** (2006)
13. Takeda, K.F., Komata, M., Takae, K., Tanaka, M., Shineha, R.: Comparative analysis of media coverage concerning the social implications on three life sciences in Japan during 1991–2020. Front. Sociol., 11 March 2025. Sec. Media Governance and the Public Sphere, Volume 10 (2025)
14. Flicker, M.: Epistemic Injustice: Power and the Ethics of Knowing. Oxford University Press (2007)
15. Cologna, V., Mede, N.G., Berger, S., et al.: Trust in scientists and their role in society across 68 countries. Nat. Hum. Behav. (2025).

Chapter 29
Discussing Emerging Science and Technology Through Science Fiction

Shingo Fujisaki

29.1 Introduction

Fujiko F. Fujio, a leading *mangaka* (manga artist) in Japan, has illustrated numerous works featuring science fiction (SF)-inspired settings, such as those seen in his signature series, "Doraemon." However, Fujiko refers to his short stories, which seem to incorporate more SF elements than his full-length works, not as "SF" but rather as stories that are "*sukoshi fushigi*" (slightly fantastic). Fujiko stated that, in this sense, they are comparable to works such as "Journey to the West" and "Arabian Nights" [1].

As a writer who has also written a number of SF novels, I feel more comfortable thinking of "SF" as an abbreviation for "*sukoshi fushigi*" rather than for the usual "science fiction." In other words, these stories are not necessarily concerned with "science" but rather intended to be enjoyed for their far-out ideas, fantastical imaginings, and shifts in perspective; nothing more, nothing less.

Nevertheless, it seems true that SF has started many researchers down their paths and inspired them along the way. I have worked as an editor of a science magazine and as a science writer, and researchers whom I have met through interviews and other activities have often told me about their experiences influenced by SF. That is, the fictional worlds depicted in the SF genre can lead to changes in the real world, at least in the field of science and technology.

Recently, there has even been a movement to actively and intentionally "utilize" SF in research and development (R&D), for formulating business strategies, training workers, and for other purposes in enterprises and other organizations. This is called "SF prototyping," and it is said to have emerged as a hot topic for discussion in Japan around 2020 [2]. The following year, 2021, I became a Journalist-in-Residence

S. Fujisaki (✉)
Fujimino Higashi, Fujimi-Shi, Saitama, Japan
e-mail: fujisaki@v001.vaio.ne.jp

S. Murata et al. (eds.), *Molecular Robotics II*,
https://doi.org/10.1007/978-981-95-2394-8_29

(JIR) in an emerging science and technology research project, called "molecular cybernetics."

In this section, I will reflect on some of the activities I have performed within the project, comparing them with popular methods such as SF prototyping. Before describing specific activities and their results, however, I would like to introduce some previous SF works I was aware of, as well as some discussions on SF prototyping.

29.2 Previous SF Works Dealing with Molecular Robots

The words "molecular cybernetics," "molecular robotics," and "molecular robots" probably appear in only a handful of SF stories other than those penned by me. However, if this is interpreted as a kind of "nanotechnology" or "nanomachine," there are many examples, and they are already a frequent element in SF stories.

Representative SF novels in which nanomachines similar to molecular robots appear include Greg Bear's "Blood Music" (1987), Michael Crichton's "Prey" (2002), and Keikaku Ito's "Harmony" (2008).

In "Blood Music," "noocytes"—lymphocytes embedded with a kind of DNA computer—possess human-like intelligence in each cell and they work in concert. Once these cells are in the body, other cells gain intelligence, and the entire body is altered to suit the noocyte's needs. Eventually, the noocytes spread across all of North America, bringing about a world in which there is no longer a boundary between living and non-living things [3].

The terms "molecular machine" and "molecular assembler" appear in "Prey." The nanomachines that appear have gallium arsenide eyes, as well as adenosine triphosphate (ATP)-based processors. They do not multiply by themselves but are multiplied by the self-organization of parts produced by specialized *E. coli* bacteria. When they are gathered in countless numbers, they acquire collective intelligence, and when they graft themselves to human physiology, they can manipulate the human host's consciousness at will and even change their appearance [4].

The "WatchMe" (constant internal surveillance system) creation appearing in "Harmony" seems to be a type of molecular robot, but it does not attack humans at its own will and is used by some powerful individuals. Although "installed" in almost all humans in the name of health care, WatchMe in the brain can extinguish human consciousness, making it susceptible to control by others [5].

What these three works have in common is that they depict the advent of a "dystopian" world that could arise from technologies such as molecular cybernetics and robotics. In the stories, fear of "self-reproducing, yet intelligent machines" and distrust of dual use are fueled.

29.3 Discussion of SF Prototyping

The "Digital Daijisen" dictionary describes SF prototyping as "making predictions about the future based on fantastical ideas like SF, rather than the accumulation of past experience, and using them as a basis for business creation and R&D."[1]

I also used to hold this general impression of SF prototyping.

However, I have read such commentaries as "Science Fiction Prototyping" (2011) by futurist Brian David Johnson, who first proposed this concept, Kyosuke Higuchi's "*Mirai ha yosoku suru mono dehanaku souzou suru mono de aru*" (The future is not something to predict, but rather something to create) (2021), and Dohjin Miyamoto's "*Furubita mirai wo dou kowasu*?" (How can we get rid of the old-fashioned future?) (2023), which offer a somewhat different take.

First, SF prototyping does not seem to involve "predicting the future." It is based on real ideas and developments in science and technology and is about "imagining" rather than predicting. Miyamoto often uses phrases like "creating an unconventional vision of the future" [2]. Whether or not this is a reasonable vision of the future is not the point. Rather, it is said that it is better to be as "outlandish" as possible. By "working backward" to identify the paths that could bring about that future, we can determine what is needed today.

In other words, it is not a method of adapting to a predicted future but a method of building an interesting or desirable future for oneself.

Another important point missing from the previous definition is that SF prototyping is a means of discussion and dialogue. If one wants to have a product, one may end up writing an SF novel or something similar. However, it is a means, not an end, and Higuchi has stated that it is not necessarily needed to complete the work [6]. The ethos is something like "Let's all have fun discussing interesting ideas together with the intention of writing some really far-out SF stories but without overthinking it."

For this reason, Higuchi's and Miyamoto's book devotes significant space to the ways of proceeding with the project, including the management of workshops and other activities. I am sure many people have experienced the difficulty of inventing the idea, product, or research topic of the century after simply being told to dream up such a project without any means or protocol.

29.4 Background on Article Publication

After becoming a JIR, the first thing I did was find a medium with as many readers as possible to write articles as a "science writer" and "author." I thought this would be advantageous in terms of outreach and would make it easier to obtain meaningful data when I wanted to know how people were reacting to what I had written. Consequently,

---

[1] Digital Daijisen. https://daijisen.jp/digital/ (Accessed Jan. 31, 2025).

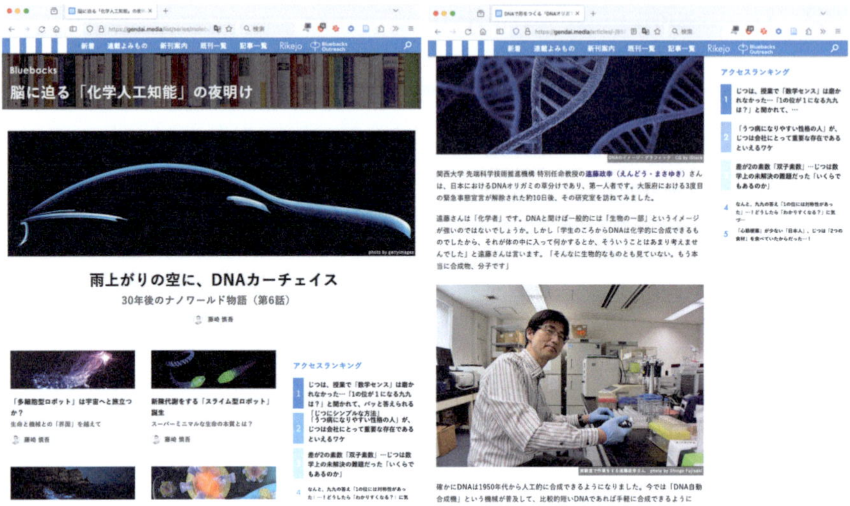

Fig. 29.1 Table of contents (left) and some of the articles (right) in the "The Dawn of 'Chemical Artificial Intelligence' and its Implications for the Brain" series

I was able to start working with Kodansha and start publishing my works on the "Kodansha Bluebacks" website.

Kodansha's "Bluebacks" was a small-sized paperback series released in 1963 and is the oldest small-sized paperback label in Japan dealing exclusively with popular science, with over 2000 titles published to date [7]. Its website also functions as a kind of web magazine, with many articles in various fields.

I wrote a series of articles titled "The Road to Life 1.0" (18 articles in total) on the site between 2017 and 2019.[2]

The content of the articles raises questions about "What is life?" based on the results of research on the origin of life, synthetic biology, and other fields. Molecular cybernetics and molecular robotics are adjacent to these fields, with many studies spanning both.

As such, I was able to begin a series of articles on molecular cybernetics, "The Dawn of 'Chemical Artificial Intelligence' and its Implications for the Brain," positioned as a sequel to "The Road to Life 1.0."[3]

The series ran from 2021 to 2023, with 18 articles in total (Fig. 29.1).

[2] Fujisaki, S.: The road to life 1.0. https://gendai.media/list/series/seimei10.

[3] Fujisaki, S.: The Dawn of "Chemical Artificial Intelligence" and its Implications for the Brain. https://gendai.media/list/series/molecular_cybernetics (Accessed Jan. 31. 2025).

29.5 Characteristics and Aims of the Article Series

The "Dawn of 'Chemical Artificial Intelligence'" series was organized as a set of three installments; the first two being non-fiction and the last being fiction (SF). There were 18 articles in total, comprising six three-article series.

Bluebacks, including its website, is essentially scientific entertainment, with little presented for express purposes of journalism, education, or general enrichment. As such, the articles in the series had to align with the overall nature of the medium. However, cases of fiction articles being published on the website are quite few.

In keeping with the above characterization, the main objectives of the non-fiction articles were as follows:

1. To inform the general public about science and technology related to molecular cybernetics in an easy-to-understand and entertaining manner and to raise the profile of the researchers and their projects.
2. To inform researchers about how their research is perceived by society (writers and the general public) as a reference case.

Regarding fiction, however, I had the following in mind:

3. Present the world featuring cutting-edge ideas and imaginings about the future not openly discussed by researchers, in the form of non-intimidating fiction.
4. By considering fantasies that go half a step further than researchers' fantasies, I hope to come upon ideas such researchers may find useful.
5. Considering the nature of the medium, addressing ethical issues head-on is difficult, so they can be raised indirectly in articles that fantasize about the state of things 30 years into the future.

To expand on number 3, I note that by the nature of their profession, researchers typically abstain from making statements that (may) lack scientific basis. However, in informal settings such as over drinks, they sometimes share (occasionally dangerously) novel ideas and wild fantasies. My aim is to recreate this situation in a dialogue based on the premise of "writing fiction" and convey to the public how fascinating the topics discussed are.

The ethical issues mentioned in number 5 are based on the project's request to JIR to "stimulate discussion on ethical, legal, and social issues (ELSI) and responsible research and innovation (RRI)" in molecular cybernetics and robotics.

29.6 Producing the Series

In each three-article portion of the series, one or two research scientists involved in or otherwise relevant to molecular cybernetics were interviewed. In total, 10 researchers collaborated (Table 29.1).

The essay topics were selected from among the elemental technologies necessary to realize the project's goal of "Chemical Artificial Intelligence" (Chemical AI) to

Table 29.1 Data on "The Dawn of 'Chemical Artificial Intelligence' and its Implications for the Brain"

Episode no	Title	Topic	Collaborating researchers (titles omitted)	Facebook	X (formerly Twitter)
1	"DNA Origami": Making Shapes with DNA	"DNA origami"	Masayuki Endo	26	42
2	Rolling DNA, Rotating DNA, Walking DNA	"DNA origami"	Masayuki Endo	17	33
3	Sprites of the DNA Playground	Fiction	Masayuki Endo	7	3
4	Like Earthworms, "Proteins" Slither and Swarm	Molecular robots	Akinori Kuzuya and Akira Kakugo	22	47
5	"Muscles" are Beginning to Undergo a New Evolution	Molecular artificial muscle	Akinori Kuzuya and Akira Kakugo	16	38
6	Armchair Love	Fiction	Akinori Kuzuya	60	26
7	The Birth of "Artificial Primordial Cells" that Proliferate even in the Absence of Life	Giant vesicles	Taro Toyota, Hironori Sugiyama	17	132
8	The Origin of Life Comes in a Tiny Mansion	Giant vesicles	Taro Toyota, Hironori Sugiyama	23	32
9	Molecular Letters and Soap Bubble Envelopes	Fiction	Taro Toyota, Hironori Sugiyama	27	17
10	It's Evolving! A Micro-robot that Moves Using "Muscle"	Cyborg insects	Keisuke Morishima	1	49
11	The Day "Cyborg Cockroaches" Became Heroes	Cyborg insects	Keisuke Morishima, Kan Shoji	94	28
12	Dad's Big Upgrade!	Fiction	Keisuke Morishima	1	6
13	Launched by Light: Man-made Microbial Rockets	Peptide actuators	Kazunori Matsuura	76	32

(continued)

Table 29.1 (continued)

Episode no	Title	Topic	Collaborating researchers (titles omitted)	Facebook	X (formerly Twitter)
14	Created with the Power of Chemistry, "Replica Viruses" May also Find Applications in Cancer Treatment	Replica viruses	Kazunori Matsuura	28	30
15	Viral "Gifts" and the Shining Sea Waves of May	Fiction	Kazunori Matsuura	37	12
16	A Regenerating "Slime Robot" is Born	Slime-like molecular robots	Shogo Hamada	12	28
17	Will "Multicellular Robots" Travel into Space?	Multicellular molecular robots	Shinichiro Nomura	16	29
18	A DNA Car Chase under the Sky after the Rain	Fiction	Shogo Hamada, Shinichiro Nomura	1	7

avoid any kind of overlap. Rather than the significance of the technologies, etc., the main selection criteria were whether they would attract the public's interest and whether they contained attractive images, photographs, videos, etc.

The non-fiction articles consisted primarily of an introduction of each researcher and a commentary on their work. Fiction, by contrast, presents, in principle, the researcher's "prediction" or "imagining" of a future (a world in which the research has been implemented in society) about 30 years ahead in the form of an SF short story.

However, what I intended to write in those fictional works was not a profound SF, as described in [2], but a (seemingly) modest and "*sukoshi fushigi*" story occurring among family members and those close to them.

There are several reasons for this, including practical limitations. First, the story cannot be too large-scale or complicated due to the necessity of summarizing it in a "short story." Therefore, we limited the story setting to "inside the home" and its surroundings. In doing so, I hoped that the audience would become somewhat familiar with the emerging science and technology to which they likely had no prior exposure.

Additionally, although not necessarily direct, the media asked me to avoid stories that were too dark or contained excessively cruel or sexually explicit scenes. It was expected that molecular cybernetics, which "supported" the serialization, would probably feel the same way. This would make it difficult to depict an obvious dystopian setting.

Furthermore, I wanted to present ELSI/RRI in molecular cybernetics and robotics in a way that did not lead readers in any direction but rather left them to make their own judgments and come to their own conclusions. If the emphasis is on entertainment, dystopian settings that evoke fear and loathing are more likely to be pleased, but for the above reasons, my intent was not to do so.

The ELSI/RRI-related themes covered in each fiction episode included "Misuse of molecular robots by terrorist organizations" (Episode 3), "Manipulation of consciousness and behavior by molecular robots" (Episode 12), and the impact of molecular robotics/cybernetics on "The concept and acceptance of death" (Episode 6), on "Communications between humans and plants" (Episode 9), on "The marine environment and biodiversity" (Episode 15), and on "The human growth" (Episode 18).

I conducted most of the interviews with these researchers face-to-face at the research sites, but also did online interviews. During the on-site interviews, I interviewed the researchers for 2–3 h, and I toured and photographed their research laboratory, experiment rooms, and other scenes depicting their work.

The manuscripts written on the basis of the interviews were first checked by Bluebacks editors, and if no problems were found, the researchers were asked to confirm the accuracy of the descriptions. Each stage of revision involved several exchanges. The articles were then posted on the website and reprinted on "Yahoo! News" and "SmartNews." They were updated once every one or two months.

Title and topic of each publication, collaborating researchers, and number of shares of the article on Facebook and X (formerly Twitter) (as of February 12, 2024). Lines in bold frames are fiction, while others are non-fiction.

29.7 A Reflection on the Series as SF Stories—General Audience Reactions

I recorded several indicators to reflect the reactions of the general readership to the serialized articles. These included the ranking in the "Magazine Access Ranking (IT/Science)" on Yahoo! News, the number of comments on the same news site, and the number of shares on Facebook and X (formerly Twitter).

As the rankings and number of comments could not be recorded in some editions (e.g., when the ranking showed only the top 10), the number of shares is used as the main indicator in the following discussion.

The average number of shares for non-fiction was 29.0 on Facebook and 43.3 on X, while the average number of shares for fiction was 22.2 and 11.8, respectively. In other words, the non-fiction works were more popular than the fiction works. In terms of ranking, non-fiction was almost always in the top 10, while fiction did not feature in the top 10 in all cases.

The main reason for this can be attributed to the fact that the media is fundamentally a platform for scientific articles and news, and fiction is not desired or expected to be published in the first place. In one instance, there was a comment that suggested, "I thought I was reading an article, but I was puzzled because it ended up being a short story" (of course, the preface stated that it was fiction, but this was apparently overlooked).

In addition, as mentioned in 6, the serialized fiction does not contain any scenes of violence or sex, nor does it have other unpleasantly stimulating elements. Furthermore, the initial aim was to depict a world about 30 years into the future rather than to pursue human drama. Therefore, overall, it was considered less entertaining than regular popular fiction, which may have been one reason for its lack of popularity.

The results of a comparison of the number of shares based solely on fiction also suggest this.

The most popular of the six short stories was "Armchair Love" (total number of shares on Facebook and X: 86) in the 6th series installment, followed by "Viral 'Gifts' and the Shining Sea Waves of May" (total number of shares on Facebook and X: 48) in the 15th series installment. This is quite remarkable considering that there were less than 10 shares in three instances (Table 29.1).

The 6th installment was a story in which the protagonist overcomes his conflict with his deceased father through the behavior of a movable armchair made of "molecular artificial muscle," and the 15th installment was a story in which a character pursuing the memory of her missing lover is healed when an "artificial virus" restores her memorial scene in the sea that was once lost. In other words, both stories are relatively "emotional" or "sentimental," which distinguishes them from the other four.

After all, as long as they are in the form of novellas, "human drama" may tend to be appreciated.

On the other hand, my plan to have readers share in the enjoyment of the researcher's "sharp ideas, future delusions, etc." may not have been very successful. The reason for this could be the author's lack of ability, the fact that the stories deal with unfamiliar, emerging science and technology, or the fact that few general readers are interested in the delusions of researchers in the first place—I believe that all of these factors may be related, but the third option is probably the most convincing. However, verifying this objectively would surely be difficult.

29.8 Reflections from the Perspective of SF Prototyping—Researchers' Reactions

Based on what was presented in [3], I would like to examine whether the fiction I wrote in "The Dawn of 'Chemical Artificial Intelligence' and its Implications for the Brain" can be called SF prototyping.

First, as mentioned in [5], the researchers and I tried to envision a future about 30 years from the present, but we did not try to predict it. We tried to think from an unconventional mindset as much as possible. In this sense, the project fulfills one of the requirements of SF prototyping.

Although it was only over a brief period, the researchers and I had numerous conversations on the premise that we would be writing fiction. To start the dialogue, I asked, "Roughly speaking, in 30 years' time, if the results of your research were to be implemented in some form in society, what kind of world do you think we would end up with? Or, what would you like it to be like? Please share your visions." These and other questions were sent to the researchers in advance.

In reality, I rarely asked these questions exactly this way. I always led them along in the flow of the conversation, but the purpose was in the direction described above.

Reactions varied across the researchers, with some "not really getting on board" with it. In such cases, I managed to draw out ideas and information by throwing around various fantasies that I had invented myself.

However, many researchers were really into it, and in some cases, they had prepared several fantasies before the interview. During such moments, the conversation would get very lively, and as one fantasy after another would pile up, epic tall tales would unfold—too big to fit into a short story. It was such fun and exciting that it seemed as if we could have continued for hours.

Although no alcohol was consumed during these talks, I believe that SF itself proved to be a sufficient substitute for lubricating our discussions. Miyamoto says that SF "increases the psychological safety" of those participating in the discussion [2], and the fact that they were free to indulge in whatever fantastic imaginings they had may have caused the researchers to drop any inhibitions or pretenses and really get into it.

A researcher who was reluctant to let the dialogue end once told me, half-seriously, that he "would like to give me a role here as an 'SF writer-in-residence.'" Rather than the access rankings or the number of reader shares, these words may have been the greatest accomplishment. It could be said that this, too, was SF prototyping.

We did not go so far as to come up with a new research theme by working backward from the numerous fantasies that emerged from the dialogues and discussions and the contents of the stories that were eventually written. This would have been a task for the researcher rather than for myself, and there was no time for it.

However, during the course of the dialogue, the researchers came up with new ideas or began to reexamine past themes and ideas that had been stuck in the corners of their minds. In some cases, they specifically mentioned such ideas or themes in their e-mails after reading the completed fiction manuscripts.

As for ELSI/RRI in molecular cybernetics and robotics, which I mentioned in a rather roundabout way in the fiction, not many responses were received from the researchers.

Although 10 researchers were interviewed, only one of them mentioned ELSI/RRI in the comments I received after reading the story. It stated, "I realized that I should pay more attention to social acceptability. I would like to make use of this in my future research." I also received a verbal comment, not from the researcher I interviewed but from another individual who stated that he liked the way I wrote the piece, which did not lead the reader toward a specific way of thinking.

In conversations with the researchers before writing the piece, I sometimes deliberately raised concerns I had in order to elicit material. For example, concerns like "Wouldn't it be dangerous if molecular robots started multiplying on their own?" Most of the responses to these concerns were either, "We can't deny such a danger, but there is no cause for concern at this point" or "It would be very difficult to make such a thing possible, and even if it were possible, it would be some time in the future."

I do not believe there was much discussion on this issue worth getting worked up over.

29.9 How SF Writers Can Be Leveraged in Research on Emerging Science and Technologies

Finally, based on the experience of serializing "The Dawn of 'Chemical Artificial Intelligence' and its Implications for the Brain," I would like to list some points to keep in mind when making similar attempts in the future.

First, although it may sound paradoxical, it may be better not to have SF writers produce works.

If a company or research institute asks a writer to write or if the writer is involved as a JIR (as in this case), the writer will be forced to be aware of the restrictions mentioned in [6]. They also tend to have an underlying intention to make the content "useful" and "informative" to the client, although in some cases, SF creativity is not influenced by these factors. However, I feel that rather than writing without any restrictions, there is a tendency to be uncontroversial, reserved, or otherwise anodyne.

If the work is used for outreach, it may not be entertaining enough to have the desired effect. Results such as the number of shares mentioned in [7] suggest this.

Contrarily, if the work is used for SF prototyping, the world depicted in the work may not be the result of the artist's maximum imagination. Then, research themes, etc., derived from them may not be as fresh as expected.

Although this is not the reason, the role of SF writers does not seem to be emphasized much in the commentaries by Higuchi and Miyamoto. Higuchi clearly states that he recommends "trying to have SF stories written by non-SF authors" [6]. He goes on to emphasize the value and insights gained when people who have never written novels discuss and create works together.

I agree with this, and if done poorly, the writers may, out of professionalism, take the wild fantasies that emerge from various "amateur" discussions and neatly organize and prune them into a presentable story. That might work for outreach, but it would not be suitable for prototyping.

However, as I mentioned in [8], I think the writer could be a catalyst for getting the discussion going. They may be suited to the role of encouraging fantasy to fuel fantasy and brewing SF to make it more palatable.

I have one regret here.

Although I was positioned as a JIR, I was not provided with a place to stay at a specific research institute. In addition, when I began my activities, face-to-face interactions with the researchers were severely restricted due to the outbreak of the COVID-19 pandemic. Consequently, I believe that I was in effect a "contract reporter" or "dedicated writer."

Owing to these circumstances, the dialogue with the researchers may have remained within the framework of normal reporting, similar to what is done in newspapers and magazines. In addition, there was little room to display playfulness, digress into unrelated topics, or engage in trivial chatter to find ideas. However, is it not often the case that wild imaginations emerge from such seemingly futile efforts?

If I ever have the opportunity to become an "SF writer-in-residence," I would certainly be overjoyed to help researchers write some far-out SF.

Taking the above into account, with regard to the ELSI/RRI, the extent to which the writers could stimulate discussion or serve as catalysts for promising solutions was relatively low, as far as I can glean from the responses of the researchers to completed works and the preliminary interviews.

Although I am not familiar with ELSI/RRI, it is said that in recent years, researchers themselves have been required to "collaborate with various stakeholders from the early stages of research and development, determine the desirable state of science and technology and measures to implement them in society, and always deal with them in a proactive manner" [8]. It is also said that if this is not done, the freedom of research will rather be lost.

If this is true in reality, my impression at this point is that many researchers in the field may not yet be fully aware of this. What can SF writers, who are essentially nothing more than "*sukoshi fushigi*" storytellers, do in response? This is an issue I would like to consider in the future.

References

1. Fujiko, F.: Fujiko Fujio Land "*Shonen SF Tanpen* (Youth SF Short Stories)" Volume 2. Chuokoron-Sha, Tokyo (1989)
2. Miyamoto, D.: Furubita mirai wo dou kowasu? (How can we get rid of the old-fashioned future?). Kobunsha, Tokyo (2023)
3. Bear, G.: Blood Music. Arbor House, New York (1985)
4. Crichton, M.: Play. HarperCollins, New York (2002)
5. Ito, K.: Harmony. Hayakawa Publishing, Tokyo (2008)
6. Higuchi, K.: Mirai ha yosoku suru mono dehanaku souzou suru mono de aru (The future is not something to predict, but rather something to create). Chikumashobo, Tokyo (2021)
7. Editorial staff of Kodansha, Bluebacks: 60th anniversary of Kodansha Bluebacks. Kodansha, Tokyo (2023)
8. Iwase, K. et al.: ELSI for natural science researchers. Center for Research and Development Strategy, Japan Science and Technology Agency, Tokyo (2021)

Chapter 30
Applying LARP to Explore ELSI in an Emerging Field of Science, Its Challenges, and Opportunities

Akihiko Mori

30.1 Introduction

The Journalist in Residence (JIR) program facilitates collaborative projects by journalists, which are undertaken in a particular professional or situation over an extended period. This enables them to pursue and achieve their journalistic objectives. Participating journalists are required to report on a designated media platform, such as online digital news media, in a specified format to achieve their proposed objectives. This may include, but is not limited to, formats such as straight news reporting, documentaries, opinion pieces, and others.

In the Molecular Cybernetics JIR program, two objectives were set: first, to enhance the diversity of Japanese science journalism, and second, to investigate the ethical, legal, and social issues (ELSI) pertaining to this field of research. To this end, two types of media were selected for coverage: global news media and the Japanese edition of *WIRED*. Although scientific achievements are disseminated and presented internationally in many formats, Japanese science journalism tends to be domestically oriented because of the potential limitations of language and a relative lack of diversity. Moreover, as Italian science journalist Matin W. Angler points out in his book *Science Journalism: An Introduction*, a large part of science journalism can be "infotainment" [1], that is, science stories that are easy to understand, quick to consume, and lack criticality. This tendency is further entrenched in the contemporary digital information age, which is characterized by the predominance of social media platforms. Addressing this imbalance and promoting responsible research and innovation (RRI) necessitate critical and creative engagement with the ELSI of science.

Owing to my involvement with the Molecular Cybernetics JIR program, the first objective has been realized, although the second objective remains unfulfilled. One

A. Mori (✉)
Kyoto, Japan
e-mail: akihico@mojimoji.co

S. Murata et al. (eds.), *Molecular Robotics II*,
https://doi.org/10.1007/978-981-95-2394-8_30

factor contributing to this failure is the potential complexity of journalism. This assertion is supported by the observation of Alan Rusbridger, former editor-in-chief of *The Guardian*, who noted that journalism often operates as a "rear-view mirror" [2]. Additionally, journalists are typically more interested in "the man who bites a dog" than the reverse, favouring exceptional events over gradual or invisible changes—and largely avoiding the realm of prediction, speculation, and uncertainty. This highlights the limitations of journalism within the contemporary media landscape, where the emphasis is predominantly on past events rather than on future prospects. During my research in the JIR program, I had the opportunity to visit several laboratories and engage in numerous discussions with scientists. However, it became apparent that the field of molecular cybernetics has not yet reached a sufficient level of maturity to be a subject of journalistic coverage. This conclusion is supported by the notion of journalistic approach, as articulated by Rusbridger, which emphasizes the examination of contemporary events and their analysis, rather than the anticipation of future occurrences. This insight is articulated in the article "Chemical AI Realizes the Future of Molecular Cybernetics," [3] which was published on Wired.jp on October 3, 2023. The article elucidates the concept of molecular cybernetics through an interview with three researchers involved in the research: Dr. Satoshi Murata, Dr. Akinori Kuzuya, and Dr. Taro Toyota, who are co-authors of "Molecular Cybernetics: Challenges toward Cellular Chemical Artificial Intelligence," published in *Advanced Functional Materials* in 2022 [4]. Researchers are attempting to integrate extremely small molecule-sized devices into artificial cells, modularize them, create multicellular (large-scale) structures, and develop advanced information processing capabilities, including AI. Moreover, their long-term objective is to facilitate the exchange of molecular information between cellular chemical AI and living cells. However, despite the numerous scientific advancements made and the technology developed to achieve molecular cybernetics, the field remains in the early and basic stages of scientific research.

Nevertheless, conventional journalistic attitudes frequently prove inadequate for addressing gradual and imperceptible, yet precipitous and profound shifts in ELSI, particularly in light of advancements in science and technology. Concurrently, as Rusbridger elucidates, news editors are disinclined to engage in prediction, speculation, and uncertainty. This proclivity is indicative of contemporary journalism's inability to address long-term concerns such as climate change. To address the aforementioned shortcomings of journalism and the current research phase of molecular cybernetics simultaneously, an alternative objective, designated as "SF LARP," has been established.

30.2 Method

30.2.1 Process of Live Action Role Play (LARP)

Live action role play (LARP) can be defined as a form of improvisational theater in which participants adopt the role of a character and engage in improvised conversations and performances. A prominent example of this genre is Nordic LARP, which has served as a foundational element in the development of science fiction live action role play (SF LARP) originally designed for this research. LARP is regarded as an open-source theatrical form, wherein there are no fixed or definitive rules; rather, there is a co-creative and collective aspect to the performance. However, as Koljonen et al. (2021) observe in their book *Larp Design: Creating Role-Play Experiences*, the term "LARP" is used to describe a variety of role-play experiences, each of which may be defined by three key elements: the LARP script, the duration of the game, and the debriefing activities [5].

The role of the LARP designer is to establish the fundamental structure and parameters of the LARP, which may be conceptualized as a comprehensive script. In their capacity as game masters, LARP designers are responsible for defining the character settings, establishing the game rules, developing the storylines, coordinating schedules, organizing workshops, and overseeing each session. Prior to the initiation of the LARP, the script is presented and explained in detail, often in the form of a briefing workshop, with the objective of facilitating the players' actions. The duration of the LARP constitutes its primary component, encompassing all on-site activities, including the aforementioned briefing workshop, rehearsals, and the post-play cool-down period [5]. The debriefing activities are not merely a conclusion; rather, they facilitate the emergence of the collective meaning of the LARP [5]. The act of recalling, debating, and analyzing the events that transpired during the LARP with the participants serves to reinforce the overall experience of the LARP. The LARP process is divided into two distinct categories: "Ingame," which encompasses the activities occurring within the game, and "offgame," which refers to events taking place apart from the game [5].

Players act in the runtime, demonstrating behaviors that are either aligned with or divergent from the intentions of the LARP designer. These actions may be perceived as appropriate or inappropriate and in line with or contrary to the designer's expectations. To ensure ethical conduct during the event, a widely utilized and pivotal rule has emerged in Nordic LARPs, known as metatechniques. During the game, the participants are required to wear a virtual mask to assume the roles of the characters delineated in the script. In the absence of the opportunity to reclaim their identity and express themselves, particularly when confronted with challenging scenarios such as sexual, violent, or discriminatory situations, the experience may prove detrimental to the players. Metatechniques are defined as a means of circumventing such unforeseen circumstances during the runtime by enabling players to communicate with the individual behind the mask throughout the runtime [5].

30.2.2 A Way to Speculate About the Future

The primary benefit of LARP is the opportunity for improvisation and role-playing, which affords a distinctive perspective that differs from one's usual outlook. This process facilitates the identification of potential issues that may have otherwise remained unperceived, and the experience of emotions that may have been previously inaccessible when operating behind a mask. The application of LARP has been diverse, encompassing domains such as the enhancement of children's cognitive abilities and the conceptualization of innovative wearable technologies.

Artists' employment of LARP is indicative of a speculative discourse concerning the future. Ai Hasegawa, an artist whose primary focus is on speculative design projects, examines the prospective ELSI of emerging scientific and technological advancements. A notable example of this is her work *(Im)Possible Baby, Case 01: Asako & Moriga*, in which she explores a potential future in which babies are born to same-sex parents. This is achieved by simulating and visualizing their prospective children using DNA data from a real-life lesbian couple. Hasegawa's most recent interactive artwork, entitled *Multiple Futures*, employs LARP to investigate a speculative multi-species concept. *Multiple Futures* is a speculative fiction piece based on the themes of faith and community, set 100 years from the present. The technology utilized in this piece combines virtual reality and artificial intelligence (a large language model [LLM]), allowing the generation of a future in real time. In this interactive artwork, the viewer assumes the role of a specific character through LARP with LLM to explore a multi-species future. The concept of multi-species, which refers to the symbiosis between multiple species, including humans, is gaining recognition for its importance for sustainability, particularly in the context of ethnographic and indigenous studies. The notion of multi-species demands acknowledgement of the intricate interdependencies between humans, their immediate and indirect stakeholders, and a multitude of species. This necessitates dismantling conventional anthropocentric perspectives. However, this complexity is regarded as a concept that can only be attained through the literal expansion of human capabilities. The *Multiple Futures* project explores the potential of utilizing a LARP-based speculative approach to investigate complex concepts of this nature.

An expert interview was conducted with Ai Hasegawa to identify the key concepts and detailed processes involved. The interview revealed that Hasegawa employs functional cards in the design of the LARP. Initially, participants are instructed to record each futuristic perspective as a setting on a designated World View card, with the option of producing as many as they see fit. Subsequently, the participants select three shuffled World View Cards and engage in a discussion to determine which future they will adopt for the purposes of the LARP. The participants are then directed to create Event Cards. These cards are divided into two categories: Happy Event Cards, which encapsulate positive occurrences, and Problem Cards, which encapsulate negative events. Following this, players determine which character they will assume within the context of the game. Participants are permitted to assume character roles based on their own preferences but are advised to adhere to the vectors of oppression and privilege, as delineated by Kathryn Pauly Morgan. During the game, each player draws an Event Card and plays the LARP in accordance with the instructions on that card. Finally, participants are guided through a series of debriefing activities.

Hasegawa's approach is characterized by a distinctive emphasis on collective and participatory design within the context of LARP. This approach empowers players to participate in the creative process, affording them the autonomy to shape the setting, the narrative arc of the event, and the characters within it. The incorporation of play-style analysis and functional cards (World View and Event Cards) facilitates player engagement in the LARP, even for those who are less accustomed to improvisation. This approach fosters a sense of autonomy and ownership among players regarding the LARP scripts.

30.2.3 The Focus and Storylines of SF LARP

In order to develop a LARP designed to identify and address the ELSI of emerging scientific advancements, a science fiction narrative framework was employed. The narrative forming the basis of this LARP was created in collaboration with the JIR and Shingo Fujisaki, a participant novelist in the JIR; Mikihito Tanaka, Professor of Waseda University School of Political Science and Economics; and Kanako Takae, Assistant Professor at the Waseda University Institute for Advanced Studies in the Social Sciences. The LARP enables participants to engage in speculative play concerning future scenarios, a feature that has led to its categorization as an SF LARP.

To provide a more precise framework for SF LARP's speculative aspect, a particular focus was defined: moral enhancement. The distinction between enhancement and therapy lies in the objective of each: The former aims to augment innate, healthy human abilities, whereas the latter seeks to acquire abilities that surpass the boundaries of human capability. Enhancement encompasses both physical and cognitive abilities. The former includes capabilities such as running and jumping, whereas the latter includes thinking, memory, and calculation skills. The concept of enhancement is endorsed by transhumanists, who seek to use technology to improve humanity. Moral enhancement aims to augment human ethical capabilities in a manner analogous to the enhancement of physical capabilities. Military organizations have demonstrated interest in fortifying soldiers' mental and emotional capabilities. Similarly, the fundamental concept of moral enhancement endeavors to exert a constructive influence on science and international politics by enhancing human ethical capabilities.

30.2.4 The Storyline

The following narrative constitutes the storyline employed in the SF LARP. The setting is situated in the future, specifically in 2055. Participants engaged in research at the Molecular Cybernetics Laboratory (MCL), a research institute dedicated to the implementation of molecular cybernetics in society. The field of molecular cybernetics, a chemical technology that has been awarded the Nobel Prize in chemistry, is poised to facilitate the advanced control of the next generation of human intelligence. The MCL is currently confronted with several social issues. Recently, researchers in a certain country have announced the successful completion of an experiment aimed at enhancing human intelligence. This experiment, which was conducted on human subjects, has prompted significant debate within the scientific community. The researchers employed molecular cybernetic technology to artificially expand human intellectual capacity. Despite substantial criticism, many scientists are pursuing analogous research avenues.

The considerable intellectual abilities attained through intelligence enhancement have the potential to intensify existing disparities and challenges faced by developing regions in terms of the provision of social welfare. To address this issue, the MCL has engaged in fundamental research on the concept of "ethical enhancement," with the objective of controlling intelligence enhancement.

In view of the MCL's pioneering position in the domain of molecular cybernetics research, the institute has now been divided into two distinct groups: those who advocate for the advancement of these technologies and those who do not. However, in light of the increasingly challenging circumstances, the MCL has established a task force with the objective of developing moral enhancement technology and is conducting interviews to recruit highly skilled researchers from within the institute. You are a researcher at the MCL. As a participant in the interview process, you may assume the role of either the interviewer or researcher (interviewee) or, alternatively, that of an expert.

30.2.5 Molecular Cybernetics SF LARP Process

30.2.5.1 Briefing

A briefing on the game of Molecular Cybernetics SF LARP is provided as follows. This game is played as a career interview, which is an active thinking process for enhancement technologies. The format of the career interview is as follows: It is conducted by an interviewer and includes the individual being interviewed, as well as two randomly selected experts. The objective of this is to obtain a diverse range of opinions. The selection of these experts is intended to ensure a range of perspectives, with the experts being individuals who hold either favorable or unfavorable positions on the subject of enhancement technologies.

30.2.5.2 Choosing Characters

Character selection constitutes the initial step in the LARP process. Participants are required to select one of the four available roles: interviewer, interviewee, or one of two experts. Following the allocation of these roles, the LARP is initiated.

30.2.5.3 Runtime

The LARP is conducted on an online platform. To create a degree of distance from the participants' actual identities (although their voices inevitably reveal who they are), each participant configures a digital avatar, and their Zoom username is modified to correspond to the designated role. Once the LARP has commenced, the narrative progresses as per the direction of the interviewer, who recruits the individuals to be interviewed. The dissemination of future news material is reserved exclusively for experts, with the objective of establishing them as such, that is, as individuals with a comprehensive understanding of intellectual and moral improvement. Ultimately, the interviewee must decide whether to accept the interviewer's proposal to develop a moral enhancement technique or to decline the offer. Furthermore, they present their rationale for making their decision.

30.2.5.4 Debrief Activities

A debriefing session is held after the LARP runtime, during which an open discussion and questionnaire survey are conducted.

30.2.6 Future News Materials

To enhance player engagement in the LARP, the Molecular Cybernetics SF LARP employs Future News Materials, the way that functional cards were incorporated into Hasegawa's approach. The Future News Materials are authored by Ryoma Komiyama, a science journalist employed by one of Japan's leading newspapers, *The Asahi Shimbun*.

30.2.7 Examples of Future News Materials

Stimulation of serotonin secretion suppresses violent behavior in mice

On January 15, a group of researchers from the Institute for Advanced Biology published a paper in the *British Journal of Science* in which they announced that they had successfully conducted an experiment to suppress violent behavior in mice by activating the function of a specific gene. The gene in question is involved in the secretion of serotonin, otherwise known as the "happy hormone," and the researchers hypothesize that this technology could be applied to humans in the future.

Monkeys share food, drastically reducing the number of deaths

On January 15, the Primate Science Research Center published a paper in the journal *Animal Behaviour*, in which the authors describe how ultra-fine robotic manipulation of the brains of rhesus monkeys has drastically reduced food fighting between members of a troop. The study also demonstrated that the nutritional status of weak individuals who had previously been unable to find food on their own was improved, and that almost all cases of death from disease were eliminated.

Strikebreaking through ethical capacity building? Some human rights concerns

In the present year, the South Asian country of T initiated a program of genetic engineering for employees of more than 100 state-owned companies, with the objective of enhancing their levels of "hard work" and "ethical capacity." The country has witnessed frequent industrial action in recent years, with workers expressing discontent with inflation and the escalating disparity between the affluent and the impoverished. The government's use of genetic engineering in this context is perceived as a strategy to mitigate these issues, though concerns regarding human rights have been raised by Western governments and non-governmental organizations.

More sick monkeys, possibly as a result of an inability to defend themselves

On November 15, the Primate Science Research Centre published a paper in the journal *Animal Behaviour*, which found that rhesus monkeys with increased functionality in the part of the brain responsible for empathy were more likely to become ill due to poor nutrition. This phenomenon is hypothesized to occur because of the monkeys' inability to evade or protect themselves from aggression by conspecifics whose brains have not undergone this particular process.

30.2.8 SF LARP Character Settings

Interviewer: Luca

Role: MCL Research Officer. Luca is a researcher and the head of the Moral Enhancement Research Team (the Task Force). His remit includes the recruitment of a talented researcher, Mr. Bessho, to the research team. The text does not specify whether the MCL Research Officer himself is in favor of or opposed to moral enhancement, leaving this decision to the LARP participants. Furthermore, any behavior that constitutes harassment during this interview will result in disciplinary dismissal.

Interviewee (researcher): Bessho

Role: MCL Researcher. Bessho is a researcher who specializes in the study and development of fundamental technologies associated with molecular cybernetics. The LARP player is at liberty to select his research area, with Bessho's position in terms of whether he is pro-enhancement or cautious not being specified in the setting (LARP players are at liberty to set this as they wish).

Experts: Nick and Fukuyama

Role: Members of the MCL Institute, they are also researchers. Their task is to provide the interviewees with insights from their own perspectives.

Nick: Being a radical transhumanist, Nick believes that humanity will evolve positively through intelligence enhancement.

Fukuyama: Being cautious and conservative in his views on transhumanism, he sees intelligence enhancement as a "recently emerging technology" and believes that its potential needs to be carefully studied.

30.2.9 SF LARP Rules

The rules below are suggested to all participants:

- Take 10 min at the beginning to write about your play.
- The LARP time is 40 min.
- After 30 min, it is time for Decision Making.
- Who will be interviewed during Decision Making?

The interviewee must choose one of the following options and give reasons for their choice:

A. Participate in the Ethical Enhancement study.

B. Refuse this offer.

Meta statement:

The act of raising one's hand in the midst of the performance is indicative of an impending "meta statement." This statement may pertain to the request for guidance on the modification of the dramatic narrative, or alternatively, if the individual perceives the trajectory of the drama to be divergent from their desired course, they may employ the term "Cut" to effect a cessation and alteration. Other participants must refrain from soliciting reasons for such actions. The reason for this may be personal and difficult to express, such as the story having touched on a traumatic event.

30.3 Result and Analysis

30.3.1 Summary of LARP Runtime Result

The SF LARP invited 12 researchers divided into three groups involved in Molecular Cybernetics.

Group A: A deep and wide-ranging exploration of two conflicting ideas: individual enhancement and collective (plural) enhancement.

The participants agreed that as members of the MCL, the promotion of research into moral enhancement should be undertaken to maintain the sustainability of humanity. In the context of the LARP, the subjects of individual and plural enhancements were thoroughly explored. The participants concurred that individual enhancement is comparatively more straightforward to cultivate in comparison to plural enhancement, because plural enhancement influences the cultural diversity of society. Consequently, the interviewee accepted the offer, as the Task Force team demonstrated an understanding of the potential problems of plural enhancement from scientific and ethical perspectives. The interviewee expressed a sense of personal motivation to engage with such a challenging issue as a researcher (Fig. 30.1).

Note

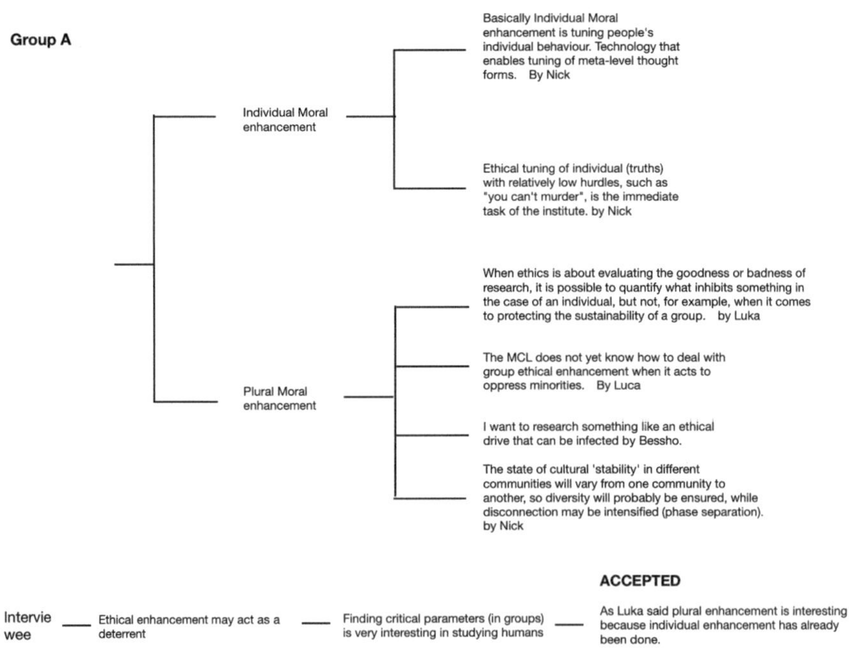

Fig. 30.1 The ELSI framings constructed by group A through LARP.

Notably, this group divided the discourse on moral enhancement into individual and collective categories. Given the uncertainty surrounding the feasibility of achieving such a categorization through a single journalist's interview with multiple researchers, the LARP research is regarded as a noteworthy accomplishment. It is also important that the interviewees cited their desire to undertake challenging tasks as their reason for accepting the offer, and these researcher-specific personalities should be confirmed. This group is also notable for its invention of digital avatar LARP. As a LARP designer, I have adopted their concept for the design of the entire SF LARP.

Group B: The importance of research lies in the people involved, and not in the research itself.

This group's representation of this act in the LARP depicted the personal transformation of a researcher, who was initially somewhat skeptical about moral enhancement research, ultimately undergoing a change of heart and accepting the offer. Bessho's assertion that "the development of the atomic bomb and the development of nuclear power are not similar" was a thoughtful and suggestive contribution, addressing the issue of moral enhancement not in itself, but in its application within society. The LARP explored how researchers can orient their own research toward the development of beneficial moral enhancement technologies. Ultimately, a democratic approach to research emerged as the most compelling factor for the interviewees in accepting the offer (Fig. 30.2).

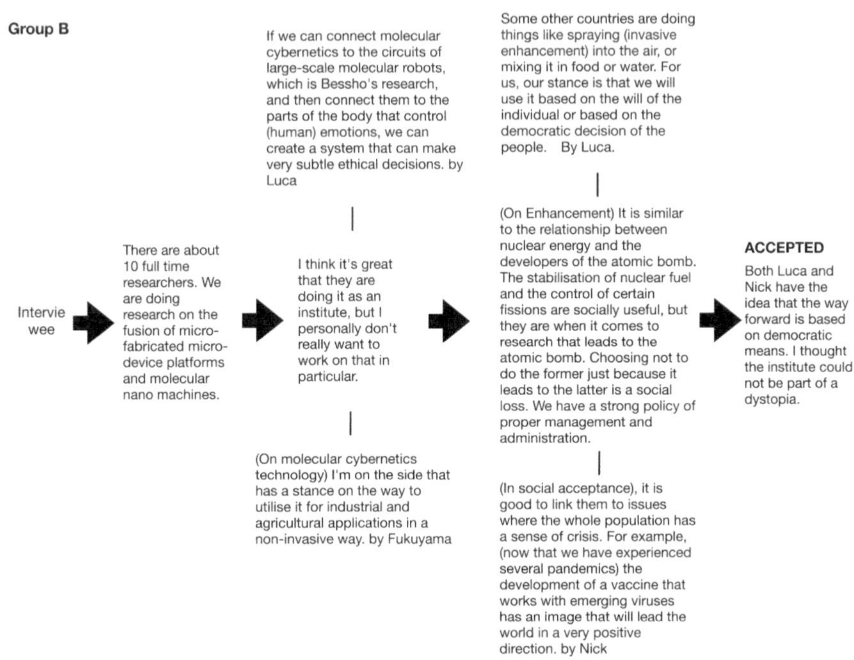

Fig. 30.2 The ELSI framings constructed by group B through LARP

Note

The process of persuading skeptical researchers in this LARP group provided insights into the elements that researchers consider ethically correct, or at least their tendency to do so. The findings of this study demonstrate that even for highly complex and problematic technology such as moral enhancement, LARP can potentially serve as a tool for understanding the ethical issues that researchers must clarify before becoming involved. These perspectives are not easily discernible through the conventional interview methods employed by journalists.

Group C: Undo technology is the key to moral enhancement.

In the LARP, the interviewer assumed a leadership role in the discussion and proposed a specific solution to the moral enhancement issue. This solution involved implementing a technology that reversed the enhancement process. Although this proposition may have initially evoked images of a dystopian future, it was nevertheless regarded as rational within the MCL framework. In this context, the promotion of moral enhancement was perceived as a hindrance, particularly in a world where the development of intelligence enhancement has become uncontrollable. As in the case with Group A, the question of individual or plural was raised. The interviewer then proposed a compromise between the individual and plural perspectives (Fig. 30.3).

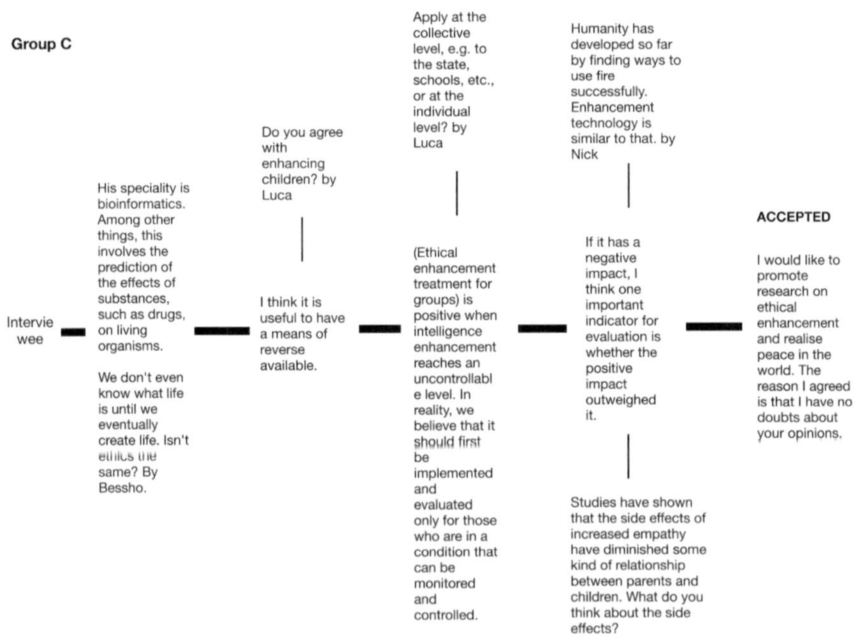

Fig. 30.3 The ELSI framings constructed by group C through LARP

Note

Notably, this group provided insights into how researchers with clearly defined intentions, beliefs, and ideas could establish themselves in the unexplored field of moral enhancement. The interviewees exhibited a high level of expertise in moral enhancement (bioinformatics) and provided unambiguous responses to the interviewers' and experts' inquiries. Analogous to Group B, the LARP can potentially serve as a valuable case study for analyzing how researchers can be persuasive and convincing.

30.3.2 Analysis of Debriefing

In the post-run debriefing, participants from each group provided positive feedback about the LARP, ranging from direct feedback such as "interesting" or "fun" to suggestive feedback such as "Being in the role gave me a perspective I would not normally have." As one researcher observes, "the term 'moral enhancement' itself has a bias" and "the actual discussion of moral enhancement raises the question of who decides the criteria for what is ethically right or wrong." Moreover, a recurring feedback was that the role of the expert is more complex and perplexing than those of the interviewer and interviewee.

Conversely, the open discussion that took place at the 7th Annual Conference on Molecular Robotics at the University of Tokyo on March 13–14, 2024 appears to have elicited less favorable feedback than that received following the runtime debriefing. This was because of the considerable difference in the amount of information between those who participated in the LARP and those who did not; this is a point that must be addressed in the future. During the open discussion, one researcher who participated in the LARP proposed expanding the LARP to include other experts and participants from diverse fields. Although the LARP's central theme was the exploration of ELSI by molecular cybernetics researchers, the suggestions put forth by this researcher merit consideration for future implementation.

A review of the feedback provided by all LARP participants following the post-runtime debriefing revealed that 63.6% of the participants rated it as significant, while an additional 36.4% rated it as "extremely significant" in terms of overall assessment. The primary reasons for this evaluation pertained to the opportunity to virtually experience a situation that necessitated deliberation on significant ethical issues in molecular cybernetics, and the ability to articulate statements that would not typically be expressed, in conjunction with observing the reactions of fellow researchers. It also provided an opportunity to "deepen my understanding of ELSI and to reflect on various matters."

In the context of the LARP setting, the interview session was evaluated as "pleasant" by 63.7% of participants, while 36.4% indicated that their perception of the session depended on the character they were portraying. The underlying reason for this observation was as follows: "Within the context of recruiting the most accomplished researchers, I perceived an element of predictability in the outcome. Although it was straightforward to draw conclusions, the setting was unlikely to engender a conflict. Additionally, it was challenging to comprehend the roles of the experts in the interview."

In response to the question, "How did you feel about this discussion compared to other cases you have experienced of discussing research ethics (i.e., ELSI) and the social acceptance of science?", 72.7% of the participants indicated that they found it useful, while 18.2% found it very useful. In addition, the following question was posed: "How does SF LARP differ from other cases you have experienced in discussing ethics of research (i.e., ELSI) and social acceptance of science?" In response to the question, 72.7% of the participants found it interesting and enjoyable, while 27.3% found it interesting to express their opinions in the discussion created by the other actors. Furthermore, all participants expressed a strong expectation, or even a very strong expectation, of further applications of the SF LARP, including its potential use in emergency response and ethics training.

The following question was posed to the participants: "What changes in attitude toward research have you experienced before and after the SF LARP? The responses to this query were as follows: "I have re-acknowledged that even seemingly good scientific and technological proposals and results can have problems of ambiguity of use"; "I have become more aware of researchers in different positions"; "I felt that I could ask bold questions because of the LARP with avatars." Conversely, the use of an avatar led to a perception of divergence between the participant's actual self and their LARP persona. This prompted reflection on the extent to which a transformation in consciousness was achieved.

The open discussion was also evaluated as significant or very significant by 81.8% of the participants, but there was one dissenting opinion: "In contrast to the participants who attended the SF LARP, those who did not attend found it challenging to participate in the discussion."

30.3.3 Future Implications

This reflective essay provides a concise overview of the merits and demerits of SF LARP. Many participants concurred that the merits of SF LARP lie in its ability to transform science fiction into a playful endeavor, thereby facilitating the exploration of ELSI from novel vantage points.

However, a significant challenge in integrating ELSI into scientific research is the gap between theoretical frameworks and their practical applications in laboratory or technological settings. Although RRI emphasizes anticipatory governance and inclusive stakeholder engagement, the operationalization of ELSI often remains abstract, posing challenges for researchers to internalize and act on these considerations in their day-to-day practice. To address this gap, LARP can be employed as a training tool to simulate ethical dilemmas and engage participants in decision-making processes.

The exploration of ELSI in the practice of LARP was spontaneous and autonomous, which is interesting and stimulating from a journalistic perspective. For instance, Group A proposed a framework for thinking in terms of individual and plural enhancement, whereas Group B presented the important perspective of considering ELSI in the context of molecular cybernetics, using the difference between the development of the atomic bomb and research into nuclear energy as an example. Obtaining such information through journalistic investigation requires considerable time and effort. Questions posed by journalists are often biased, whereas in LARP, important perspectives in the field of research are explored spontaneously, autonomously, and comprehensively by people who share the same scientific interests, which is of great value. The positive outcomes observed are attributed to the JIR program.

Conversely, LARP design has certain limitations. From the perspective of LARP designers, defining the precise outcome of the LARP is challenging, which often results in SF LARP yielding outcomes that the designers cannot anticipate or predict. This is because of the inherent nature of LARP, which fosters autonomous and spontaneous discussions and facilitates perspectives that cannot be achieved through conventional Q&A or discussion. Consequently, controlling the outcomes is difficult. In this instance, an attempt was made to clarify the outcome by designing an interview session as an ethics improvement session. However, the design proved to be complex, and as the responses above suggest, this should be noted as a difficult aspect for participants.

In future iterations, it would be advantageous to utilize the outcomes of this LARP to create tangible achievements or artifacts outside the LARP, such as in the JIR program, where further actions will be taken to produce journalistic works using the results of the LARP. In summary, LARP should be designed and used as a tool to achieve a specific goal.

30.4 Conclusion

In conclusion, the use of SF LARP to explore ELSI, particularly in an emerging scientific domain, is advantageous for scientists engaged in ELSI and journalists who require an alternative investigative approach. As demonstrated by the Molecule Cybernetics SF LARP, ELSI can be explored from various perspectives that would not typically be considered. Notably, the exploration in this context is characterized by greater spontaneity and autonomy compared to conventional interview methods employed in journalistic investigations and other research methodologies. Although SF LARP has proven advantageous in exploring potential issues or problems in emerging scientific domains, its design requires refinement to ensure precise control over outcomes. To enhance the precision and efficacy of SF LARP, the designer is recommended to engage in research on alternative applications to refine the design.

References

1. Angler, M.W.: Science journalism: an introduction. Routledge, Abingdon (2017)
2. Rusbridger, A.: Climate change: why the guardian is putting threat to earth front and centre, the guardian. https://www.theguardian.com/environment/2015/mar/06/climate-change-guardian-threat-to-earth-alan-rusbridger (2015). Accessed 06 May 2024
3. Mori, A.: Chemical na AI ga jitsugen sugu, Bunshi Cybernetics no mirai (The Future of Molecular Cybernetics reclaimed by the Chemical AI) [Japanese], WIRED.jp. https://wired.jp/article/chemical-ai/ (2023). Accessed 06 May 2024
4. Murata, S., et al.: Molecular cybernetics: challenges toward cellular chemical artificial intelligence. Adv Funct Mater **32**(37) (2022)
5. Koljonen, J., et al.: LARP design: creating role-play Experiences. https://trepo.tuni.fi/handle/10024/136330 (2021)

Chapter 31
A Perspective from the Science Newsroom

Ryoma Komiyama

31.1 Scientific Articles are Not Being Read

My apologies for bringing this up seemingly out of nowhere, but let me begin with a tweet posted on Twitter (now called X) in January 2014 by a prominent researcher.

> There is no mention of her being a woman, let alone wearing an apron. And yet, the article impressively conveyed just how incredible the discovery was. I thought the article was a great breath of fresh air, and, of course, it turned out that it was written by a foreign journalist.
>
> A new method to create pluripotent cells in 30 minutes, developed by Japan's RIKEN, has the world raving!–
>
> Post on X (formerly Twitter) by Shunji Yamanaka, Project Professor at The University of Tokyo, Jan. 30. 2014.

Many of you may recognize the words "apron" and "pluripotent cell," but this is a post by Haruko Obokata about "STAP cells," which was a hot topic in Japan at the time. It was right after the announcement by RIKEN, when the allegations of research misconduct had not yet surfaced, and the media coverage was unquestionably positive.

However, the tweet concerned not the STAP cells themselves but how the Japanese media covered the topic. "I thought it was a good article, but it was written by a foreign journalist. Japanese reporters only wrote uninspiring articles that had nothing to do with the discovery itself, such as the fact that Obokata was wearing an apron or that

Note: The following manuscript is based on a lecture given by Mr. Komiyama at the annual conference of Molecular Robotics held on November 13, 2022, in the session titled "When Science Meets the Headline (see 7.1.3.3)." Therefore, slides and photos used on the day of the lecture have been omitted.

R. Komiyama (✉)
The Asahi Shimbun, Chuo City, Japan
e-mail: ryomakom@gmail.com; komiyama-r@asahi.com

S. Murata et al. (eds.), *Molecular Robotics II*,
https://doi.org/10.1007/978-981-95-2394-8_31

she was a woman," he lamented, which can be interpreted as an indirect criticism of Japanese media delivered by praising foreign reporting.

The post was reposted 2,500 times and liked 1,200 times, and it garnered considerable sympathy in the online space. In other words, many people agreed with the assertion that the Japanese media, as usual, was putting out drivel. That is how I interpret it.

In fact, the domestic media are focusing on the person herself, Obokata, and writing articles about what her apron is all about. For example, the Asahi Shimbun, where I work, published this article.

(photo from the society page article).

The headline was "Cultivating competitive spirit, a big discovery at age 30," and it featured a large photo of Obokata in an apron. In the article (written in Japanese), which is over 1,200 characters long, there is not a single word about the research. Instead, it is full of gossipy little tidbits, such as the yellow color of the walls in the lab, the Moomin stickers on the walls, and the fact that she thinks about her research even while in the bath or on a date.

At the time, I was in charge of life sciences in the Science and Healthcare Department at the Osaka head office, so I was the reporter directly in charge of the STAP cell report. When I attended the press conference, Obokata's remarks were full of catchy statements that could easily be quoted in an article. Even in the laboratory where we were shown around after the press conference, she appeared in her apron and introduced her pet turtle. I got a strong impression that this was a great character.

So, the newspaper is indeed writing articles that have nothing to do with science. That is what the earlier post is pointing out.

But I feel I still have to say that we also write articles about science. The same day the article on the apron appeared, the following also appeared on the front page of the newspaper:

(Photo of the article on the front page).

The second page and the science page looked like this.

(Photo of the second page and the science page).

Most of these articles explain the research and its scientific significance, with the total number of words exceeding 6,000. In other words, more than five times the length of the article on the society page, "This is who Ms. Obokata is," was devoted to explaining the science.

The prominent researcher whose tweet I mentioned above and the many people who agreed with him were either unaware of the existence of these articles or ignored them, focusing only on the gossip articles and criticizing the media.

Even outside of topics such as STAP cells, this kind of thing is not uncommon. A typical example is when a Japanese researcher wins a Nobel Prize. Criticisms are often made that the article is full of nonsense and talks about things that have nothing to do with research, such as the recipient was like this when he or she was a kid, friends back home are happy, or the recipient was doing like this when the Nobel Prize committee contacted him or her.

But in fact, we also write about research. For example, in the case of Syukuro Manabe, who received the Nobel Prize in Physics in 2021, the first and second pages looked like this.

(Photos on pages 1 and 2).

All the articles only touched on the scientific significance of the research.

> Another prominent science writer posted the following when the 2023 Nobel Prize was announced:
>
> "Most contributors to the Japanese mass media devote virtually zero reporting on the prize unless it is awarded to a Japanese. This is because they are not interested in science in the first place. I think it's an unfortunate attitude."
>
> —Post on X (formerly Twitter) by Kaoru Takeuchi, A science writer, Oct. 1st, 2013.

This also received over 2,500 likes, but, quite simply, the assertion is false. The Asahi Shimbun writes articles of around 1,000 words every year, regardless of whether a Japanese national wins the prize (moreover, many reporters are interested in science).

Nevertheless, why is the Japanese media's coverage of science criticized in this way? One possible answer is that the more gossipy articles are more likely to be shared on social media sites. Users skim these kinds of "fluff articles" that appear on their personalized feeds and criticize them as "nonsense."

In short, it is not that articles introducing science are not being written. Rather, they are being written but are not being read.

31.2 In the Past, It Was More...

Let me tell you a little about an old story that happened more than 40 years ago, which I heard from a former Asahi Shimbun reporter named Akira Ozeki, a senior colleague of mine.

Kenichi Fukui of Kyoto University, known for his "frontier molecular orbital theory" of chemical reactions, won the Nobel Prize in Chemistry in 1981. On the night in October when the prize was decided, the Asahi Shimbun's science department received the first news about it via foreign media. However, the reporters did not know who "Fukui" was and what kind of research he was conducting. They did not even know how to write the Kanji characters for his name. This apparently caused a great deal of panic.

However, through an unexpected stroke of luck, the science department had a "scheduled article" on Fukui. It was not a draft of the Nobel Prize but of the Order of Cultural Merit that Mr. Fukui was to receive. Although the government had not yet announced the award's recipient, it had set a date for releasing the information to the public, and the news media had been informed in advance. Therefore, the manuscript had already been written, and the reporters at the head office breathed a sigh of relief when they found it.

So, at the time, Mr. Ozeki was a first-year reporter assigned to the Kyoto bureau. He had earned a master's degree in physics at graduate school, but he was not yet a science reporter and was in charge of covering incidents, accidents, and police work. The reporter who was supposed to be covering Kyoto University had gone out for a drink in the city's Gion district, leaving only Mr. Ozeki at the bureau. So, he was ordered to go to Dr. Fukui's home, and he headed there in a cab while reading a scheduled article of the Order of Cultural Merit.

When he arrived, reporters from other newspapers were already surrounding Dr. Fukui. All of them were what we call in the business "*satsukap*," or reporters overseeing the team in charge of the prefectural police and criminal cases. They were asking Dr. Fukui questions like "Who do you respect?" and "What did you learn from your mother?"—none of which had anything to do with science. When Mr. Ozeki, drawing on his background, asked a question related to the tradition of theoretical physicists at Kyoto University, who think with "pencil and paper," Mr. Fukui's eyes suddenly lit up, and he began to talk. Mr. Ozeki happily shared this anecdote with me, a junior staff member.

What I wanted to convey through the above story is not that Mr. Ozeki is a suave, savvy newsman, nor that knowledge of science is important. There are two points. First, it has always been the case that when someone wins a Nobel Prize, their "personality" and other aspects, which have nothing to do with science, are reported extensively. Second, things are still much better now than they used to be. Nowadays, most potential Nobel Prize winners are listed in advance, and a variety of drafts are prepared (with the only exception being Koichi Tanaka of Shimadzu Corporation, who received the 2002 Nobel Prize in Chemistry). It is also standard practice to send reporters with a reasonable understanding of science to the press conferences of Nobel Prize winners.

31.3 What is a "Science Reporter?"

In the first place, what is the role of a "science reporter" in this context?

The Asahi Shimbun's Science Department was established in 1957, and the following address was given at a symposium held in 2007 to commemorate the 50th anniversary of the department's establishment.

"The year 1957 was marked by a succession of major scientific news events, including the opening of Showa Station in Antarctica, atomic fire lit in Tokai-mura, and the launch of Sputnik by the Soviet Union." Three years earlier, in 1954, the Daigo Fukuryu Maru ship was exposed to radiation from a hydrogen bomb test at Bikini Atoll.

The Science Department was created in response to the need for reporters who could cover such stories. What was sought was not so much "science" but rather news related to science and "technology."

Later, the scope of coverage expanded to include earthquake prediction and organ transplants. However, the most important job was ultimately to follow up on the activities of the government's Science and Technology Agency. As a newspaper publisher, reporters in charge of covering government agencies are, for whatever reason, the "stars of the show." I understand that while they did cover basic science, that was not their focus.

An examination of Asahi Shimbun's database shows that the number of articles concerning basic science has been increasing since at least the late 1980s, with a particularly noticeable rise since around the year 2000.

For example, in the 10 years leading up to 2000, the number of times the word *Nature* (the science journal) appeared in articles other than local pages was 123 times. However, in the decade leading up to 2010, this number increased to 460, and in the decade leading up to 2020, it increased further to 482. A similar check in the *Science* journal shows that the number of times has increased from 54 to 341 to 236.

However, this is seemingly not merely due to a growing genuine interest in science. One thing that can be noted is that the number of Japanese Nobel Prize winners has increased.

Before 2000, the last recipient of three natural sciences prizes was Susumu Tonegawa in 1987, when he received the Nobel Prize in Physiology or Medicine. However, since 2000, it is no longer uncommon, with Hideki Shirakawa (chemistry prize) receiving in the same year, Ryoji Noyori (chemistry prize) receiving in 2001, and Masatoshi Koshiba (physics prize) and Koichi Tanaka (chemistry prize) receiving in 2002.

The top of the front page of every issue features a bold headline accompanied by a brief explanation of the achievement. This is followed by a more detailed description of the scientific significance of the achievement on the next page. In the society section, an article is dedicated to introducing the "personality" of the laureate as described by acquaintances and friends—a format established during this period.

Writing a large number of articles requires a certain amount of preparation in advance, and each year, the reporters in charge of the awards conduct various interviews, necessitating a dedicated team of reporters. Although the task is tedious for individual reporters, it serves as good training for them as they gain an understanding of what basic science is all about, and what they learn through this training can be naturally applied when covering topics other than the Nobel Prize. I believe this process has contributed to the increase in the number of science articles.

Another reason is the government's university reforms, particularly the incorporation of national universities in 2004. This led to an increase in the number of cases where researchers approached journalists, which apparently also led to an increase in the number of articles.

One of my science reporter seniors, who is now in his 60 s, told me that when he was assigned to the science section from the society department in the late 1990s, he felt that researchers did not take him seriously enough. He further added that the researchers had little desire in communicating anything to the world. If he wanted to collect material for coverage, he had to gather English-language papers and study them by himself.

For various reasons, he eventually moved on to the society department and returned to the science department in the early 2000s. At this time, he had more opportunities to attend press conferences and lectures. He felt as if researchers has changed their attitudes drastically.

Of course, individual reporters have their own way of making connections, and some reporters were probably in close communication with researchers. However, it is certainly possible that since the incorporation of national universities, researchers have been increasingly required to promote their achievements and engage in public outreach in various ways, which may have led to an increase in the amount of contact with reporters. This has also likely resulted in an increase in the number of opportunities for journalists to be exposed to basic science.

31.4 Science Reporting is Once Again on the Decline

To offer a glimpse of my own experience, I was assigned to the science and healthcare department in 2010, seven years after I started at the paper, where I was in charge of science reporting. I came to cover various science-related topics, including space, astronomy, biology, and nuclear accidents.

Major scientific journals such as *Nature*, *Science*, and *Proceedings of the National Academy of Sciences (PNAS)* would email lists to the media in advance of publication, informing them of the articles that were coming out. At the time, all articles in major journals were assigned to reporters, and each reporter had the task of evaluating each article to gauge whether it should be published.

However, as you know, business for legacy media outlets such as newspapers continues to be extremely difficult. When I joined in 2003, the Asahi Shimbun had a circulation of over 8 million, but it had already dropped below 4 million. Of course, advertising revenue has also decreased accordingly. Consequently, the number of reporters has been decreasing, with the number of science reporters reduced considerably.

Naturally, the number of staff members available for covering the news has also been on a downward trend. The myriad effects of this are being felt in many places, one of which is that people are no longer able to take the time to carefully read academic papers. The structure where someone is in charge of all articles in major scientific journals can no longer be maintained. The current practice is to assign someone to handle only those topics deemed absolutely essential, where the consensus is "There is just no way we can go without putting out an article on this," as long as an editor or someone else is somehow aware of this.

Following the previous example, the number of times the proper noun *Nature* appeared in an article was 116 in the three years from 2018 to 2020 and only 54 in the period from 2021 to 2023. The number of articles in science also decreased from 69 to 39. Although the overall number of scientific articles does not necessarily seem to have decreased, the number of articles written based on press releases issued and press conferences held by Japanese research institutes may have increased.

31.5 How Should We Write to Increase Readership?

With all of the above in mind, how should we go about writing scientific articles, when resources are limited? How can we transform the "unread" articles, which struggle to capture the public's attention, into those that "everyone is reading?" This is where the skill and wits of the scientific journalist come in.

And, disappointingly (I have to say), the way in which the wits of the press have been drawn out and applied has not been well received by researchers and people interested in science.

One issue on this point, as is often the case when reporting on the Nobel Prize, is that the articles focus on individual researchers who have nothing to do with science. I have already touched on this earlier.

The second is taking advantage of nationalism. This is also evident when reporting on the Nobel Prize, where a big fuss is made over the fact that "a Japanese won!" The Nobel Prize is not a competition between countries, and nationality has nothing to do with science. Furthermore, YoichiroNambu, Shuji Nakamura, and Syukuro Manabe, for example, have obtained American citizenship and are therefore technically no longer Japanese "nationals," but journalists still try to attract the interest of Japanese readers by describing them ambiguously as "from Japan."

Another thing is that even if the research is about basic science, journalists add statements such as "useful for ____" to hint at possible applications. This is also common in Nobel Prize reporting. I am sure there are many researchers, not just Nobel Prize winners, who have been put off by a reporter asking them, "What use is that?".

When Shinya Yamanaka, the scientist who led the development of iPS cells, was awarded the Nobel Prize, there was extensive publicity about the possible applications of these cells to medicine. Although the Nobel Prize committee evaluated the basic part of "induced cell pluripotency," this case is still highly convincing since Yamanaka is a doctor and aims to apply iPS cell technology. Moreover, if it were an achievement purely in engineering, such as Akira Yoshino's lithium-ion battery invention, I think it would only be natural to write about the many uses of such a product.

But what about in the case of Yoshinori Osumi's research on autophagy, for example? The article on the front page of the Asahi Shimbun stated, "It is expected to contribute to deepening our understanding of various diseases, including cancer, and the development of treatments." Although this is not a lie, Dr. Osumi is strongly oriented toward basic science as a researcher. Moreover, no one can say that he has a clear roadmap for applying his research to specific treatments. Given such circumstances, the description seems to be quite a leap.

Still, if the question "What is this research for?" remains unexplained, readers will be left unsatisfied and may lose interest. At least, that is what reporters believe. They know that it is interesting as basic science, but they still, despite themselves, feel a compulsion to raise the question, "What can it be used for?" They also write it in their articles. There are many such reporters. At the very least, I was one of them.

What about molecular cybernetics, in which we are participating as Journalists in Residence? The goal is to create a "micron-sized artificial brain," and it would seem that this research has applications in mind. However, currently, they are only at the stage of creating something that could be used as a component of such a brain. It still seems far from the point where one can say with conviction and specificity, "This is what it can be used for!".

The research was conducted by a group comprising primarily Japanese researchers, so appealing to nationalism may be possible in that sense. But it has not won any major awards that anyone would know about. The truth is that introducing such research within the framework of a newspaper article is not easy in many ways.

31.6 To Discuss ELSI

One of the purposes of the JIR is to discuss Ethical, Legal and Social Issues.. It is part of the job of science reporters to consider the possible negative effects of science on society at large, and I have written some of these articles.

For example, in 2010, a colleague and I wrote a series of articles on the increasing number of grants provided by the US military to Japanese research sites. Drawing a clear line between military and civilian technology is difficult. For researchers struggling to secure funding, the relatively free use of research funds from the US military is attractive. While receiving grants from the US military is not immediately considered "involvement in military research," some researchers are concerned that their basic research will eventually be used for military purposes. This article introduced this situation.

The Asahi Shimbun's science section also ran an irregular series of articles from 2017 to 2018 titled "Considering 'the unforeseeable'." The project aimed to enable readers to think about "risks that could cause major harm." One article presented a future scenario in which an AI-generated deepfake video of the US president would surface on the Internet and cause widespread discord. Another article suggested the possibility of unexpected health risks from genetic manipulation of human fertilized eggs.

However, I think these projects fall into a rather specialized category. The focus of our work is only to present "facts." We report on events that have already materialized and are happening right in front of our eyes. Of course, we sometimes publish editorials about something that may occur in the future, but I think they are often supplemented by interviews with outside "experts" and the like. With a few exceptions, such as editorial board members, the articles are required to be specific. In fact, the series on US military funding was not so much a discussion of future risks as it was an introduction to the current situation of researchers who have received grants from the US military.

The "Unforeseeable" series would have been difficult to present to readers in articles if we had not taken almost forced approach of presenting future scenarios in a concrete manner as if they had already occurred. It began as a reflection on the Great East Japan Earthquake and the nuclear power plant incident, events that were called "unforeseen" at the time but ended up becoming reality. Several articles written before 2011 pointed out problems with government regulations regarding nuclear power plants, but they failed to garner sufficient public attention. Consequently, the crisis could not be prevented. After the incident, a veteran journalist familiar with nuclear power plants commented, "I've written so many articles, but no one reads them."

Researchers are human beings, and their preference for "evidence over argument" is probably not so different from that of newspaper readers. If this is the case, then further discussion of ELSI will face the same challenges with which we journalists have been struggling. Moreover, unlike the public, researchers are also players in this issue. The explanatory material on ELSI prepared by the Japan Science and Technology Agency (https://www.jst.go.jp/crds/pdf/2021/XR/CRDS-FY2021-XR-02.pdf), explains that ELSI "does not stifle free research and development," yet the fact that an entire chapter is devoted to explaining this paradoxically indicates that many researchers consider ELSI a "restraint."

31.7 Sometimes, There Are Science Articles People Do Read

There are times when basic science attracts significant public interest. As far as I have experienced, for example, an announcement was made in 2019 that "a photograph of a black hole had been taken for the first time." This article appeared on the front page of the Asahi Shimbun (unfortunately, it did not make the top story on the front page because the minister in charge of the Tokyo Olympics resigned over a gaffe), and I remember that it was also featured quite extensively on TV.

(Photo from the article).

Of course, photographing a black hole does not immediately lead to the development of some useful tool. In terms of practical application, this research had limited potential. Moreover, an American-led international group conducted the research, and Japanese researchers, although involved, comprised only a part of the group. It was an unlikely subject for an article that could be given any sort of nationalistic angle.

Nevertheless, why did it attract so much attention? To be honest, I do not really understand it, but the word "black hole" has a mysterious and catchy ring to it, and many people, not only space enthusiasts, have at least heard of this astronomical phenomenon. People become inspired and excited, sharing sentiments such as "This cosmic mystery has been revealed to us at last!" In this sense, I think a relatively straightforward and approachable "wow" factor was at play in this case.

31.8 Japan and the Ig Nobel Prize

Another example is the Ig Nobel Prize, not the Nobel Prize. Many journalists would likely prefer writing about the Ig Nobel Prize than reporting on the Nobel Prize, which is "impure" in many ways, as I have already mentioned.

For example, over the past 15 years, about six Asahi Shimbun science reporters have been stationed in the US. As far as I could confirm, at least three of them directly covered the Ig Nobel Prize award ceremony and interviewed Mark Abrahams, the Ig Nobel Prize chairperson. They traveled several hours all the way from Washington, D.C., where they work, to Boston, where the award ceremony was held.

I love the Ig Nobel Prize. I was studying abroad in Boston about five years ago, and I did not want to miss the opportunity to interview Mark and write an article about it. I visited his home, where he also does his editorial work. The experience was highly enjoyable.

What surprised me when I interviewed him was learning that Americans have no interest in the Ig Nobel Prize. In his hometown of Boston, the prize is hardly known, and they are struggling to raise funds for it. One of my classmates at the time was an American with a science background, but she, too, had never heard of the Ig Nobel Prize.

The unpopularity of the Ig Nobel Prize in the US is evident from the varying volumes of media coverage in the US, UK, and Japan, as well as the number of articles related to the Ig Nobel Prize available on the websites of different media outlets in these countries (see Table 31.1).

Compared to the UK and Japan, the number of articles written about the Ig Nobel Prize in the US is extremely small. As the US is a pragmatic society, it may be that the "fun of science itself," which the Ig Nobel Prize embodies, is not highly valued.

By contrast, in Japan, interest in the Ig Nobel Prize is extremely high. The number of Japanese laureates is quite large, and the prize has been awarded every year for the past 18 years. When I asked Mark, "Why are there so many?" he replied, "I ask the same question to you." Of course, the number of American laureates is not small, but their presence is relatively overshadowed when compared to the proportion of Nobel Prize winners.

One of the reasons for the large number of reports in Japan is probably also the large number of Nobel laureates. But that is not the only reason.

Table 31.1 Number of online articles covering Ig Novel prize in US/UK/Japan

US	UK	Japan
The New York Times: 27	Telegraph: 43	Asahi Shimbun: 152
The Washington Post: 15	The Guardian: 289	Yomiuri Shimbun: 267
USA Today: 4	Times: 51	Mainichi Shimbun: 185
Wall Street Journal: 16	Financial Times: 3	Nikkei Shimbun: 223

Articles found via Google search. Search conducted on Aug. 6. 2024.

In 2023, a Japanese received the Ig Nobel Prize, and my colleague wrote an article about it. I, who was not in a position to be directly in charge of Ig Nobel Prize articles, wanted to do something about it. So, I decided to write a long article introducing all the achievements of non-Japanese Ig Nobel Prize winners, adding a few words from each of them.

Initially, the article was purely a hobby or experimental piece and was published on Asahi Shimbun Digital, where writers could write as much as they wanted without worrying about space limitations. However, it was well received, and a slightly condensed version was later published in the paper version as well. Although it was published in the evening instead of the morning edition, it took up almost the entire front page of the paper. Ultimately, the article ended up being more voluminous than the one on the Japanese laureate.

In short, the popularity of the Ig Nobel Prize does not seem to have much to do with jingoistic "A Japanese won!" excitement. Another interesting thing about the Ig Nobel Prize is that "you won't win it by trying to win it." Mark says, "The prize is rarely awarded by self-nomination. The fun, surprise, and absurdity are merely secondary effects, and research that aims for these things in themselves is not eligible for the award" (Asahi Shimbun, morning edition, November 16, 2013).

I think this statement is really telling. The Ig Nobel Prize is not given to those who seek it. If you purposely *try* to be funny, it will fall flat. Americans are not highly interested in the Ig Nobel Prize in the first place, but even if they are interested and invest a large amount of money, the number of winners will not increase in proportion to the amount of money. In Japanese academic circles, the pros and cons of "selection and concentration" in allocating research funds are often discussed, but if the Ig Nobel Prize were to be used as an evaluation indicator (which is unlikely), I think this measure would be clearly harmful. This is because diversity is important.

31.9 If We Strive to Do What We Want to Do

I assume that the reason why molecular cybernetics set its sights on the "artificial brain" was probably that it is advantageous to identify future applications to obtain research funds. But who knows—perhaps there is another way to appeal to the science lovers out there.

Japan's culture appreciates basic science. In fact, many people in Japan can experience intellectual joy from the pursuit of science itself (at least compared to the US). I believe this is the case.

Whether or not it attracts as much attention as the Ig Nobel Prize article, my colleagues and I regularly write articles on basic science. It does not matter whether it is useful. We write because we think it is interesting and want as many people as possible to know about it. Despite the bumps along the way, I keep writing articles, hoping that this time, surely this time, many people will read them.

Researchers who pursue their research while following their passions and letting their muses guide them end up producing many astonishing results. Some of them even win the Ig Nobel Prize. Such a thing may happen in the field of molecular cybernetics.

I am looking forward to it. Of course, it is not a prize one sets out to win, though.

MIX
Papier aus verantwortungsvollen Quellen
Paper from responsible sources
FSC® C105338

If you have any concerns about our products,
you can contact us on
ProductSafety@springernature.com

In case Publisher is established outside the EU,
the EU authorized representative is:
Springer Nature Customer Service Center GmbH
Europaplatz 3, 69115 Heidelberg, Germany

Printed by Libri Plureos GmbH
in Hamburg, Germany